ACS SYMPOSIUM SERIES **317**

Biogeneration of Aromas

Thomas H. Parliment, EDITOR
General Foods Corporation

Rodney Croteau, EDITOR
Washington State University

Developed from a symposium sponsored by
the Division of Agricultural and Food Chemistry
at the 190th Meeting
of the American Chemical Society,
Chicago, Illinois,
September 8–13, 1985

D1441356

American Chemical Society, Washington, DC 1986

Library of Congress Cataloging-in-Publication Data

Biogeneration of aromas.
(ACS symposium series; 317)

"Developed from a symposium sponsored by the
Division of Agricultural and Food Chemistry at the
190th meeting of the American Chemical Society,
Chicago, Illinois, September 8–13, 1985."

Bibliography: p.
Includes index.

1. Flavoring essences—Congresses. 2. Odor—
Congresses. 3. Biosynthesis—Congresses.

I. Parliment, Thomas H., 1939–   . II. Croteau,
Rodney. III. American Chemical Society. Division of
Agricultural and Food Chemistry. IV. Series.

TP418.B56   1986   664'.06   86–17334
ISBN 0–8412–0987–1

Biogeneration of Aromas

FOREWORD

The ACS SYMPOSIUM SERIES was founded in 1974 to provide a medium for publishing symposia quickly in book form. The format of the Series parallels that of the continuing ADVANCES IN CHEMISTRY SERIES except that, in order to save time, the papers are not typeset but are reproduced as they are submitted by the authors in camera-ready form. Papers are reviewed under the supervision of the Editors with the assistance of the Series Advisory Board and are selected to maintain the integrity of the symposia; however, verbatim reproductions of previously published papers are not accepted. Both reviews and reports of research are acceptable, because symposia may embrace both types of presentation.

CONTENTS

PREFACE

THE RAPID EMERGENCE OF BIOTECHNOLOGY and its potential application in agricultural and food production has demonstrated the need for an interdisciplinary exchange in the area of biological generation of flavor and aroma compounds.

The previous ACS symposium, Biogenesis of Flavor Compounds, was held in 1972 in New York City. In the intervening decade, much new information has been developed on biologically generated aromas. This information includes knowledge of biosynthetic pathways, development of methods for augmenting specific desired compounds, and the formulation of techniques for producing higher quality, more intense aromas through manipulation of precursors, enzymes, and producing organisms.

The symposium upon which this book is based was intended to provide participants with an overview on flavor isolation as well as a description of new advances in the enzymatic, fermentative, and molecular biological approaches to the generation of aromatic chemicals. Interaction between participants during the meeting helped spark numerous discussions and deepen friendships (new and old) between American and foreign scientists. We were pleased to be able to provide a number of graduate students the opportunity to make presentations before an international audience.

Sections of this book begin with one or more review chapters to provide the reader with a general background. Subsequent chapters in each section present original research in these areas. In addition, we have included chapters which deal with both the legislative aspects as well as the industrial aspects of biologically generated aromas. We have attempted to make this book useful to researchers in both the chemical and biological sciences.

I (T. H. Parliment) acknowledge Cynthia Parliment for her assistance in proofreading and editing manuscripts. We dedicate our own efforts to our mentor, I. S. Fagerson, whose influence stimulated our interest in flavor research. In addition, we acknowledge with sincere appreciation the following corporate sponsors: Campbell Soup Company, The Coca-Cola Company, General Foods Corporation, McCormick & Company, Inc., and Pepsico, Inc.

THOMAS H. PARLIMENT
General Foods Technical Center
General Foods Corporation
Tarrytown, NY 10591

RODNEY CROTEAU
Institute of Biological Chemistry
Washington State University
Pullman, WA 99164–6340

PERSPECTIVES

1

Legislative and Consumer Perception of Biologically Generated Aroma Chemicals

Jan Stofberg

PFW Division, Hercules Incorporated, Middletown, NY 10940

Many of the approximately 5000 known volatile flavoring substances are generated in biogenetic processes in natural foods, or during the subsequest processing of these foods. In both cases, they are perceived as natural flavorings by both consumer and legislator. Many of these same flavoring substances can be produced by chemical processes, in pure form. In that case they are perceived as artificial flavoring substances, both by consumer and legislator in the United States. It is, however, in many cases, also possible to manufacture exactly these same flavoring substances by biogenetic processes, at a usually much higher cost. We will review how the origin of flavoring materials influences their regulatory and safety status, and how it determines consumer perception of flavorful foods.

Consumers and regulators all over the world, but particularly in the more developed countries, are becoming increasingly involved in and concerned about the source and composition of the flavorings in food. In many cases, especially on the side of consumers, there is an increased preference for what are called: "natural flavors".

The consumer concept of natural flavors

What consumers in general mean by "natural flavors" appears to be rather a vague concept. It is based on the archetype of the apple on the tree, with all its connotations of the Garden of Eden, perfection, purity and safety. More exactly, the concept certainly covers all flavoring materials biologically generated during the growing and ripening of vegetables and fruits. Indeed, many flavoring materials are being generated that way in the living cells. They are derived, as a kind of by-product, from amino acids and fatty acids, particularly in flowers, vegetables and fruits, at the time of ripening. Examples are:

0097–6156/86/0317–0002$06.00/0

1. the esters in many fruits, formed from alcohols and acids which are in turn derived from common amino acids such as valine, leucine, cysteine and methionine;

2. the saturated, as well as the mono- and di-unsaturated aldehydes, formed from unsaturated fatty acids, in particular linoleic and linolenic acid, in many vegetables;

3. the terpene hydrocarbons, alcohols and aldehydes formed in conjunction with the carotenoids and their oxidative breakdown products, particularly in citrus fruits.

It seems, however, that the ill-defined consumer concept of natural flavorings extends far beyond this limited area of biologically formed flavorings. Certainly the flavori substances generated as a result of minimal processing, such the cutting or chewing of fruits and vegetables are b considered natural. In these cases, the flavor formation place when enzymes in one part of the vegetable material c contact with flavor precursors in other tissues after the have been ruptured. Cucumbers and tomatoes have very litt which has been developed in the course of the normal meta the plant. Their "natural flavor" that consumers are with is developed by enzymatic processes after chewing. Another particularly well known example of flavor formation is the flavor of onions which is cutting. S alkyl cysteine sulfoxides are broken d disulfides, particularly dipropyl disulfide, under of alliinase. Simultaneously, the lachrymat sulfoxide is formed. (1)

In many cases, even more natural flavoring m formed by continued action of the plant's o flavor precursors, during further processing tea, cocoa beans and vanilla pods generates materials characteristic for those products further enzymatic reactions during wi fermenting processes.

For many traditional foods, furthe enzyme systems is needed. The flavorin of bread and wine as well as of yogurt in biological processes under the i yeast, starters and rennet respect which often take place during aging as perfectly natural by the average

More surprising, however, is include in his everyday con preparation by physical processe degrees of thermal processing chemical reactions takes pla parboiling of rice and c aldehydes by Strecker degr temperatures, the amino degradation recombine to flavoring materials in meat. Still higher temperatures cause the decomposition of sugars to furan and pyrone derivatives. In addition the combined reaction of sugars and amino acid degradation products, via the Amadori rearrangement, leads to dozens of furan and pyrazine derivatives. Introduction of low molecular sulfur and nitrogen compounds, resulting from the breakdown of sulfur containing amino acids, leads to large numbers of thiophenes, pyrroles and thiazoles. Since coffee, cocoa, baked bread and roasted meat are ally included in the average consumer's concept of natural their characteristic flavoring ingredients are considered ural also. (2)

000 flavoring materials have been identified in od so far. Many of those are not present in the ate of the food. They are generated by the ood by enzymatic or thermal processes. Often, s generated this way are more dependent on han on the specific food which has been o at which these different flavoring etermines the characteristic flavor

nized that many forms of ed to develop the flavors traditional food. In the lations defines natural s natural flavor the extractive, protein roasting, heating or fir constituents derived hydrtable juice, edible though ar plant material, natural or fermentation structur food is flavoring formed P deserve a little decisive f nong experts on regulatory only hydrolysis on. The words Nature-identit tuted in the generic word It should be not (4) Even two kinds of flav action, the ntained the ologically been the s in the

es only re is no

category "nature-identical", as is the case in most other developed countries. That term is used to designate flavoring substances that have undergone chemical processing, but which are identical in all chemical aspects to flavoring materials identified in natural foods, either raw or processed for human consumption.

In those countries where it is recognized, nature identity is by regulation equivalent to recognition of safety. The basis for this assumption is that occurrence in traditional food means that their consumption has been accepted as safe for humans. From a toxicological viewpoint, this may be questionable. The fact that traces of allyl caproate have been identified in pineapple (5) does not necessarily mean that its use in large quantities as a predominant ingredient in pineapple flavors is safe. Whether it is safe or not can only be determined by a proper safety evaluation, as has been carried out in the U.S. for certain applications and use levels. This has been done for all flavoring materials that have been Generally Recognized as Safe. (6)

We should not, however, ignore the importance of the consumption of large quantities of biologically formed flavoring materials, large compared to the quantities consumed of the same materials produced by chemical processes. The ratio of the total annual poundage of a flavoring material, unavoidably consumed as a component of traditional foods, and the annual poundage of the same flavoring material used and consumed as a food additive, has been called Consumption Ratio. (7)

If this ratio is greater than 1, the average consumption via the traditional diet is larger than the quantity of added synthetic flavoring material. Such a flavoring material has been called "Food Predominant", since it is consumed predominantly as a natural component of food. To illustrate the meaning of Consumption Ratio values: if the Consumption Ratio of a flavoring material is greater than 10, then the consumption of that material as a food additive adds no more than 10% to the unavoidable consumption via food; certainly an insignificant increase.

So far the Consumption Ratio has been calculated for close to 350 flavoring substances. 80% of them have a Consumption Ratio of more than 1, and are therefore Food Predominant. 60% even have a Consumption Ratio of over 10, which means that their intake as artificial flavoring materials is insignificant compared to that as ingredients of traditional food. (8)

The consumption of flavoring materials with a high Consumption Ratio cannot be managed by controlling their use as additives by regulations since their uncontrolled use would be much larger than the controlled use. This should have significant impact on the priority to be set for the safety evaluation of their use as a flavoring material.

Even though nature-identical does not have a regulatory status in the U.S., the FDA is fully aware of the significance of the Consumption Ratio, and is taking the data into consideration in setting its priorities.

The consumer perception of "natural" versus "artificial" and "chemical"

It has been explained earlier that many flavorings are generally accepted as "natural" both by consumer and legislator because they are characteristic for the flavor of traditional foods. Even though many are formed as a result of complicated chemical reactions that take place during food processing, these flavors still maintain, in the consumers mind, the image of pure and safe associated with the apple on the tree. The consumer does not see them as a completely different and inferior class of materials: chemicals!

The Consumer Research Department of the Good Housekeeping Institute (9) conducted a study on, among other aspects, the consumer attitude toward food flavorings. From the tabulated results of a consumer survey it appears that, with the exception of fresh fruits, the most desirable food flavors - the mouthwatering type - are processed foods, with meat, chocolate and baked goods ranking very high. "Natural" flavorings are, according to this survey, considered very important for many foods and are considered better quality than artificial flavorings.

Many consumers, again according to this survey, admit that they sometimes buy artificially flavored foods, but a large majority is of the opinion that artificial flavors consist of chemicals, whereas only 40% of them are aware that there are chemicals in natural flavorings. Artificial flavors are thought to leave an aftertaste and to be bitter. They are perceived to have been made of chemicals, in a laboratory, and that they are not found in nature. In the background is also the thought that they are harmful with, of course, the fear of cancer leading the list. This consumer opinion largely disregards the protection provided by the FDA regulations, and the publication of lists of flavoring materials that are permitted food additives, or are Generally Recognized as Safe by FDA. (10-11)

This attitude of the consumer in favor of natural flavorings is, in particular, promoted by the marketing strategy for newly developed food products: the natural status of the food and its flavoring is stressed and stretched as far as possible.

The creation and production of natural flavorings

This consumer preference for natural flavor has created a demand for natural (biologically rather than chemically produced) flavorings which still have all the desirable properties of flavorings containing both "natural" and "artificial" flavoring substances. Such desirable properties are, depending on the type of flavor and application: strength, heat stability, microbiological stability, uniform quality and unlimited availability.

Therefore, part of the research effort in the flavor industry has been directed towards the natural formation of flavoring substances, using thermal processes. This approach has been facilitated by the fact that FDA has recognized that processing of natural ingredients under endothermic conditions, by distillation and fractionation, leads to flavoring materials which still can be considered natural.

Another way of obtaining natural flavoring materials is by recovery of volatiles, even if they occur only in small quantities, from the processing of large quantities of flavorful foods such as orange juice and other fruit juices. The search for the presence and possible isolation of natural flavoring materials from the processing of natural raw materials is resulting in the availability of many more flavoring materials of biological origin.

Another part of the research effort of the flavor industry is being directed towards the generation of natural flavoring substances from natural ingredients by enzymatic processes. In addition to the traditional enzymatic hydrolysis, specific enzymes such as esterases, oxidases and isomerases provide possibilities for modification of natural starting materials to flavoring substances without using chemical methods.

The economic aspect of natural flavorings

All flavoring materials obtained in the ways described above are truly natural chemicals, chemically identical to, and of the same food grade quality as many traditionally known artificial flavoring materials. Because of the considerably higher cost of manufacturing involved, their cost will be many times higher than that of their synthetic counterparts. This is in line with the opinion expressed by a previous Commissioner of Food and Drugs, A.M. Schmidt that artificial flavor is no less safe, no less nutritious and not inherently less desirable than the natural flavor; that the sole purpose for distinguishing between natural and artificial flavors is for economic reasons. (12)

Distinction of synthetic and natural chemicals

Pure flavoring substances, single chemicals, cannot be distinguished, as regards their synthetic or natural origin, by traditional chemical methods. There is, however, one aspect in which the artificial and natural versions of the same flavoring substance may be different. If the artificial material has been synthesized exclusively from chemicals of petrochemical origin, no radioactive carbon-14 isotope will be present. Carbon from recent agricultural sources is in equilibrium with carbon-14 present in the atmosphere. A widely accepted interlaboratory standard for natural carbon-14 specific radioactivity is oxalic acid from the National Bureau of Standards. 95% of the 1950 specific activity of NBS oxalic acid (13.56 disintegrations per minute per gram of carbon, dpm/g) has been defined as the "modern activity" of the carbon in atmospheric CO_2. Ever since 1960, due to nuclear weapons testing, atmospheric ^{14}C activity has gone up, but it started declining again a few years later after the limited test ban treaty went into effect. Even if only part of the raw materials for the synthesis were from petrochemical origin, the content of carbon-14 in the artificial flavoring material will be significantly lower than that of the naturally-derived materials which should give values close to 100% of "modern activity".

Carbon 14 analysis has been successfully used in the distinction between natural wine- or cider vinegars and less expensive white vinegar made from acetic acid. (13)

It is to be expected that this analytical technique, though far from generally available, will have an impact on enforcing the economically justified distinction between natural and artificial flavoring materials. The method can only prove absence of materials from recent agricultural sources. It cannot distinguish between a natural flavoring and a material from natural origin which has been chemically modified without addition of carbon atoms from petrochemical sources.

In such cases, determination of the ratio between the stable isotopes carbon-12 and carbon-13 may be a way to differentiate between different sources of botanical origin, reflecting differences in the CO_2 assimilation pathways. This procedure has only been used with some degree of accuracy in a limited number of cases such as the distinction between vanillin from vanilla beans and vanillin from wood lignin. (14)

I would like to stress that, as far as the functionality in food is concerned, there is no difference between the flavoring materials from natural or synthetic origin. It is up to the marketing policy of the food industry, and the educated choice of the final consumer, which class of flavoring materials will be chosen for a certain application.

Enzymatic flavorings and Process flavorings

In addition to enzymatic processes aimed at the production of pure, single chemicals, there is a growing interest in enzymatic processes which duplicate, in more concentrated and faster form, the entire flavor complex which is being formed during aging, ripening and fermentation of traditional foods. Since their natural flavor has been formed this way, dairy products obviously present themselves as candidates for this approach, which may well lead to the most cost effective way of developing natural flavors. Butter acids, butter esters, modified cheeses are by now well established, highly effective natural flavorings.

The flavorings that are biologically prepared this way are aiming for the reproduction of the entire flavor formation occurring during traditional food preparation, rather than duplicating them by identifying the individual components of the natural food flavor and combining the identified individual aroma chemicals. By this procedure of developing the entire flavor complex in one process, it is sometimes possible to obtain a much closer match of the original flavor, since identification of many flavoring ingredients in the parts per trillion area is still a problem. Such trace ingredients are important for obtaining the characteristic natural flavor, in particular the highly odoriferous sulfur- and nitrogen-containing substances.

A similar procedure, aiming for a total flavor complex, has been followed in duplicating the flavor of many kinds of baked and roasted products, such as nuts and meats. Combinations of natural ingredients, analogous to those expected to play a role in the flavor formation of roasted foods, in particular amino nitrogen sources, sugars and fats, are subjected to thermal processes of similar temperature and duration as kitchen cooking procedures. The procedure to create this type of flavors, usually called Process Flavors, reproduces in concentrated form the processes that take place during the cooking, baking and roasting of food. If only natural ingredients are used in such a process flavor, it rightfully enjoys the same status of natural flavor as traditional foods, even though complex chemical reactions are taking place in the preparation of both foods and process flavors. To make sure that the reactions taking place remain within the range acceptable for food, the International Organization of the Flavour Industry (IOFI) has included in its Code of Practice for the Flavor Industry a set of Guidelines for Process Flavors, limiting the kind of ingredients used, and in particular the main processing conditions, time and temperature. (15)

The future: natural or functional?

The distinction between naturally and artificially flavored food used to be as simple as the difference between raspberries and hard candies flavored with amyl acetate. In the future the choice is more likely to be between genetically engineered botanical material, physically and enzymatically processed, and a manufactured food with added natural flavoring complexes, prepared by thermal and enzymatic processes, or their precursors.

The consumers will have to be educated about the true meaning of "natural". It does not just stand for "the ideal source of nutrition for humans", in contrast to other products made with so-called "chemicals". Many natural foods contain ingredients that are useless, or even harmful, such as oxalic acid in spinach and solanin in potatoes. The consumer has to understand that all substances are chemicals, that life is a chemical process, and that many new processes will have to be developed to provide sufficient food with high nutritional value per dollar for a hungry world. This is not going to be achieved by stretching the word "natural" to fit in easily with preconceived consumer ideas on that subject, but by creating understanding for the need for functional flavorings as safe and normal food ingredients, either traditionally present, or deliberately added.

Literature Cited

1. Buttery, R.G. In "Flavor Research", Teranishi, R. et. al. Ed.; Marcel Dekker, Inc.: New York, 1981; pp 175-216
2. Stofberg, J. Perf. Flav., 1985, 10, 1.
3. Code of Federal Regulations 21, 101.22.a.3.
4. Federal Register 38, August 2, 1973, p 20718.
5. Nitz, N.; Drawert, F. Chem. Mikrobiol. Techn. Lebensm. 1982, 7, 148.
6. Hall, R.L.; Oser B., Food Tech., 1965, 19, 253.
7. Stofberg, J. Perf. Flav. 1981, 6, 69.
8. Stofberg, J.; Grundschober, F. Perf. Flav. 1984, 9, 53.
9. "Food Attitude Study", the Consumer Research Department, Good Housekeeping Institute, April 1985.

10. Code of Federal Regulations 21, 172.515.
11. Code of Federal Regulations 21, 182.60.
12. Federal Register 38, December 3, 1973, p 33284
13. Krueger, D.A.; Krueger, H.W. J. Ass. Off. Am. Chem., in press.
14. Krueger, D.A.; Krueger, H.W. J. Agr. Food Chem. 1985, 33, 323.
15. "Code of Practice for the Flavour Industry", IOFI, 2nd ed., Geneva, Switzerland, 1984.

RECEIVED January 3, 1986

Biogeneration of Aromas
An Industrial Perspective

Joaquin C. Lugay

General Foods Technical Center, 555 South Broadway, Tarrytown, NY 10591

Consumer concern about "synthetic" components of foods, including aromas and flavors, although unfounded, gives the food industry reason to be responsive. Natural ingredients carry positive implications in the minds of many consumers. The industry perspective is presented in this article: the need to have natural flavors and aromas that are readily available, reasonably priced, and compatible with contemporary foods. Biogeneration of aromas is one way of satisfying these needs.

Our food supply contains naturally occurring chemicals and frequently synthetic chemical compounds (1-5). There is no chemical distinction between a flavor compound synthesized by nature and the same compound prepared in the laboratory. Yet the perception that anything natural is "good for you" and that which is synthetic is harmful, is prevalent among consumers.

This situation has not escaped the attention of the food industry. This symposium, the Biogeneration of Aromas, is indeed very timely. It addresses a need, an increasing trend in the past several years -- the consumer demand for natural ingredients and additives in food formulations. This trend runs parallel to the increasing consumer awareness of health issues. Consumers will continue to put pressure on the food industry to seek natural sources of food additives such as colors, flavors, enhancers, and other key components of formulated foods. Fortunately, recent developments in biotechnology give us reason to be optimistic that efficient and reliable methods will allow us to make natural additives -- for example, natural aromas biogenerated from microorganisms (6,7).

In order to appreciate and understand the rationale behind the increased interest in biogeneration of aromas in foods, it is important to examine the changing lifestyle of the consumer. Against this backdrop, it is easier to see the challenges and opportunities presented to the food scientist and technologist, in particular, and to the food industry in general.

0097-6156/86/0317-0011$06.00/0

Changing Lifestyles and Demographic Patterns

The food industry has entered the threshold of a new age -- the social, demographic, economic, and technological changes taking place around us signal the beginning of an era that presents challenges and opportunities, as well as pitfalls for the industry (8,9). Although no dramatic changes are expected in the total population of the U.S., distribution by age group shows that America is getting old (10-12). Forty per cent of the working population is in the 25-39 years age group, compared to 30% in 1970. The number of Americans 65 years and older has doubled since 1955.

The aging of Americans will be reflected in the buying and eating patterns of the general population. For example, the cookie industry has shown a drop in the cookie eating group, 5 to 19 years old. However, there is an increase in the 20 to 50 year old group, which creates a demand for cracker products (13). The increase in the number of women in the work force, as well as the decrease in the size of families, suggests we analyze the manner in which food is prepared and consumed in the home. There is reason to believe that the traditional approach to meal preparation and the customary family sit-down dinner may have gone the way of the dinosaur. In the last year alone, more than 500 different prepared entrees and complete meals have been introduced into the freezer section of grocery stores by food processors (13).

The consumer is not willing to spend appreciable time in the food preparation, but will not settle for less than premium-quality prepared meals. Meals requiring minimum preparation are in demand. This explains the increased penetration of microwave ovens across American households. The freezer case is now the new field of battle in the food industry. From the freezer to the table in ten minutes or less is not an unreasonable expectation from the consumer's viewpoint. No trade-off in quality is accepted.

What does this all mean to the food scientist or technologist? It means that these products must retain high quality while they are processed, stored, prepared, and served. This is a challenge that should not be taken lightly; those delivering less than the best will find themselves outside the freezer case!

Consumer Perception and Health Concerns

More than ever, consumers are expressing concern and being very selective about food products as well as the ingredients they contain. There is a perception that processed foods are less healthy than "natural" foods. Consequently, the term "natural" on the label becomes to consumers a stamp of approval that a particular food is good for you. The term "natural" means different things to different people. In actuality there is a precise definition of the term "natural" as it applies to flavors and flavorings. The Code of Federal Regulations (14) delineates not only the source of flavors, but also the processing involved that permits the use of the "natural" terminology. This should provide the consumer a better understanding of what natural flavors really are.

Another area that attracts the consumer's attention is weight control. It is estimated that 34 million people in the U.S. weigh 20% in excess of what is ideal (13). It is not surprising, therefore, to see a rising tide of diet beverages, sugar-free snacks, desserts, cereals, low calorie frozen foods, low-fat processed meats, and a host of calorie-controlled prepared meals. This trend is expected to continue -- a challenge and an opportunity for the food industry.

This emphasis on health and well-being appears to be causing many consumers to adopt the slogan "less is good." Less sugar, less salt, less fat, fewer calories, and fewer additives carry a "good for you" message. The recent pronouncements from the NIH on the relationship of diet to cancer, the role of fat and fiber, and the relationship of cholesterol to heart disease, have raised the consciousness of consumers on the type, quality, and pattern of their food consumption (15,16).

In spite of all this information, the average healthy consumer has no reason to worry. A prudent diet, applying the principle of balance, variety, and moderation -- using the range of foods from the four basic food groups, and engaging in moderate regular exercise -- is a regimen that is consistent with a rational and healthy lifestyle. In the absence of hard data to the contrary, a better pattern of food consumption and behavior cannot be recommended.

Trends in the Food Industry

The pattern of food consumption and the response of the food industry reflect the consumer's increased awareness of the health attributes of food and emphasis on general health and well-being. In addition, there is a general shift in the direction of high-quality prepared meals (13,17). In 1984, there were 575 prepared entrees introduced, an increase of 23% from 1983. Frozen meals are a $1.4 billion business that is increasing at a rate of 18% annually. Accompanying the frozen meals explosion is the increase in the use of microwave ovens. Twenty per cent of major appliances delivered in 1984 were microwave ovens -- consistent with the fast paced lifestyle that allows little time for food preparation. Microwave ovens are now in 44% of U.S. households.

Fruit juices, fruit blends, and fruit based beverages have gained popularity. Fruit juices promote a healthy and nutritious image to the consumer, which is carried over into the snack area -- where fruit rolls, considered to be a wholesome snack, have shown increased consumption.

There is a decrease in the consumption of red meat in response to concern about health effects of fat and cholesterol -- which are implicated in coronary heart disease, hypertension, cancer, and obesity. Parallel to this is the increase in the use of poultry and fish. The dollar value of poultry in 1984 was $15.4 billion, an increase of 37% from 1983. Seafood per capita consumption was 13.6 pounds in 1984, an all-time high. The use of surimi, a restructured fish product, is on the rise and could reach a billion pounds in 1990. A simulated form of crab legs using surimi technology is gaining adherents.

There is a trend in the changing American taste -- from the
basic meat, potato, and vegetables -- toward exotic international
cuisines (18). The changing population mix in the U.S. has contributed to this shift in no small way. A recent survey of ethnic
food consumption showed that over 80% of those polled have eaten
Italian, Chinese, and Mexican food more than once. The affluence
of American consumers and their increased exposure to these international foods is providing the food processor with the incentives
to satisfy the growing appetite for ethnic dishes. The end is not
in sight.

Challenges and Biotechnology Implications

In light of prevailing attitudes toward foods, the shift from traditional lifestyles, as well as current industry trends, how might
participants in this symposium address the challenges and opportunities in the biogeneration of aromas? The key challenge is to deliver biogenerated aromas that are as good as, if not exactly identical to, the traditional aromas that the consumer is familiar
with. The consumer's basis for comparison is past experiences with
aromas -- coffee brewing, bacon frying, cookies in the oven, or
steak being grilled over charcoal. In short, the quality of biogenerated aromas cannot be less than what the consumer expects.
The complexity of many natural aromas will make it very difficult
to duplicate exactly the aromas of specific foods. Therefore, the
artistry of the flavorist combined with the technical expertise of
the biotechnologist is critical for delivering high quality natural
aromas and flavors.
    Biogeneration of aromas is not new. The use of microorganisms
in foods occupies a prominent role in the history of traditional
foods. The different types of cheese aromas depend on the kinds of
microorganisms used for inoculation. Alcoholic beverages, bread,
yogurt, pickled vegetables, and certain cured meats have characteristic aromas mainly as a result of microbial action. A wide range
of odors of microbial origin have been reported (19-21). Many have
potential application in foods and beverages. Fungi are especially
interesting because of the various types of fruity aromas that have
been attributed to this class of microorganism (22). Fermented
oriental foods -- such as miso, soy sauce, natto, tempeh, and
sufu -- all have characteristic aromas brought about by the type of
bacteria, yeast, or mold used in the fermentation.
    What is different today is our ability to manipulate the ge-
netic material in living cells, giving us reason to be hopeful that
we will be able to make quantities of different types of biological
materials, including natural aromas for food application (23). Industry recognizes this opportunity has challenging technical difficulties. Biogeneration of aromas from microbial, plant, or animal
tissue will require precise understanding of the chemical nature of
the aroma components (an aroma comprises of many distinct chemicals), the biochemical processes leading to the formation of aromas, as well as the nonenzymatic reactions that could enhance or
destroy the desired aromas. Understanding the nature and chemistry
of precursor formation should provide application opportunities

where aromas could be generated at the time of food preparation.
This would result in a stronger impact with minimal loss during
processing and storage.

With all this information in hand, we have to direct the living cell -- e.g., via genetic engineering -- to synthesize the desired aromas in commercial quantities and at reasonable cost. This
is a tall order, but the rewards of success make this a worthwhile
endeavor.

There are some obvious opportunities where tailor-made natural
aromas or precursors can find immediate application. These include
direct application in formulated foods and beverages, as well as
processes where precursors release the desired aromas (e.g., high
temperature/short time operations). There are many other opportunities, especially with regard to natural aromas as replacement for
synthetic ones. But the biggest opportunity is to give consumers
what they want -- a high quality natural aroma, readily available,
at reasonable cost.

Proteases, lipases, and carbohydrases are just a few of the
enzymes that can produce precursors of flavors and aromas. Bioengineered organisms could be made to produce more efficient enzymes,
providing greater substrate specificity and higher specific activity. Such organisms may yield enzymes that are effective over a
wider range of processing conditions -- for example, enzymes stable
at higher temperatures.

Biogenerated aromas for such applications as baked goods,
processed meats, fruit based products, beverages, desserts, and
breakfast cereals, are examples of the potential market for this
technology. The fermentation industry, the forerunner of present
day biotechnology, could be the leading edge in this effort.

The Question of Safety

The safety of bioengineered microorganisms and plant tissue has
been addressed recently (24,25). While molecular biologists believe that genetically engineered organisms are acceptably safe and
that the hazards remain hypothetical, ecologists continue to express concerns about potential damage to man and the environment.
This debate will, no doubt, continue. In the meantime, FDA Commissioner Frank E. Young, in remarks to the NIH Advisory Committee,
presented FDA's position on the regulation of products from biotechnology (26). The FDA does not recommend new regulations, laws,
or regimens to address products from biotechnology. Rather, the
FDA will employ procedures now in place to assess and evaluate
foods, drugs, biologicals, and all products derived from biotechnology. Since this is a product based assessment, not a process
driven evaluation, the products will be reviewed on a case-by-case
basis. This is a rational approach for evaluation of biogenerated
aromas, since it makes use of established procedures. The result
will create a minimum of confusion.

Should toxicological testing be required to assess the safety
of biogenerated aromas, the procedure recommended by Stofberg and
Kirschman will facilitate the setting of priorities (27). With

hundreds -- if not thousands -- of possible chemical entities being generated by biochemical processes, the toxicological assessment of all chemical components becomes an almost impossible task.
Stofberg and Kirschman proposed the use of consumption ratios
(C.R.), which relates the amount of a specific chemical compound
found in a natural or traditional food to the amount added by the
food processor. A consumption ratio greater than one means that
the chemical component consumed is predominantly derived from natural or traditional foods. High consumption ratios will raise our
confidence about the safety of such chemical components in biogenerated aromas, while low consumption ratios will set a high priority for toxicological testing. (For further discussion of consumption ratios, see chapter 1.
     With the above procedure in place, the assessment of the
safety of biogenerated aromas should be timely, rational, and cost
effective. Sometimes, this stage could spell the difference between success or failure.

Conclusion

Without a doubt, there is enormous potential for the application of
biotechnology for the generation of aromas for food application.
The utilization of more natural additives in processed foods is
consistent with the consumer's desire to have less artificial components in food formulations. At the same time, the trend towards
high quality prepared meals and a wider range of ethnic foods
(Mexican, Chinese, Italian, etc.) create a demand for more flavors
and aromas that satisfy this range of taste. Biotechnology appears
to be on the threshold of delivering against these needs.

Literature Cited

1. Code of Federal Regulations, 1985, 21CFR.
2. Heath, H.B. "Source book of Flavors"; AVI Publishing Co.:
   Westport, CT, 1981.
3. Stofberg, J. Perf. Flav. 1985, 9, 2-6.
4. Furia, T.; Bellanca, N. "Fenaroli's Handbook of Flavor Ingredients"; Vols. I and II, CRC Press: New York, 1970.
5. Ames, B. Science 1983, 221, 1256-1264.
6. Outstanding Symposia in Food Science and Technology -- Genetic
   Engineering in Food Production, Food Tech. Feb. 1984, 38,
   65-127.
7. Science, Biotechnology Special Issue, February 11, 1983, 219
   No. 4584
8. Ogilvy, J.A. "Social Issues and Trends: The Maturation of
   America"; Business Intelligence Program, SRI International Report #697, 1983-84.
9. Anonymous, Business Week, 2 July 1984, 52-56.
10. Barney, G.O., "Global 2000 Report to the President"; Council on
    Environmental Quality and Department of State, U.S. Government
    Printing Office, 1980.
11. Przybyla, A. Prepared Foods June 1984, 153, 108-110.

12. Lemaire, W.H. <u>Food Engineering</u> May 1985, <u>57</u>, 90-101.
13. <u>Prepared Foods</u>, New Products Directory, Summer 1985, <u>154</u>.
14. "Code of Federal Regulations" 21 CFR 101.22a.3. 1985.
15. "Diet, Nutrition, and Cancer"; Committee on Diet, Nutrition, and Cancer, National Academy Press: Washington, D.C., 1982.
16. "Lowering Blood Cholesterol to Prevent Heart Disease"; National Institute of Health Consensus Development Conference Statement, 1984, <u>5</u>.
17. Staff, <u>Food Engineering</u> State of the Industry Overview, August 1985, <u>57</u>, 67-87.
18. Anonymous, <u>Food Engineering</u> America's Cuisine, April 1984, <u>57</u>, 66-68.
19. Schindler, J.; Schmid, R.D. <u>Process Biochem.</u> 1982, Sept./Oct. <u>17</u>, 2-6.
20. Rutloff, H. <u>Die Nahrung</u> 1982, <u>26</u>, 575-589.
21. Margalith, P.A. "Flavor Microbiology"; C.C. Thomas Publisher: Springfield, IL, 1981.
22. Maga, J.A. <u>Chem. Sens. Flavor</u> 1976, <u>2</u>, 255-262.
23. Prentis, S. "Biotechnology A New Industrial Revolution"; George Braziller Inc.: New York, 1984.
24. Brill, W.J. <u>Science</u> 1985, <u>227</u>, 381-384.
25. Kolata, G. <u>Science</u> 1985, <u>229</u>, 34-35.
26. Young, F.E., Press Release, FDA Commissioner's Remarks to the NIH Director's Advisory Committee, 24 June 1985.
27. Stofberg, J.; Kirschman, J.C. <u>Fd. Chem. Toxic.</u> 1985, <u>23</u>, 857-860.

RECEIVED January 27, 1986

3

Legal and Scientific Issues Arising from Corporate-Sponsored University Research in Biotechnology

Considerations for Negotiating Research and Licensing Agreements

S. Peter Ludwig

Darby & Darby P.C., 405 Lexington Avenue, New York, NY 10174

A discussion and review of legal and scientific issues that arise in negotiations for corporate/university patent license agreements and corporate sponsored university research arrangements in the area of biotechnology is presented. Some approaches for negotiation of disputed issues are also considered. The topics discussed are:

      I.   Goals of the Parties in a Corporate
           Financed University Reseach Project
     II.   Arrangement of Terms
    III.   Financial Terms
     IV.   Fundamental Agreement Terms
      V.   Confidential Information.
     VI.   Patent Considerations
    VII.   Impact of Federal Funding
   VIII.   Preference For United States Industry
           (35 U.S.C. 204)
     IX.   Product Liability
      X.   Diligence to Commercialization
     XI.   Division of Royalty
    XII.   Termination Provisions

The field of Biotechnology has been largely developed by University scientists. Investigators from the academic world, confronted with the increasing difficulties of locating financial support for their research, have increasingly turned to corporate sponsors for research funds.

Sponsored university research in the biotechnology field is usually initiated by a corporate research or product development department to avail itself of the unique expertise available in the academic world. A research agreement outlines the responsibilities and

expectations of the parties. Often, a patent license
agreement (covering the inventions that may result
from the research) is simultaneously arranged between
the parties. This paper will highlight some of the
legal and scientific issues that arise in negotiations
for corporate/university patent license agreements and
corporate sponsored university research arrangements in
the area of biotechnology. However, many of the same
issues arise in other fields of chemical research.
   Some issues that arise in arranging for
   (A) license agreements under which a corporation
       obtains the right to use university owned
       patents and inventions, and
   (B) corporate sponsored university research agree-
       ments will be considered.

I. Goals of the Parties in a Corporate Financed University Reseach Project

The primary goal of a university is to advance the
boundaries of basic scientific investigation. University
investigators are traditionally interested in furthering
"basic" scientific principles or knowledge as opposed to
developing existing (known) products or processes.
Industry sponsored research projects involve a coinci-
dence of interest between a university investigator and
a commercial sponsor. The commercial sponsor (corporation) is interested in prospecting for a novel product
(or process) in a specific field (e.g. expression of
natural aroma compound by a recombinant organism) and
joins forces with a university researcher working in the
same area (or acquires rights to an existing invention
owned by an institution).
   The goal of the university researcher will be to
acquire new knowledge, obtain funding for laboratory
research and secure material for publication. The
profit motive and the development and acquisition of
exclusive rights in new products and processes are
usually the objectives of an industrial research sponsor
or licensor.
   The university/corporate research collaboration,
and a subsequent patent license, should also benefit the
public at-large by making possible the rapid development
and effective use of inventions. The public also
benefits when a commercial entity licenses a university
owned invention (perhaps created with government funding)
and develops it into a viable commercial product.

Suggested Approach For Negotiations. In the current
financial climate (marked by reduced federal funding for
university research) university administrators and
researchers are actively seeking alternative sources of
funds for research. The fact that a corporate sponsor
is willing to fund a research project does not always

guarantee acceptance of the project by university administrators. The project must also appear to fulfill the
university's basic mission of advancing scientific
knowledge. The university investigator must balance his
academic interests (and those of his university) with the
commercial interests of an industrial research partner.
Industrial collaboration and research support are desirable to stimulate rapid progress and to allow the public
to benefit quickly from the results of academic research.
By focusing on the potential benefit to the public from
the proposed collaboration (or license) both parties
are more likely to achieve their objectives.

II. Arrangement of Terms

Initial contact is usually made between the university
scientist who is to conduct the desired biotechnology
research (or the inventor of the product or process to be
licensed) and a representative of the commercial sponsor
(or licensee). In most instances, the scientist is
generally an employee of the university and will not be
empowered to make any commitments of facilities, personnel
or patent licenses. Such arrangements are generally the
province of university administrators. The financial
terms of a research (or license) agreement will generally
be negotiated between a university administrator (with
the input of the scientist who is involved with the
project) and representatives of the commercial sponsor.
   The corporate sponsor (or licensee) who begins to
negotiate the terms of a biotechnology research project
or license agreement with the principal university investigator, may be wasting time. Research scientists are
often unaware of the prevailing institutional policy
regarding such arrangements. Corporate personnel are
well advised to ascertain at an early stage the identity
of the university official empowered to negotiate
financial terms for research and licensing agreements.

III Financial Terms

Two financial elements are generally involved in funding
university research projects. Direct Costs involve
payments for the actual out-of-pocket and operating
expenses of the sponsored research project. The salary
of scientists' and technicians, cost of research materials, and related travel (by investigators), are examples
of Direct Costs.
   Many universities will also require payment of the
Indirect Costs of a sponsored research project. This
payment (generally a fixed percentage of the Direct
Cost) is designed to reimburse the institution for rent,
electricity, capital improvement and new equipment.
The Indirect Cost varies widely from one institution to
another and may be as much as 70% of the direct research

cost. Most institutions have a federal "indirect cost" rate, i.e. the amount the federal government pays the institution for the indirect cost of federally sponsored research projects.

In some instances, an institution will request an additional (institutional support) payment. Such payments generally take the form of grants for diverse educational projects including faculty training, support of graduate education or advanced technical library facilities.

Suggested Approach For Negotiations. When arranging the terms of a research agreement, the university (and the sponsor) should attempt to avoid overlapping contract research payments with federal grants. The research agreement should specify that funds received from the commercial sponsor will not be used to pay for scientific activities that are funded by government grants (or another commercial sponsor).

Before accepting a license agreement under a university owned patent, check to see if the invention was made with federal funding (see Section VII - Impact of Federal Funding).

IV. Fundamental Agreement Terms

(a) A corporate sponsored university research agreement should allocate responsibility for supplying personnel and resources and performing specific tasks. If special personnel or technology are required, the agreement should spell this out. If access to clinical facilities is an important consideration for the joint research project, these arrangements should be identified in the agreement. Failure to do this can result in inadvertent misunderstandings and delays.

(b) Even if it is not anticipated that the research will lead to patentable inventions, some understanding should be reached regarding the maintenance of research records. Many academic researchers in the biotech field do not maintain research notebooks that would pass muster in an industrial laboratory. Industrial research sponsors should not be lulled into a false sense of security on this point. Even though an academic laboratory looks like an industrial laboratory, it does not function the same way. Academicians may be penalized for failing to publish, but are not subject to any sanctions if they do not maintain contemporaneous notebook records that are witnessed and dated at periodic intervals.

(c) The sponsored research agreement should contain a specific policy regarding the scientific records that will be kept in the course of the project. A suitable policy might include the following points:

(i)   all laboratory research procedures
and experiments should be recorded in writing in a
research notebook by the investigator at the time
of performance;
        (ii)   the research notebook should be a
permanently bound (preferably hardcover) volume with
prenumbered pages;
        (iii)   all notebook entries must be signed and
witnessed by qualified scientific personnel at periodic
intervals (at least twice a month is recommended);
        (iv)   the date on which each experiment or
procedure is performed must be recorded at the time
of performance;
        (v)   the university will be responsible
for custody of the research notebook (or a suitable
copy e.g. microfilm). This obligation should survive
termination of the agreement.

Suggested Approach For Negotiations. Maintenance of
adequate research records is necessary even though no
"inventions" are contemplated to arise under the research
arrangement. Negotiate for and obtain a guarantee of
adequate record keeping. These records may be required
for patent and non-patent purposes, including:
        (a)   establishing priority of invention;
        (b)   determining inventorship;
        (c)   fulfilling requirements under government
funding agreements (applicable to projects that are
jointly sponsored by industry and the federal government
e.g. periodic reporting requirements).
        (d)   domestic and foreign health or product regis-
trations.
        (e)   product liability claims.
        For patent license agreements, the licensee should
determine (in advance) the nature and custodian of the
scientific records which are available to support the
conception and reduction to practice of the patented
invention. The information in these records may become
important later on if the resulting patent must be
litigated.

V.   Confidential Information

The competing interests of confidential information and
academic freedom can raise a difficult issue in an
industry/university research agreement. If an academic
institution is not free to publish the results of its
research findings, it has failed to fulfill a part of
its basic mission. Many universities have by-laws or
regulations which guarantee faculty members complete
academic freedom to publish research findings. At the
other end of the spectrum, the corporate research
sponsor is justifiably concerned about the loss of
potentially valuable technical information through

publication prior to applying for patent, or even worse, public disclosure of corporate trade secret information for which patent protection is not available. Resolution of the publication issue is often the key to concluding an acceptable research or license agreement.

Three types of information (outlined below) are generally involved in university/industry research collaborations.

(a) Background Information is information that is in the public domain at the inception of the research collaboration. Such information usually forms the basis of the "prior art" and may be the starting point for the collaboration. (E.g. vectors useful for cloning in specific organisms).

(b) University Technical Information is information developed (by the university) as a result of the research collaboration. This may include essential know-how or research findings that form the basis for patent protection. (E.g. the recombinant product or a fusion protein).

(c) Corporate Technical Information embraces confidential technical information developed solely by corporate scientists prior to the collaboration and disclosed to university personnel to expedite work on the sponsored research project. Such information is not usually available in the literature or to the general public. (One example might be the composition of a flavor used in a beverage developed wholly by corporate scientists for use in a corporate project.)

To fulfill its basic academic mission, the university will insist upon retaining for its researchers the right to publish articles containing Background Information and University Technical Information. If the corporate research sponsor (or licensee) insists on maintaining all University Technical Information in strict confidence, it is doubtful that an agreement can be reached between the parties.

The policy of academic freedom is strongly embedded in university by-laws and university regulations. In most instances institutional administrators are not empowered to waive the right of faculty members to publish research findings. Although the industrial sponsor may not object to publication of Background and University Technical Information, it will generally wish to avoid any disclosure of Corporate Technical Information. The corporate party will want to review proposed publications by University researchers, in advance, to insure that Corporate Technical Information is not inadvertently revealed or patent rights lost through publication. This type of understanding can provide the

means for avoiding the disclosure problems imposed by
the academic freedom issue.

Suggested Approach. The corporate sponsor should ask
the university to maintain Corporate Technical Information in confidence. Since it was not generated by
university personnel, there is little justification for
university administrators to insist that such information
must be available for publication.
     In most situations the industrial sponsor should
try to agree to publication of Background and University
Technical Information (subject to prior review by both
parties). To circumvent the "academic freedom" obstacle
(and protect valuable research developments), corporate
sponsors of university research should request the right
to review papers or articles by university scientific
personnel pertaining to the Sponsored Research field
sufficiently in advance of the time they are to be
submitted for publication. The advance review period
should afford the corporate party adequate time to
evaluate the proposed publication for the presence of
patentable subject matter or Corporate Technical Information. Publication of an invention prior to filing a
patent application can bar most foreign patent protection
(e.g. European Patent Convention Article 54) and
disclosure.
     If the preliminary review by the industrial sponsor
of a proposed publication reveals the presence of (i)
patentable subject matter, or (ii) Corporate Technical
Information, the agreement should provide that the
sponsor has the right to request that publication be
delayed (e.g. for 30 days) during which time a patent
application can be prepared and filed, or Corporate
Technical Information deleted from the article to be
published.

Tangible Products of Research. Distribution of tangible
products of corporate/university research can also be
troublesome, particularly when the information involves
living materials (e.g. cell cultures generating useful
aromas). University research personnel will generally
provide biological research materials to their counterparts at other institutions, upon request, and without
requiring any restraints on the use or distribution of
such materials. Information and materials resulting from
corporate sponsored research are not immune from such
transfers. Exchanges of information and materials are
often made by academicians on an informal basis, without
legal counsel.
     An innocent release of biological specimens to
fellow academic researchers can result in the loss of
patent or trade secret rights to a valuable product
(e.g. cell line). Recovering biological materials that
have been disseminated (free of any restraint) may be

virtually impossible. University research agreements should clearly specify the right of the parties to tangible biotechnology research results. This includes products (e.g. genes that express compounds which emit a desired aroma), intermediaries (e.g. products that are used to enhance the "aroma" compound by the gene) and reagents. Any corporate/university agreement which contemplates the availability of tangible research results should consider the following points;

(i) Who will have title to tangible research results?

(ii) Does the industrial sponsor obtain only license rights (if so, what are these rights, e.g., do they include the right to commercialize, use for research and development, etc.?)

(iii) Is the tangible material patentable?

(iv) Can the anticipated research results realistically be held as trade secrets.

(v) Distribution. Will university personnel have the right to distribute the tangible product to other reseachers; if so, under what conditions?

(vi) Who will have access to the tangible research results; will each person having access be required to sign a confidentiality agreement?

(vii) Publication rights (oral and written). Consideration of these issues in advance can avoid serious loss of rights after the sponsored research has begun.

VI. Patent Considerations

Two separate situations must be addressed under the heading of patent rights.

A. Patents and applications on inventions made in the course of corporate sponsored university research projects.

B. Straight patent licensing (in which a corporation is seeking a license to use existing university owned inventions, patents and applications).

The situations discussed above are considered separately.

A. Corporate Sponsored University Research Agreements

(a) Ownership of Patent and Application. The corporate/university research agreement should specify the party that will have title to any patents or patent applications covering inventions made as a result of the sponsored research. In most instances, universities will insist on retaining title to patents on inventions made by university personnel. The corporate sponsor may secure an option or other right to acquire an exclusive license under such patents and inventions.

(b) <u>Decision on Patent Filing</u>. Because the corporate research sponsor is likely to be responsible for commercial exploitation of the underlying invention, it is imperative that the corporate party retain some input on deciding which inventions should be patented and, when and where patent applications should be filed for such inventions. If the university is reluctant (or financially unable) to bear the cost of patenting, the corporate sponsor may wish to advance these costs and receive a corresponding credit against future royalties that may become due under a subsequent license agreement on the invention.

To avoid the loss of patent rights it is imperative that some mechanism be established that will enable scientists to promptly disclose inventions to the parties and permit a rapid decision on patent filings. Incorporation (in the sponsored research agreement) of a time table for invention disclosure and patent filing is one way to achieve this goal (e.g., all inventions will be disclosed in writing to the corporate sponsor within 30 days after they are disclosed to the university by the inventor).

(c) <u>Responsibility for Prosecution and Maintenance of U.S. and Foreign Patent Applications and Patents.</u> The license or research agreement between the parties should identify the party responsible for prosecution and maintenance of patent applications (and the resulting patents). There are benefits (and disadvantages) in allocating responsibility for such matters to either the university (licensor) or the industrial party (licensee).

Commercial considerations favor allocating responsibility for patent prosecution to the corporate sponsor (licensee). A commercial entity is more likely to be acquainted with the nature of the product or material that will be sold and therefore may have a better insight as to the type (scope) of protection that would be most helpful in the marketplace. Such insight can be particularly important in connection with foreign patents. University personnel will ordinarily have no means available to determine if there would be any benefit in obtaining, maintaining or discontinuing a foreign patent application (or patent).

There are also arguments in favor of allocating responsibilty for patent filing and prosecution to the university. In the case of a sophisticated invention, it may be faster, easier, and less expensive for patent attorneys located at the institution to communicate with and receive instructions from university inventors. They may also be more familiar with the technology in the field of the invention. This can be important in responding to the numerous issues that arise when patent applications are filed in a multitude of foreign jurisdictions.

Corporate personnel should be alert to the possible conflict of interest that may arise when a commercial party (licensee) is responsible for filing and prosecuting the patent applications (owned by the University) on the licensed invention. This is particularly the case where the amount of royalty to be paid by the licensee is measured by the scope of the granted patent claims. What will result if the licensee does not secure the grant of patent claims that are sufficiently broad to embrace the product he is selling? If the agreement calls for payment of a royalty on "products covered by a patent claim", the university may be disappointed to learn that no royalties are due to them under the agreement. This type of situation can become the focal point for a dispute and may be averted by allocating responsibility for patent prosecution to the licensor (university).

Another approach is to give primary responsibility for obtaining patents on university owned inventions to the corporate licensee. This should include the proviso that he will inform patent counsel for the licensor (university) of the issuance of all official actions, and also inform the counsel prior to making any substantive response to a patent office action or discontinuing prosecution or maintenance of any patent or application. Although this practice avoids the conflict of interest problem discussed above, it also involves a duplication of effort and expense. This may not be justified in many situations.

(d) <u>Faculty Consultants</u>. As an employee of the university, a faculty member is generally required to assign his inventions to his employer. Hence, difficulties may arise when a firm engaged in collaborative research with the university hires the principal investigator/inventor (a university faculty member) as a private consultant in the field of the research collaboration. In such situations, the commercial entity may contend that inventions or patents arising during the research collaboration become their property (as a result of the consulting agreement). Sorting out the rights of the parties may be difficult due to the overlapping nature of their research efforts. Research work performed at the institution on an industry sponsored research project is likely to be identified as the property of the university if a research agreement is in effect.

To avoid difficulties, university faculty members (particularly those involved in the sponsored research project) should not be engaged as private consultants to the corporate sponsor/licensee in the field of the license or the sponsored research. This makes it imperative for the agreement between the parties to define the scientific field of the research collaboration or patent license with some specificity. A comprehensive discus-

sion of such rights is obviously beyond the scope of
this paper.

B.   Patent Licensing Agreement

In patent licensing agreements, several issues (beyond
those discussed above) must be considered. For the most
part, these issues involve the impact of federal funding
and are considered below in item VII.

VII.   Impact of Federal Funding

The United States Government has rights to inventions
that are made with federal assistance under the provisions of 35 U.S.C. §200 through 212.
    When accepting a patent license from a university
(or any other educational or non-profit institution), it
is imperative for corporate researchers and businessmen
to determine if the underlying invention was made with
federal funding (usually a research grant from a federal
agency). The importance of making this determination,
at a preliminary stage in the agreement process, cannot
be overemphasized.
    Observance of the requirements imposed by the U.S.
Patent Laws (Title 35 U.S.C. §200 et seq.) for inventions
made with federal funding is fairly straightforward.
However, failure to observe the formalities can have
serious repercussions, particularly for the licensee
(including loss of the license). Also, certain requirements of the Patent Law may affect the commercial value
of the license. The message here is fairly straightforward. At a preliminary stage of the negotiations for
a patent license on a university owned invention (or for
a research collaboration) inquire to determine if the
investigator/inventor is the recipient of federal
funding. While an affirmative answer should not be
considered a "death knell" for the proposed arrangements,
the facts should promptly be brought to the attention of
patent counsel for the parties.

VIII.   Preference For United States Industry
        (35 U.S.C. 204)

A non-profit organization which receives title to any
invention made with federal funding may not grant an
exclusive license to use or sell the invention in the
United States unless the licensee agrees that any
products embodying, or produced through the use of, the
subject invention will be manufactured substantially in
the United States (See section 204 of the U.S. Patent
Law - Title 35 U.S.C. §204).
    In individual cases, the requirement for such an
agreement may be waived by the sponsoring federal
agency. A waiver requires a showing that reasonable but

unsuccessful efforts have been made to grant licenses on similar terms to potential licensees that would be likely to manufacture substantially in the United States or that under the circumstances domestic manufacture is not commercially feasible.

Failure to obtain the undertaking required by §204 from the licensee can result in exercise by the government of its rights to take control over the patents and the underlying invention (so-called "march-in rights"). Non-U.S. corporations should be alert for compliance with this provision of the U.S. Patent Law. If the University licensor must seek a waiver to comply with the strictures of Section 204, they should try to build up a "paper" record to show that considerable effort has been made to locate a licensee that will manufacture in the United States, or that a qualified licensee is not otherwise available.

IX. Product Liability

Universities have no means to monitor activities in the commercial world. They are reluctant to assume any responsibilities for commercial products or processes embodying inventions made in university laboratories by university personnel. Responsibility for commercial products and processes is perceived as a risk to be undertaken by the licensee (under whose auspices commercialization of the invention is undertaken).

Because inventors and their employers may become involved as defendants in product liability suits, university patent licensors will generally demand indemnification from a corporate licensee against any claim or suit for injuries or property damage arising from use or sale of the licensed product or process.

It is not uncommon for university license agreements to call for the licensee to provide adequate product liability insurance, naming the university as an insured party. (For an interesting consideration of the issues of product liability for injuries caused by recombinant DNA bacteria, see "Strict Product Liability for Injuries Caused By Recombinant DNA Bacteria", 22 Santa Clara Law Review, 117.)

X. Diligence to Commercialization

Due to its inability to monitor activities in the commercial biotechnology or aroma field, the University licensor may not be satisfied with an undertaking by the licensee to "use best efforts" in commercializing products embodying the licensed invention. As with any other licensor, a university will be interested in deriving maximum income from a patent license arrangement. However, because it has no means at its disposal for monitoring commercial market activities, a university

licensor has no mechanism to assess a licensee's compliance with a "best efforts" arrangement and will generally
require a minimum performance guaranty that embodies
some objective criteria. The approaches that may be
taken will be familiar to those conversant with patent
licensing of more conventional chemical products and
processes. One important distinction is that many
universities will insist that a product be brought to
market within a defined time period, failing which the
license is forfeited.

Any one (or more) of the following examples
of minimum performance guarantees may be adopted.

(a) Minimum Annual Royalty; the licensee
agrees to pay a predetermined amount on a periodic
basis to maintain exclusivity of the license (or to
avoid loss of the license). The minimum royalty may
be credited against royalties due on actual sales of
product. Failure to make minimum royalty payments can
result in having the license become non-exclusive, or
in termination of the license agreement.

(b) Requirement for Commercialization;
failure to have a product embodying the patented invention on the market within a fixed time period results in
loss of exclusivity, or the license can be terminated by
the licensor. This type of provision can be coupled
with a minimum royalty arrangement or loss of exclusivity
as in (a) above.

(c) Minimum Guaranteed Sales; the licensee is
required to sell a minimum quantity (or wholesale/retail
dollar value) of product embodying the invention in
order to maintain his exclusive license (or to avoid
termination).

Suggested Approach For Negotiations. When confronted
with a demand by a university licensor for a minimum
performance guaranty, bear in mind that part of the
motivation may be the University's concern with exercise
of march in rights in a federally funded invention (and
consequential loss of the revenue stream provided by the
patent). Try to create an arrangement that will allow
both to show that concrete steps are being taken to
advance the licensed invention to a commercial product
that will afford some benefit to the public.

If the issue comes up in conjunction with a combined
research/license arrangement, consider the possibility
of offering some concession in the research agreement in
exchange for a reduced performance guaranty in the
license (e.g agree to support a post doctoral student at
the University in a research project related to the
licensed invention in return for a reduced minimum annual
royalty).

XI. Division of Royalty

Unlike commercial entities, university licensors do not always retain 100% of the royalty received from licensing an invention. Royalty payments received by university licensors from commercial firms may have to be divided with the inventor pursuant to the requirements of Federal law or the university's internal patent policy.

(a) Impact of Federal Law. If the university is licensing an invention made with federal funding, the U.S. patent laws (35 U.S.C. §202(c) (7)(D)) require that the university share royalties with the inventor. No mention is made of the sharing arrangement that will be deemed satisfactory under the law. Specific provisions for division of royalty are generally set forth in the university's by-laws or patent policy.

(b) Revenue Sharing. Many universities have adopted a policy of sharing revenues on all inventions with faculty inventors. The policy may be articulated in the university by-laws, patent policy or faculty employment agreement. If such a policy is in effect, the institution will be required to allocate royalties on all inventions made by employees (even if such inventions were not made with federal funding). Some institutions believe the policy of sharing revenues with inventors is counterproductive and will not do so (except when required for compliance with the applicable provisions of the U.S. Patent law).

XII. Termination Provisions

More so than a commercial licensor, a university is interested in achieving a level of "certainty" in its patent license agreements. Many institutions rely on the revenue from licensed inventions as an integral part of the funds available for support of educational and scientific activities. If the original licensee backs out of the license after one or two years, it may take some time to locate a new licensee and work out an acceptable arrangement. Academic institutions understand that the value of an invention tends to diminish as time passes. Hence, the university will be interested in an arrangement that only permits termination of a license agreement on the happening of narrowly defined events.

On the other side, a licensee does not wish to become locked into an agreement, particularly where further development work may prove that the licensed invention is unmarketable, or that commercialization is unrealistic from the economic standpoint.

To avoid a stalemate on this issue, a flexible approach is required.

Suggested Approach For Negotiations. Consider a termination provision that will allow the licensee to terminate if the university technology does not meet certain
predetermined goals. There is probably little value in
having a licensee that is not optimistic about the
commercial prospects of the licensed invention. Thus, if
the licensee encounters problems in development due to
loss of enthusiasm, change in business plans, etc. that
are not related to the university technology, the agreement should permit him to terminate by making a predetermined payment to the licensor.

The following references may be useful to obtain
further insight and information on research relations
between industry and the University.

Suggested Reading

1. Baldwin, D.R.; Green, J.W. Soc. Res. Adm. J. 1984,
    15(4), 5.
2. Battenburg, J.B. Soc. Res. Adm. J. 1980, 11,3.
3. Carey, S. Wall St. J. East. Ed , 9 Feb. 1982, p. 33.
4. Crittenden,A. New York Times 22 July 1981, p. D1.
5. Lepkowski,W. Chem. Eng. News 1983, 61, 8.
6. National Science Foundation, University/Industry
    Research Relationships Government Printing Office,
    Washington, D.C. 1980.
7. Smith, K.A. Phys. Today 1984, 37, 2.
8. The Society of Research Administrators and the
    National Science Foundation, Industry/University
    Research Relations: A Workshop for Faculty
    Government Printing Office, Washington, D.C.
    1983.
9. Tatel, D.S.; Guthrie, R.C., Educ. Res. 1980, 64, 2.
10. Varrin, R. D.; Kukich, D. S. Science 1985, 237, 385.

RECEIVED January 27, 1986

ANALYTICAL METHODOLOGY

Sample Preparation Techniques for Gas–Liquid Chromatographic Analysis of Biologically Derived Aromas

Thomas H. Parliment

General Foods Technical Center, 555 South Broadway, Tarrytown, NY 10591

Sample preparation is a critical step in the analysis of bio-generated aromas. This paper will review techniques necessary to isolate and concentrate these volatile components. Particular emphasis will be placed on procedures which are relatively simple and which permit high sample throughput.

The techniques to be discussed can be classified as follows:

1. Headspace Sampling
2. Headspace Concentrating
3. Distillation/Extraction
4. Direct Analysis of Aqueous Samples
5. Direct Adsorption of Aqueous Samples
6. Direct Extraction of Aqueous Samples

The purpose of this presentation is to review procedures for the analysis of volatile compounds generated through biological processes. Numerous techniques have been proposed to separate the volatile chemicals from the nonvolatile materials and water, and to concentrate them. After sample preparation, the complex aroma sample can be separated into its individual components by high resolution gas chromatography and the aroma chemicals then structurally identified by spectral techniques.

Sample preparation of biologically generated aromas is complicated by a number of factors:

1. Concentration Level

The level of these volatile chemicals is frequently low--typically, at the parts per million (mg/kg) level or less. Thus, it is frequently necessary not only to isolate the volatile materials, but also to concentrate them by several orders of magnitude.

0097–6156/86/0317–0034$06.00/0
© 1986 American Chemical Society

2. Matrix

Volatile chemicals are frequently intracellular and compartmentalized, thus requiring disruption to liberate these aromas. In biological systems, water is generally the component in greatest amount. In addition to water, there are also lipids, carbohydrates and proteinaceous materials present which compound the isolation problem. The presence of insoluble materials normally precludes direct gas chromatographic analysis of the aroma bearing biological materials.

3. Complexities of Aromas

Biologically generated aromas are frequently quite complex and include a wide range of polarities. For example, strawberries have been shown to possess more than 350 volatile compounds (1). In addition, the classes of compounds frequently encountered in biological materials includes alcohols, aldehydes, ketones, esters, ethers, sulfides, mercaptans, amines, aromatic and heterocyclic compounds and hydrocarbons. Representative classes of compounds identified in strawberries are indicated in Table I.

Table I. Classes of Compounds in Strawberries

Class	Quantity	Class	Quantity
Hydrocarbons	34	Bases	1
Alcohols	56	Acetals	20
Aldehydes	18	Phenols	3
Ketones	20	Furans	8
Acids, aliphatic	40	Mercaptans	2
Acids, aromatic	6	Sulfides	3
Esters	130	Thio esters	2
Lactones	10	(Ep) Oxides	5
		Total	358

4. Variation in Volatility

Aroma chemicals encompass a wide range of volatiles. For example, tomatoes contain both methyl mercaptan (bp 6°C) and vanillin (bp 285°C). Techniques appropriate for the analysis of low boiling components are quite different from those for higher boiling constituents.

5. Instability

Many aroma compounds generated biologically are
unstable. Examples of this are mercaptans which
can be oxidized to sulfides, and terpenes which
can be thermally degraded.

There is no universal technique which can be used for the
preparation of all samples under all conditions. One must be aware
of the limitations of each sample preparation technique, and design
the technique to the objective of the study.

A number of years ago, Weurman (2) reviewed techniques of
aroma isolation; more recently, a number of additional reviews on
aroma research have appeared (3-11).

An interesting paper comparing several sample preparation
techniques was published by Jennings and Filsoof (12). In this
work, equal amounts of ten aroma chemicals (Table II) were combined
and then subjected to various isolation techniques.

Table II.   Composition of Model System

Compound	Bp,$^{\circ}$C	Wt. %
Ethanol	78	9.6
Pentan-2-one	102	9.5
n-Heptane	98	8.2
Pentan-1-ol	138	9.7
Hexan-1-ol	157	9.8
n-Hexyl formate	178	10.5
Octan-2-one	174	9.8
d-Limonene	176	11.3
n-Heptyl acetate	192	10.3
γ-Heptalactone	84.8 (5 mmHg)	11.5

(Reproduced in part from Ref. 12, Copyright 1977,
American Chemical Society.)

The samples were separated by capillary gas chromatography and peak
areas compared. The results of that study which are shown in
Figure 1 demonstrate that no single isolation and concentration
technique is uniformly satisfactory. Rather, the choice of technique is determined by the information desired. They did conclude,
however, that distillation-extraction gave results which most
nearly agreed with direct injections of the neat mixture.

The goal of this article is to review current sample isolation
and concentration techniques which have value in the analysis of
biologically generated aromas. Relatively simple and straightforward techniques will be emphasized since the researcher frequently
wishes to analyze a number of samples, e.g., cell cultures, fermentation broths and plant materials, in a short period of time.

Headspace Sampling

Manual Procedures Direct analysis of volatiles above an equilibrated sample contained in a sealed system is a technique which has

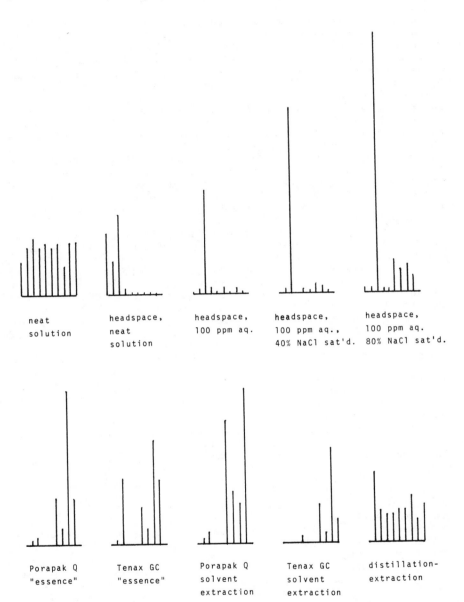

Figure 1.   Relative integrator response (arbitrary units) for several methods of sample preparation. (Reproduced from Ref. 12. Copyright 1977, American Chemical Society).

great appeal. Arduous sample preparation is eliminated; one simply homogenizes the sample, transfers it to a vial, seals the vial with a septum and equilibrates it for a period of time at a constant temperature. Hypovials (Pierce Chemical Company, Rockford, IL) are quite satisfactory for this type of analysis. Gas-tight glass syringes should be used for sampling the vapors, since disposable syringes have adsorptive properties. Syringe needles with Huber points are useful since they do not cause coring of the septum. Occasional analysis of water vapor should be made as a blank to en- sure no residual components are contaminating the syringe or the injector of the gas chromatograph.

Salt may be added to the aqueous samples which will increase the amount of aromatic materials in the headspace; however, this may cause disproportionation of the chemicals in the headspace. For example, addition of salt to apple juice increased the concen- trations of alcohols more than aldehydes, and the latter more than esters (13).

Reproducibility can be a significant problem with manual head- space techniques; therefore, rigorous efforts must be made to stan- dardize all times, temperatures, and procedures.

An improvement on the manual syringe technique is shown in Figure 2. In this case, two syringe needles are permanently af- fixed to two ports of a three-wave valve (3). A gas-tight syringe (1) is attached to the third port. The needle at (4) is placed into the injector of the gas chromatograph and the apparatus is clamped into place. The biological sample is placed in a vial (2) and held for a prescribed period of time at a constant temperature to permit equilibration to occur. The vial is then punctured by the needle (2). When ready to inject, the three-way valve is turned so that vapors in the sample can be transferred back and forth between vial (2) and syringe (1). This should be done sev- eral times to ensure equilibration of the sample. Finally, the valve is rotated so that all of the vapors can be injected directly from the syringe into the gas chromatographic injection port which is located at position (4). After injection, the valve is closed. The apparatus is left in place between gas chromatographic runs so that coring of the septum and concurrent leakage is avoided.

It is our experience that these techniques are applicable to components approaching 200°C in boiling point such as ethyl heptanoate (bp 187°C). Compounds with higher boiling points do not have sufficient vapor pressure to be sampled and detected reproducably by this technique unless the sample temperature is raised. Another difficulty arises from the fact that the largest component normally injected using this technique is water vapor. Water can cause deterioration of polar gas chromatographic liquid phases and precludes chilling the head of the column to focus the organic sample. The presence of water also prevents the use of high resolution gas chromatographic columns unless special precau- tions are taken.

These techniques are simple to perform, and they are fre- quently able to analyze the lower-boiling components in a satisfac- tory manner. A distinct advantage they possess is that insoluble and cellular material do not cause interference. Another advantage of headspace sampling is that the sample represents, most accurately, the composition which we perceive by odor, since no disproportionation due to adsorption or selective solvent extraction occurs.

An American Chemical Society Symposium was devoted to headspace techniques, and the proceedings have been published (7).

Automated Procedures   Some of the difficulties associated with manual procedures can be eliminated with an automated headspace sampler. Such a device has been described in the literature (14), and is commercially available. The schematic diagram of such a semi-automatic headspace analyzer is shown in Figure 3. Precise control of times and temperatures, as well as the capability to hold samples at high temperatures, produces better chromatographic reproducibility. We have been able to analyze ppm levels of ethyl dodecanoate in aqueous solution with this system. Significant amounts of water vapor are introduced into the gas chromatographic column under these conditions. For this reason, columns with non-polar liquid phases or bonded Carbowax-type liquid phases should be used. Two versions of these automated headspace samplers are commercially available (Perkin-Elmer Corp., Norwalk, CT; Hewlett-Packard Corp., Palo Alto, CA). Both can be interfaced with fused silica capillary gas chromatographic columns.

The advantages of automated headspace analysis are:

1. Sample preparation is minimized.
2. Various types of samples (liquids, solids, sludges) can be analyzed.
3. A significant concentration factor may be achieved.
4. Solvent peaks associated with extractions are eliminated.
5. GC columns last longer since only volatiles are introduced onto the column and excessive column temperatures are not required.
6. Reproducibility is good.

If increased sensitivity is required:

1. Increase the sample temperature.
2. Use a larger sample.
3. Add an inorganic salt (Na_2SO_4) to an aqueous sample to increase the amount of organic compound in the headspace.

Headspace Concentrating by Condensation

In cases where the preceding techniques do not possess sufficient sensitivity, concentrating techniques may be employed. Trapping of volatile organic compounds may be accomplished directly on the gas chromatographic column if the boiling point of the chemicals is sufficiently high, or if the gas chromatographic column is cooled. Such techniques have been described by Jennings (15).

Alternatively, headspace vapors may be concentrated by an apparatus similar to that designed by Chang et al, (16) and shown in Figure 4. Purified nitrogen is swept through the sample, which is contained in vessel (B). The nitrogen gas is disbursed through a sparger (D), and aroma and water are condensed in the flask (G). In addition, the system includes subsequent traps cooled at progressively lower temperatures. The various condensates are then extracted with a low-boiling solvent such as methylene chloride, which is concentrated and injected in the gas chromatograph.

Several pounds of material may be treated at one time using a large-scale apparatus of this type. This technique is non-destructive, since temperatures are kept low. Extended periods of time are required for the sweeping, thus this technique is not appropriate for routine analysis.

Headspace Concentrating by Contact with Solvents

Another technique for concentrating volatile aroma compounds involves sweeping the vapors at a low rate into a low-boiling refluxing solvent (17) as shown in Figure 5. In this way, the volatiles are retained in a water-immiscible solvent which can be subsequently analyzed by gas chromatography. A continuous variation of the nitrogen entrainment/Freon extraction technique has recently been described (18) and is shown in Figure 6. The sample is placed in vessel (7) and purified nitrogen is passed through it at a rate of 50 ml per minute. The volatile materials are transferred to an extractor (9). The upper phase in extractor (9) is 10% alcohol in water, and this is continuously extracted with Freon 11.

This system has been refined and evaluated (19) using a series of chemicals representative of fruits or fermented beverages. These researchers found the technique to be quite reproducible and effective for terpene hydrocarbons. It is not quantitative for polar, water-soluble compounds. Extraction time for this apparatus may be as long as 24 hours, hence it cannot be used for rapid screening of aromas generated from biological processes.

These headspace concentrating techniques have the advantage that the volatile chemicals are removed and concentrated under very gentle conditions which reduces artifact formation.

Headspace Concentrating by Adsorption

Over the last several years, much research has involved the use of porous polymers and carbon for concentrating aroma chemicals. Materials such as Tenax GC, Porapack and Chromosorbs .possess the valuable characteristic of retaining a wide variety of volatile aroma compounds while permitting water and low molecular weight alcohols to elute rapidly.

Manual Procedures   Prior to use, these packings should be conditioned by placing the polymer in a trap at an elevated temperature for an extended period of time under a flow of oxygen-free nitrogen. If this is not done, difficulties with impurities will be encountered. Williams et al (20) recommended conditioning traps at 180°C for 15 hours at 30 ml/min nitrogen flow.

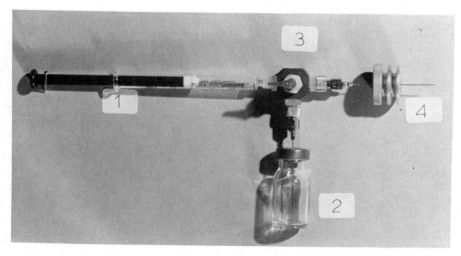

Figure 2. Manual headspace sampling device.

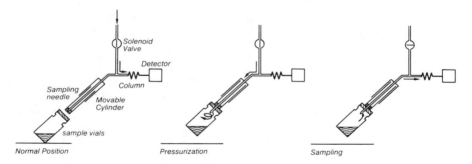

Figure 3. Semiautomatic headspace sampling device.
(Reproduced with permission from Ref. 14. Copyright 1977,
Dr. A. Huethig Publishers).

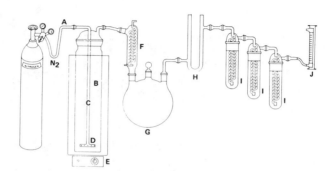

Figure 4. Apparatus for the isolation of trace volatile
constituents in headspace gas of foods. (Reproduced from Ref.
16. Copyright 1977, American Chemical Society).

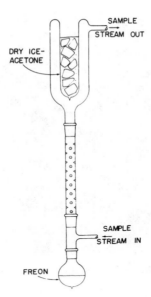

Figure 5. A Freon co-condensation unit. (Reproduced with permission from Ref. 17. Copyright 1979, Dr. A. Huethig Publishers).

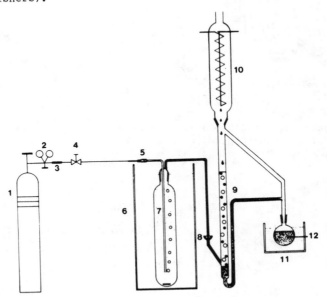

Figure 6. Apparatus for the enrichment of headspace components by nitrogen entrainment Freon/extraction. (Reproduced with permission from Ref. 18. Copyright 1980, Friedr. Vieweg & Sohn).

Figure 7 shows a relatively simple apparatus which was used for a study on citrus aromatics (21). The amounts of packing materials which have been employed range from 30 mg (21) to greater than 1 gm (22). Desorption of the desired aromas is generally accomplished by heating and may be performed directly into the gas chromatograph. Alternatively, the sample may be desorbed into an intermediate cooled trap and sampled via a micro syringe or evaluated by organoleptic techniques (22). The most commonly used polymers are Tenax GC and Porapack Q.

Solvents may also be used for desorption of porous polymer traps as discussed by Schaefer (23). Recently Parliment (24) described a micro solvent extraction apparatus for desorbing volatile organics from C-18 or Tenax traps. Less than a milliliter of solvent is used to desorb the volatiles from the trap, and the solvent is recycled to achieve complete desorption.

The apparatus is shown in Figure 8. The tube containing the adsorbent (3) is placed in adaptor (1), and microfunnel (2) is used to direct solvent into the adsorbent tube. A 5 ml pear-shaped flask is placed at the lower end of adaptor (1), while a microspiral condenser with drip tip is placed at the upper joint. Solvent is allowed to percolate through the trap until about 0.5 ml collects in the pear-shaped flask. Then this solvent is brought to reflux and the resin trap is continuously extracted for 10-15 minutes. Finally, the apparatus is disconnected and distillation used to reduce the volume to about 75 microliters.

Automated Procedures At least four commercial automated instruments exist which will concentrate headspace volatiles on adsorbents and thermally desorb them into a gas chromatograph. Times, temperatures, and gas flow rates are accurately controlled. Manual handling steps which can introduce variability are eliminated. Since these instruments are automated, sample throughput is enhanced; they can be interfaced to high resolution gas chromatographs.

A major disadvantage of these instruments is their cost, which is several thousand dollars.

Distillation/Concentration Techniques

If the volatiles can be removed from the sample via steam distillation without decomposition, then several procedures involving distillation/extraction are available.

Manual Steam Distillation/Extraction Samples can be prepared using the apparatus shown in Figure 9. The sample, containing an internal standard, is placed in pear-shaped flask (2). Steam is generated externally and introduced through inlet assembly (1) into the bottom of flask (2). Steam-distillable organic compounds are condensed in the spiral condenser (3) and collected in a receiver. The aqueous condensate is extracted with a few hundred microliters of a solvent (such as methylene chloride) and the organic phase analyzed by gas chromatography. Preparation times are quite short, e.g., five to six minutes for distillation, one minute for extraction. Using a system such as this, methyl anthranilate can be

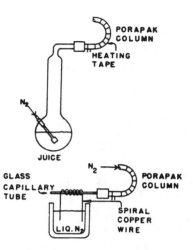

Figure 7. Porous polymer trap for stripped volatiles.
(Reproduced with permission from Ref. 21. Copyright 1978,
Academic Press).

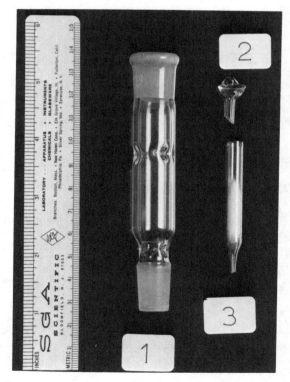

Figure 8. Micro solvent extraction apparatus for desorbing
aromatics from porous polymers. (Reproduced with permission
from Ref. 24. Copyright 1985, Walter de Gruyter & Co.).

quantitatively distilled and concentrated from grape juice in as
little as ten minutes. If the sample is heat sensitive, the aqueous receiver can be connected to a vacuum source, and the distillation performed under vacuum.

<u>Simultaneous Steam Distillation/Extraction</u>   An elegant apparatus
was described by Nickerson and Likens (<u>25</u>) for the simultaneous
steam distillation and extraction (SDE) of volatile components.
This device has become one of the mainstays in the flavor field.
In this apparatus, both the aqueous sample and water-immiscible
solvent are simultaneously distilled. The steam which contains the
aroma chemicals and the organic solvent are condensed together, and
the aroma compounds are transferred from the aqueous phase to the
organic phase. Typical solvents used are diethyl ether, pentane or
a mixture thereof; normal extraction times are one to two hours.

Advantages of the SDE Apparatus are as follows:

1. A single operation removes the organics from an
   aqueous material and concentrates them.
2. Recoveries of volatile, water-immiscible organic
   compounds are generally quite high.
3. Only a small amount of organic solvent is required, thus one doesn't have to concentrate
   large volumes of solvents.

A number of modifications of this apparatus have been proposed.
One such improvement is described by Schultz <u>et al</u>, (<u>26</u>) and is
shown in Figure 10. In this paper, the authors describe the effects of pH, time, pressure and extracting solvent on the recovery
of typical aroma compounds. Results showing percent recovery at
various initial concentrations of aromas are shown in Table III.

Table III. Recovery of Components by SDE from the Model Mixture at
            Various Degrees of Dilution[a]. (Recovery as Percent
            of Initial Amount)

	Conc. of each component, ppm			
	210	21	2.1	0.21
Ethyl butyrate	95	98	95	93
Ethyl hexanoate	100	104	100	96
Ethyl octanoate	101	98	90	89
Ethanol	0			~0.01
1-Hexanol	99	98	91	86
Linalool	99	101	97	95
Carvone	99	97	90	83
Limonene	99	93	80	85

[a]At pH 5.0 and atmospheric pressure, with 125 ml of
hexane; SDE time 1 hr.

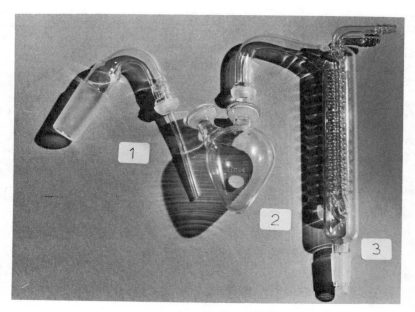

Figure 9. Micro steam distillation apparatus.

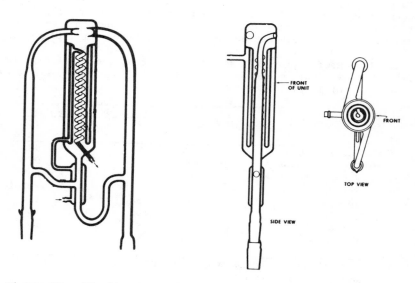

Figure 10. Simultaneous steam-distillation extraction (SDE) head. (Reproduced from Ref. 26. Copyright 1977, American Chemical Society).

It is apparent that compounds which are water immiscible and steam distillable are recovered quite effectively.

A micro version of the distillation extraction apparatus has been described by Godefroot et al, (27). This apparatus uses heavier than water solvents, e.g., methylene chloride or carbon disulfide as the extractant. Because only one milliliter of solvent is used, no further concentration of solvent is required. The authors found 15 minutes distillation/extraction time sufficient for recovery of nonpolar compounds, e.g., mono and sesquiterpenes, while one hour was required for oxygenated and higher boiling compounds. This apparatus was evaluated by Nunez and Bemelmans (28) for low levels of aroma compounds in water. They reported that results were satisfactory for volatile levels greater than 1 ppm.

Both heavier and lighter than water versions of the steam distillation apparatus are commercially available (Alltech, Deerfield, IL; Chrompack, Bridgewater, NJ). Diagrams of the two versions are shown in Figure 11.

Simultaneous Distillation and Adsorption   Jennings and Nursten (29) suggested steam distilling dilute aqueous solutions, followed by passing the condensate over activated carbon. The volatile aromatic compounds were desorbed with carbon disulfide. A simultaneous distillation and adsorption (SDA) version of this procedure has recently been published (30), and the apparatus is shown in Figure 12. Ten mg of activated carbon (60-80 mesh) are used in the trapping tube and typical distillation times are 40 minutes. The researchers found methylene chloride and carbon disulfide to be the most effective desorption solvents. The procedure was applied to both a model system as well as Perilla Leaves. The chromatograms obtained compared favorably with those generated by an SDE apparatus.

Direct Analysis of Aqueous Essences

The most simple sample preparation technique is direct gas chromatographic injection of an aqueous essence on a bonded fused silica column. This technique may be employed when aqueous distillates are available. For example, Moshonas and Shaw (31) described a method for the analysis of aroma constituents of natural fruit essences. The essences were collected from the first stage of an evaporator and the essence injected directly into a capillary gas chromatograph.

They stressed that the availability of fused silica capillary columns coated with cross-linked non-polar liquid phases permitted development of this technique. Such columns resist the deterioration which was previously encountered with aqueous samples. These authors applied this technique to several citrus essences as well as to fruit essences such as grape, apple and strawberry.

Adsorption of Organics from Aqueous Solutions

If particulates can be removed from an aqueous sample, it is possible to concentrate the volatile aromatic compounds directly on

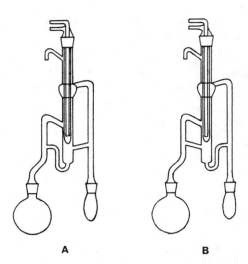

A   **B**

Figure 11. Micro steam distillation-extraction apparatus.
A: For solvents lighter than water. B: For solvents heavier
than water. (Reproduced with permission from Chrompack, Inc.,
Bridgewater, N.J.)

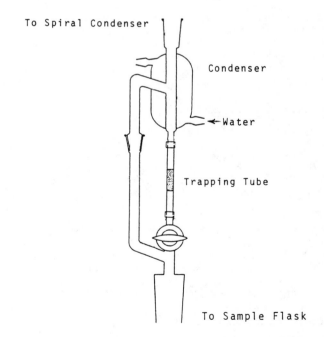

Figure 12. Simultaneous steam-distillation adsorption (SDA)
head. (Reproduced with permission from Ref. 30. Copyright
1981, Walter de Gruyter & Co.).

adsorbents. Clarification can be achieved by centrifugation, ul-
trafiltration or reverse osmosis. The clarified aqueous material
can be concentrated on mini columns of macroreticular resins, e.g.,
XAD-4 or on activated carbon as described by Tateda and Fritz
(32). These investigators used disposable pipets as their sorption
columns, with column dimensions of 1.2 mm by 2.5 cm. Typical aqueous sample volumes are about 100 ml and flow rates are 1 ml/min.
Desorption is achieved with less than 100 microliters of an organic
solvent. They found acetone to be an effective solvent for XAD-4
desorption, and carbon disulfide to be most satisfactory for desorption from activated carbon. The purpose of their study was to
investigate typical organic contaminants found in drinking water,
but they also investigated alcohols, esters and carbonyl compounds,
which are typical aroma chemicals.

In a similar vein, Parliment (33) used reverse phase C18 adsorbents to concentrate and fractionate low levels of volatile organic compounds from dilute aqueous streams. The aqueous phase
must be particulate free to prevent fouling of the adsorbent, but
soluble solids are not necessarily a problem. For example, a commercial cola beverage containing caramel color, caffeine, phos-
phoric acid and sweetener was passed over a reverse phase column.
Desorption with acetone produced an aroma concentrate which could
be analyzed by gas chromatography.

Extraction of Volatile Organics from Aqueous Solutions

Certain aqueous materials can be extracted directly with a solvent
to prodice an aroma concentrate. This technique has found utility
in studies on wine aroma (34). The investigators studied model
wine systems containing 12% ethanol and 4.4% sucrose. Extractions
were performed manually, and the organic solutions were concentrated under vacuum. They concluded that Freon 11 (bp 24°) was
the solvent of choice; if the organic essence is to be stored for
an extended period of time, they recommended methylene chloride as
the solvent. The presence of sugar did not create any difficulties.
More recently, a 2:1 mixture of pentane:methylene chloride has
been used to extract the aromatics from white wines (35). These
investigators used a continuous liquid/liquid extractor for eight
hours to remove the aromatics; the organic phase was then concentrated to 1 ml volume using a Vigreux column. Such a procedure
should have application in studies on fermentation broths.
We have found direct extractions of biologically-generated
aromas can be performed quite rapidly by a procedure which has
recently been described (36). The extraction device is a Mixxor
(Lidex Tech., Bedford, MA), and is shown in Figure 13. A quantity
of aqueous sample (e.g., 10 mls) is placed in the lower vessel, and
a low density solvent is added. The sample is mixed by moving the
piston up and down. Phases are allowed to separate, and the piston
is slowly depressed forcing the solvent into the upper collecting
chamber. We have found that if only a few hundred microliters of
solvent are used it is possible to force the solvent into the axial
chamber (C). The organic sample can be removed from (C) with a
syringe for gas chromatographic analysis. This extraction technique is rapid and produces an efficient extraction. The procedure

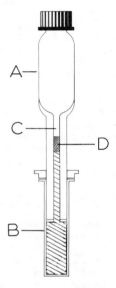

Figure 13. Mixxor extractor, used to extract aqueous
solutions with minimal amount of solvent.

was applied to a model system of ethyl esters in water. Detection limits were less than 0.2 ppm and recoveries were generally over 90%. The procedure was applied to several plant materials such as fennel, rosemary and celery seed with satisfactory results.

Conclusion

There are many sample preparation techniques which one can choose. Choice will be determined by the requirements of the study. Questions to be considered include: Is high sample throughput required? Is high precision desired? How much time is available for the study? Does the sample survive $100^{\circ}C$? What instrumentation and apparatus are available?

Each method described has advantages and disadvantages. No single method is ideal, and the choice of technique should be based on informed decisions.

Acknowledgment

The author wishes to thank Judy Schinkel for secretarial assistance.

Literature Cited

1. van Straten, S.; Maarse, H. "Volatile Compounds in Food." Division for Nutrition and Food Research TNO, Zeist, The Netherlands, 1983.
2. Weurman, C. J. Agric. Food Chem. 1969, 17, 370.
3. Schreier, P. In "Chromatographic Studies of Biogenesis of Plant Volatiles"; Huthig: New York, 1984; pp. 1-32.
4. Maarse, H.; Betz, R. In "Isolation and Identification of Volatile Compounds in Aroma Research"; Akademie - Verlag: Berlin, 1981; pp. 1-59.
5. Teranishi, R.; Flath, R.; Sugisawa, H. In "Flavor Research - Recent Advances"; Marcel Dekker, Inc.: New York, 1981, pp. 11-51.
6. Morton, I.; MacLeod, A. In "Food Flavors. Part A. Introduction"; Elsevier: New York, 1982; pp. 15-48.
7. Charalambous, G. "Analysis of Foods and Beverages. Headspace Techniques"; Academic Press: New York, 1978.
8. Jennings, W. In "Gas Chromatography with Glass Capillary Columns"; Second Edition. Academic Press: New York, 1980; pp. 183-200.
9. Reineccius, G.; Anandaraman, S. Food Sci. Technol., 1984, 11, 195.
10. Leahy, M.; Reineccius, G. In "Analysis of Volatiles. Methods. Applications"; Schreier, P., Ed.; de Gruyter: New York, 1984; pp. 18-47.
11. Schreier, P.; Idstein, H. Z. Lebensm. Unters. Forsch. 1985, 180, 1.
12. Jennings, W.; Filsoof, M. J. Agric. Food Chem. 1977, 25, 440.
13. Poll, L.; Flink, J. Food Chem. 1984, 13, 193.
14. Kolb, B.; Pospisil, P.; Borath, T.; Auer, M. J.H.R.C.&C.C. 1979, 2, 283.

15. Jennings, W. In "Gas Chromatography with Glass Capillary Columns"; Second Edition. Academic Press: New York, 1980; pp. 54-56.
16. Chang, S.; Vallese, F.; Hurang, L.; Hsieh, A.; Min, D. J. Agric. Food Chem. 1977, 25, 450.
17. Jennings, W. J.H.R.C.&C.C. 1979, 2, 221.
18. Rapp, A.; Knipser, W. Chromatographia 1980, 13, 698.
19. Guichard, E.; Ducruet, V. J. Agric. Food Chem. 1984, 32, 838.
20. Williams, A.; May, H.; Tucknott, O. J. Sci. Fd. Agric. 1978, 29, 1041.
21. Lund, E.; Dinsmore, H. In "Analysis of Foods and Beverages. Headspace Techniques."; Charalambous, G., Ed.; Academic Press: New York, 1978; pp. 135-186.
22. Withycombe, D.; Mookherjee, B.; Hruza, A. In "Analysis of Foods and Beverages. Headspace Techniques"; Charalambous, G., Ed.; Academic Press: New York, 1978; pp. 81-94.
23. Schaefer, J. In "Flavor '81"; Schreier, P., Ed.; de Gruyter: New York, 1981; pp. 301-314.
24. Parliment, T. In "Semiochemistry: Flavors and Pheromones"; Acree, T.; Soderlund, D., Eds.; de Gruyter: New York, 1985; pp. 181-202.
25. Nickerson, G.; Likens, S. J. Chromatogr. 1966, 21, 1.
26. Schultz, T.; Flath, R.; Mon, R.; Eggling, S.; Teranishi, R. J. Agric. Food Chem. 1977, 25, 446.
27. Godefroot, M.; Sandra, P.; Verzele, M. J. Chromatogr. 1981, 203, 325.
28. Nunez, A.; Bemelmans, J. J. Chromatogr. 1984, 294, 361.
29. Jennings, W.; Nursten, H. Anal. Chem. 1967, 39, 521.
30. Sugisawa, H.; Hirose, T. In "Flavor '81"; Schreier, P., Ed.; de Gruyter: New York, 1981; pp. 287-299.
31. Moshonas, M.; Shaw, P. J. Agric. Food Chem. 1984, 32, 526.
32. Tateda, A.; Fritz, J. J. Chromatogr. 1978, 152, 329.
33. Parliment, T. J. Agric. Food Chem. 1981, 29, 836.
34. Cobb, C.; Bursey, M. J. Agric. Food Chem. 1978, 26, 197.
35. Moret, I.; Scarponi, G.; Cescon, P. J. Sci. Food Agric. 1984, 35, 1004.
36. Parliment, T. Perfum. Flavorist 1986, in press.

RECEIVED January 15, 1986

Advances in the Separation of Biological Aromas

G. R. Takeoka, M. Guentert, C. Macku, and W. Jennings

Department of Food Science and Technology, University of California, Davis, CA 95616

The potential for achieving improved gas chromatographic separations of biological aroma mixtures has been enhanced by developments in column technology, and in methods of sample injection and sample preparation. Both smaller and larger diameter columns, coated with a variety of bonded stationary phases in a range of film thicknesses, have become available. New and simplified methods for the direct injection of headspace samples can eliminate the errors that may accompany conventional methods of purge and trap. Combining these advances to follow the production of volatile compounds by the ripening banana indicates smooth, progressive increases in the concentrations of those volatiles. There was no evidence of the cyclic phenomena that had been reported earlier.

Aroma compounds frequently exist as trace components, dispersed in complex systems that complicate their isolation and identification. While many fields benefitted from the invention of gas chromatography, it has been especially valuable to those in flavor chemistry. Gas chromatography is not a mature science; in many ways, it is still an art, and developments in the field continue. While these developments offer some exciting new opportunities, they can also be employed improperly with adverse effects on the analytical results.

Three of the more recent developments that have alleviated many of the aroma chemist's analytical handicaps are:

1) developments in new stationary phases;
2) developments in specialized columns; and
3) developments in sample introduction.

Although not offered in every combination, today's state-of-the-art columns are available in a variety of lengths, and with internal diameters ranging from 0.05 mm to 0.75 mm, and with cross-linked chemically bonded films of stationary phase that range from 0.1 to 8 um in thickness.

0097–6156/86/0317–0053$06.00/0
© 1986 American Chemical Society

Theoretical Considerations

The driving force behind the creation of these many choices emerges from the resolution equation, which in the following form can be used to estimate the number of theoretical plates required for a given degree of separation of any two solutes (1):

$$n_{req} = 16 \ Rs^2 \ [\alpha/(\alpha-1)]^2 \ [(k+1)/k]^2 \qquad [1]$$

Rs represents the degree of separation desired (1.5 corresponds to baseline separation);

α represents the relative retention of the two solutes and is largely governed by the degree of their interaction with the stationary phase;

k is the partition ratio of the second solute, which for a given stationary phase varies inversely with temperature and directly with the stationary phase film thickness.

Columns whose efficiencies approach or equal the theoretical limit (1):

$$h_{theor. \ min} = r[(11k^2+6k+1)/3(1+k)^2]^{1/2} \qquad [2]$$

are not unusual today; as a result, there are essentially two choices for increasing n: longer columns, or columns of smaller diameter. For a given set of conditions, both require a higher pressure drop through the column; higher pressure drops result in steeper van Deemter curves, and diminish the prospects for maintaining good separation over a broad range of the chromatogram. For most of us, long columns should be avoided, not only for this reason we have just explored, but also because both the optimum, and the optimum practical gas velocities vary inversely with column length (2). Shorter columns of smaller diameter find some utility in producing separations that are roughly equivalent to those obtained on "standard" columns, but in shorter times. Figure 1 shows a separation of a lemon oil on fused silica columns measuring (numbering from the top), 1) 30 m x 0.25 mm; 2) 10 m x 0.1 mm; 3) 5 m x 0.1 mm; 4) 2.5 m x 0.1 mm; and 5) 4 m x 0.05 mm; all columns were coated with DB-1 (dimethyl polysiloxane) with film thicknesses as detailed in the figure legend.

Column diameters

Small diameter columns require some modification of our normal analytical procedures. The minimum split ratio is usually about 1:1000; because the flow rate through the column is so limited (ca. 50 uL/min). At lower split ratios, the sample resides too long in the injector, depositing an elongated band on the column and generating widened peaks. Figure 2 explores the result that this has on the resolution achieved between 2-methyl-1-butanol and 3-methyl-1-butanol. The best results were achieved with a split ratio approaching 1:2000, clearly demonstrating the effect that a short starting band can exercise on component separation.

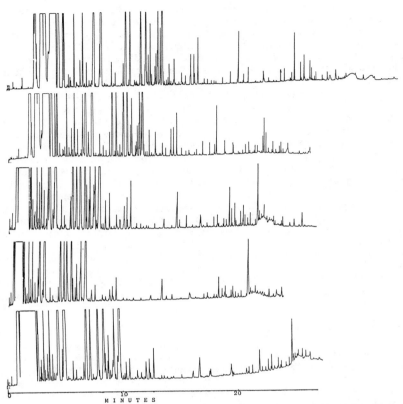

Figure 1. Chromatograms produced by split injections of a lemon oil on columns of different dimensions. All columns coated with DB-1. Column dimensions (from top to bottom): 1) 30 m x 250 um; 2) 10 m x 100 um; 3) 5 m x 100 um; 4) 2.5 m x 100 um; 5) 4 m x 50 um.

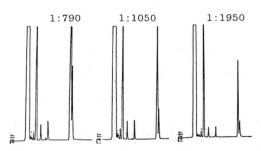

Figure 2. Effect of split ratio on the resolution of isoamyl and active amyl alcohol in hexane. Column, 6 m x 50 um DB-1, d_f 0.2 um; 40°C isothermal; hydrogen carrier at 43.5 cm/sec (51.3 uL/min).

There have also been developments in the reverse direction, i.e. larger diameter columns (3). Small diameter open tubular columns are restricted to lower flow volumes of carrier gas; low flows require good instrumental design, and improved operator techniques. The continued popularity of packed columns is largely due to the fact that they offer a more "user friendly system", although at the price of inferior separations in longer times that are qualitatively and quantitatively less reliable. While larger diameter open tubular columns are not new (e.g. 4), it is now possible to obtain large diameter open tubular columns of fused silica, that reflect state-of-the-art deactivation procedures, whose stationary phases are both cross-linked and surface-bonded. In almost every case, these can be directly substituted for the packed column, and operated under the same conditions of temperature, carrier flow, and injection size to produce a superior analysis in a shorter time (Figure 3). Under these packed-column carrier flow conditions, the column is operating far above its optimum velocities, and the chromatographic results can be further improved by decreasing the gas velocity. At velocities approaching u_{opt}, the addition of a make-up gas may be necessary (1). The 0.53 mm diameter column represents what might be called "the point of diminishing returns" for large diameter open tubular columns. If the diameter is increased still further (which has been done in conventional glass, but not in fused silica columns), the efficiency loss become excessive (see Equation [2], above).

Stationary Phase Developments

There have also been developments in stationary phase formulations. Wines can serve as good examples; Figure 4a shows a Freon 11-CH_2Cl_2 (9:1) extract of wine injected on a 30 m x 0.25 mm fused silica column coated with DB-1. In spite of several areas dominated by atypically shaped peaks, this is an acceptable separation for a real-world sample. In this case, the leading fronts on the malformed peaks testify not to interaction between the column and those solutes, but rather to a lack of interaction between the solutes and the stationary phase, or a solute-stationary phase incompatibility. The offending peaks represent free acids, which are essentially insoluble in the dimethyl polysiloxanes. As a result, the acids are forced to chromatograph under overload conditions; denied access to the stationary phase, their movement along the film surface is vapor-pressure limited (5). Figure 4b shows this same extract on a bonded polyethylene glycol column. Compound identifications are shown in Table I.

This particular stationary phase, the DB-Wax, maintains its chromatographic properties at temperatures as low as 0°C. By starting at a temperature of 40°C, the front end is separated reasonably well; if better separation in this area were required, there would be no problem in starting at a still lower temperature. The free acids now look acceptable; there are at least two areas that could be improved, around the bases of the 2-phenylethanol (peak 6) and monoethyl succinate (peak 12). It is not clear from this chromatogram whether the problem is an interaction of those solutes with the

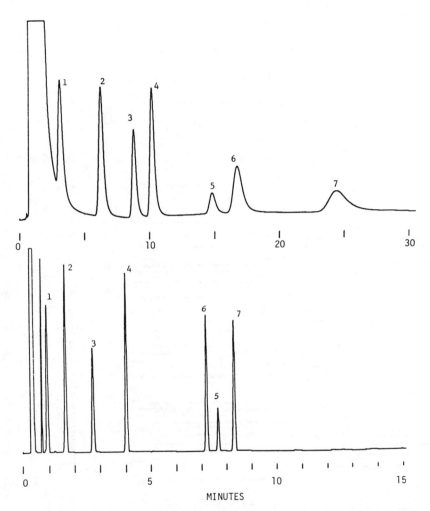

Figure 3.   Comparison of packed and Megabore columns, EPA Method
602. FID, and identical injections in both cases.   Top, 6 ft x
1/8 in stainless steel column packed with 5% AT-1200 + 1.75%
Bentone 34 on 100/120 Chromosorb WAW; 50°C (4 min), 8°/min to
100°C; 20 min hold.   Helium carrier at 50 mL/min.   Bottom, 30 m
x 530 um Megabore column, coated with a 1 um film of DB-Wax;
45°C (2 min), 8°/min to 100°C.   Helium carrier at 30 mL/min.
Solutes: 1) benzene; 2) toluene; 3) ethylbenzene; 4) chloroben-
zene;   5) 1,3-dichlorobenzene;   6) 1,4-dichlorobenzene;   7) 1,2-
dichlorobenzene.   Chromatogram courtesy of J & W Scientific,
Inc.

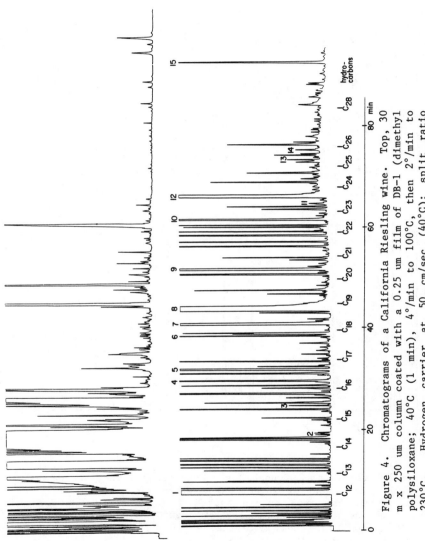

Figure 4. Chromatograms of a California Riesling wine. Top, 30 m x 250 um column coated with a 0.25 um film of DB-1 (dimethyl polysiloxane; 40°C (1 min), 4°/min to 100°C, then 2°/min to 230°C. Hydrogen carrier at 50 cm/sec (40°C); split ratio 1:30. Bottom, 30 m x 250 um column with a 0.25 um film of DB-Wax (bonded polyethylene glycol). See Table I for solute identification and DB-Wax operating conditions.

column, or incomplete separation of other solutes on the backsides
of those peaks.

Table I. Compounds Identified in California Riesling Wine (Fig. 4b)

1) 2-methyl-1-butanol + 3-methyl-1-butanol (1208)	9) octanoic acid
2) furfural (1451)	10) decanoic acid
3) linalool (1550)	11) 4,5-dihydroxyhexanoic acid
4) ethyl decanoate (1633)	γ-lactone (2335)
5) diethyl succinate (1671)	(sherry lactone)
6) 2-phenylethyl acetate	12) monoethyl succinate
(1791)	13) N-(2-phenylethyl)acetamide (2543)
7) hexanoic acid	14) ethyl pyroglutamate (2571)
8) 2-phenylethanol (1886)	15) 2-(4-hydroxyphenyl)ethanol (tyrosol)

Conditions: 30 m x 320 um DB-WAX column, 40°C for 1 min, then
2°C/min to 200°C. H_2 @ 50 cm/sec; split ratio 1:20. Parenthetical
numbers are Kovats retention indices.

In other cases, stationary phases have been tailored to achieve
specific separations. In one case, a new stationary phase was designed to achieve the separation of a particular mixture of volatile
priority water pollutants whose separation has posed a real problem
(6,7). A serendipitous finding was that this new stationary phase,
DB-1301, also promises to be very useful for the separation of some
chlorinated pesticides. Nor have we reached the end of this road;
an immobilized form of 2330, with utility for those interested in
dioxins, in positional isomers of the fatty acids, and in other
challenging separations, will soon be available. Fused silica
columns with bonded particulate materials, reminiscent of the old
PLOT-type columns, are also available. The primary utility of the
PLOT-type columns currently available is for fixed gas analysis, but
newer types on the horizon will permit a choice of adsorptive-type
separations, partition-type separations, or a combination of both.

Effect of Film Thickness

Another development, occasionally misrepresented and offering opportunities for confusion, is that of the thick-film columns. Commercially available thick film columns were formerly limited to about
1 um films, but today we have columns with stationary phase film
thicknesses of 3, 5, and 8 um. According to Equation [1], above,
the magnitude of the partition ratio k influences the number of
theoretical plates required to separate two solutes. The magnitude
of the multiplier $[(k+1)/k]^2$ is approximately 10,000 for k = 0.01,
about 36 for k = 0.2, and 1.4 for k = 4. Under a given set of
conditions, k increases in direct proportion to the increase in
stationary phase film thickness, d_f. Hence, the great utility of
the very thick film columns is for the separation of very low k
solutes; the number of theoretical plates required to separate two
higher k solutes is affected only slightly, and that benefit is
usually offset by the other disadvantages conferred by these very

thick film columns (8). Figure 5 shows Muscat wine headspace injections; although not clearly evident in the figure, peaks 6, 7, and 8 were (barely) baseline resolved on the standard 0.25 um film column shown in the top chromatogram. In an effort to increase the resolution of those solutes, a 1.0 um column was substituted, resulting in co-elution of solutes 6 and 7 (bottom); with partition ratios, k, of this magnitude, the decrease in the $[(k+1)/k]^2$ multiplier failed to compensate for the decrease in n. This sample would fare even worse on an ultra-thick film column. One of the more novel uses of these latter has been in the detection of water contaminants by direct water injection, where the ultra-thick film has been used to retain halogenated contaminants until the electron capture detector recovered from the rapidily eluted water peak (9,10). The technique, which is of course not limited to this mode of detection, should also have utility for the detection of volatile aroma compounds in aqueous solution.

It is interesting that while h_{min} is smaller for low-k solutes in open tubular columns of conventional film thickness, in these very thick film columns the reverse is true: h_{min} is now smaller for high-k solutes (Figure 6).

Column Bleed

Another point of contention relates to column bleed; in general, thick film columns have higher bleed rates, and some argue that the bleed rate is directly proportional to the stationary phase film thickness (11). This generalization was once correct, but columns prepared with high purity polymers that are extremely clean, and bonded to properly deactivated tubing, do not exhibit this limitation unless the columns are abused. Some bleed problems can be associated with residues from "dirty" samples that remain on the column, while others are attributable to

1) low molecular weight fragments in the polymer;
2) low molecular weight fragments resulting from phase depolymerization caused by residual traces of catalyst; and to
3) stationary phase degradation due to reaction with an improperly deactivated support; or
4) stationary phase degradation due to reaction with silanols produced by high temperature exposure of the stationary phase to molecular oxygen (12).

Bleed attributable to the first two causes would correlate directly with the mass of the stationary phase, but bleed attributable to the latter two would not. The bleed rates from columns containing 5 um films of bonded stationary phases from polymers selected from a number of in-house syntheses can be roughly equivalent to those of "standard" film columns, and do not correlate with the mass of stationary phase in the column (13).

Sample Introduction

The other area of interest to the aroma chemist is that of sample introduction. One of the major advantages cited for the large

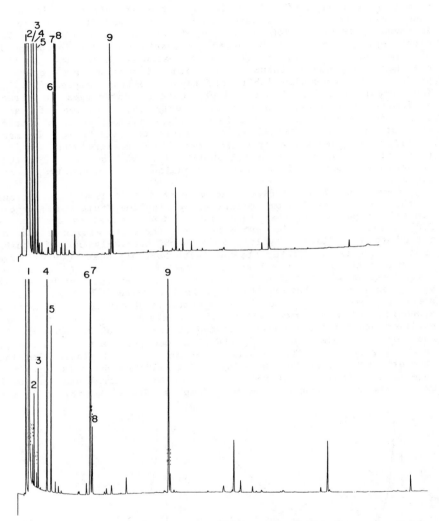

Figure 5.   On-column headspace injections (500 uL) of a California Muscat wine; effect of stationary phase film thickness. Top, 30 m x 250 um column, 0.25 um film of DB-1. Bottom, identical column, but with a 1.0 um film of DB-1. The columns were run under identical conditions: 30°C (2 min), 2°/min to 40°C, then 4°/min to 170°C; hydrogen carrier at 50 cm/sec (30°C). Note that separation of the low-k solutes (peaks 1-5) are improved in the lower chromatogram, but resolution of higher-k solutes (note especially peaks 6-8) have been adversely affected. In addition, elution times are much longer (note peak 9).

diameter open tubular column is that the injection hardware is
essentially the packed column injector, although results can be
improved by incorporation of an "anti-flashback" liner (14,15). The
simple split injector can be quite satisfactory for the introduction
of some samples, including many of the essential oils, into smaller
diameter open tubular columns. On-column injections are generally
preferred for thermally labile and higher boiling samples (15,16).
The on-column injector also offers certain advantages in direct
headspace injections, where a relatively large volume of vapor
sample is injected inside the fused silica column, while a section
of that column is chilled in a suitable coolant (17). In many
cases, direct headspace injection can yield results that are, in
terms of both quantitative reliability (17) and sensitivity (10),
superior to those obtained with conventional purge and trap techniques.

     Some years ago, studies that attempted to follow fluctuations
in the production of aroma volatiles by ripening fruit were largely
limited to purge and trap techniques. One such study on bananas
(18) indicated that the production of these volatiles could be
fitted to two diurnal cycles, which were precisely out of phase.
Similar results were later obtained with Bartlett pears (19). Purge
and trap techniques are subject to a variety of errors (20), and we
have long been interested in better techniques that might either
verify or correct that earlier hypothesis. Figure 7 shows an air-
swept ripening chamber designed to permit inclusion of internal
standards.   The internal volume of the chamber was ca. 10 L, and
purified breathing air was introduced at the rate of 250 mL/min. In
the initial trials, massive fluctuations occurred in the area of the
internal standard peak.   Insertion of an elongated glass "blending
chamber" preceding the sampling point eliminated this problem;
injection-to-injection reproducibility of the internal standard peak
was in the neighborhood of ± 3%. Samples were withdrawn at 30 min

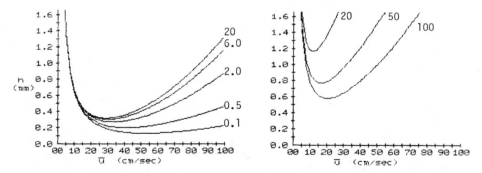

Figure 6.   Computer-generated van Deemter curves for solutes of
the indicated partition ratio (k) on (left) a "normal" 30 m x
320 um column, with a 0.25 um film of stationary phase, and
(right), the same column with a 5 um film of stationary phase.
Note that h varies directly with k on the left, and inversely
with k on the right.

intervals, and analyzed by direct headspace injection, using liquid nitrogen to achieve a distribution constant focus (15). All volatiles appeared to exhibit steady and progressive increases, although at different rates, to a plateau level; the earlier conclusions were very probably due to sampling errors engendered by the purge and trap methodologies employed.

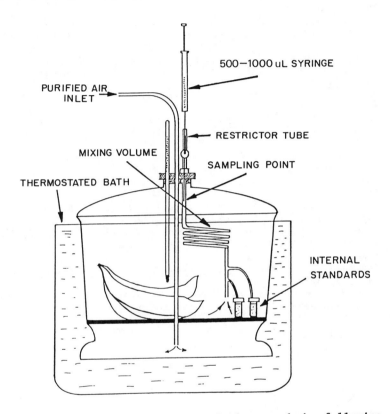

Figure 7.   Headspace sampling chamber used in following the production of aroma volatiles by ripening fruit.   See text for discussion.

Literature Cited

1.   Jennings, W.   In "Gas Chromatography With Glass Capillary Columns", second edition; Academic Press: New York, 1980.
2.   Ingraham, D. F.; Shoemaker, C. F.; Jennings, W.   J. High Res. Chromatogr. 1982, 5, 227.
3.   Ryder, B. L.; Phillips, J.; Plotczyk, L. L.; Redstone, M. Paper 497, Pittsburgh Conference on Analytical Chemistry and Applied Spectroscopy, 5–9 March 1984; Atlantic City, NJ.

4. Mon, T. R.; Flath, R. A.; Forrey, R. R.; Teranishi, R. In "Advances in Chromatography"; Zlatkis, A., Ed.; Preston Technical Abstracts, Evanston IL. 1967; pp. 30-33.
5. Yabumoto, K.; Ingraham, D. F.; Jennings, W. J. High Res. Chromatogr. 1980, 3, 248.
6. Mehran, M. F.; Cooper, W. J.; Jennings, W. J. High Res. Chromatogr. 1984, 7, 215.
7. Mehran, M. F.; Cooper, W. J.; Lautamo, R.; Freeman, R. R.; Jennings, W. J. High Res. Chromatogr. 1985, 8, (in press).
8. Jennings, W. Amer. Lab. 1984, 16, 15.
9. Grob, K.; Grob, G. J. High Res. Chromatogr. 1983, 6, 133.
10. Mehran, M. F.; Cooper, W. J.; Jennings, W. J. Chromatogr. Sci. 1986, 24 (in press).
11. McNair, H. M.; Ogden, M. W.; Hensley, J. L. Amer. Lab. 1985, 17, 15.
12. Jennings, W. J. Chromatogr. Sci. 1983, 21, 337.
13. Guthrie, E. (J & W Scientific, Inc.) Personal communication, 1985.
14. Freeman, R. R. (J & W Scientific, Inc.) Personal communication, 1985.
15. Jennings, W.; Mehran, M. F. J. Chromatogr. Sci. 1986, 24 (in press).
16. Jennings, W. In "Sample Introduction in Capillary Gas Chromatography"; Sandra, P., Ed.; Huethig: Heidelberg, 1984.
17. Takeoka, G.; Jennings, W. J. Chromatogr. Sci. 1984, 22, 177.
18. Tressl, R.; Jennings, W. J. Agric. and Food Chem. 1972, 20, 189.
19. Jennings, W.; Tressl, R. Chem. Mikrobiol. Technol. Lebensm. 1974, 3, 52.
20. Jennings, W.; Rapp, A. "Sample Preparation for Gas Chromatographic Analysis"; Huethig: Heidelberg, 1983.

RECEIVED February 19, 1986

Changes in Aroma Concentrates during Storage

G. R. Takeoka[1], M. Guentert[1], Sharon L. Smith[2], and W. Jennings[1]

[1]Department of Food Science and Technology, University of California, Davis, CA 95616
[2]IBM Instruments Inc., P.O. Box 332, Danbury, CT 06810

Kiwifruit (Actinidia chinensis Planch.) concentrates were prepared by vacuum distillation, followed by continuous liquid-liquid extraction and concentration of the extracts. The concentrates underwent various changes when stored at -10°C. The artifacts produced were analyzed and characterized by capillary GC, GC/MS and GC/FTIR. Carboxylic acids, probably produced by free-radical oxidation, comprised the major portion of the artifacts. Possible mechanisms of artifact formation and methods to minimize their production are discussed.

The formation of artifacts during sample preparation has been addressed by various researchers (1-4). A less frequently examined problem is the formation of artifacts during storage of aroma concentrates. DeMets and Verzele (5) noted the decrease of myrcene and the production of new volatiles, mainly attributable to the degradation of bitter acids, during the storage of hops. Badings (6) reported that storage at -40°C was required to keep a concentrate of cold-stored butter volatiles unchanged for 12 hours. The influence of storage temperature was also observed by Kepner et al. (7) who found that while the amount of germacrene D in Douglas fir oil did not change for one year when stored at -20°C, it disappeared within 6 weeks when the oil was stored at 5°C. They postulated that the loss was not a thermal isomerization but rather interaction with another component in the oil. The autoxidation of caryophyllene in Cannabis oil has been reported by Paris (8). Similarly, a marked decrease in caryophyllene level along with a corresponding increase in caryophyllene oxide level was observed in rough lemon (Citrus jambhiri Lush.) leaf oil after storage at 9°C for 7 months (9). The instability of the distilled oil was attributed to the removal of nonvolatile natural antioxidants which are normally present in most cold-pressed citrus oils.

The formation of artifacts in aroma concentrates is influenced by a number of factors including storage temperature, atmosphere, the presence of pro-oxidants and/or antioxidants, the concentration

0097-6156/86/0317-0065$06.00/0

and stability of the volatiles present in the concentrate, and
time. During our recent investigation of the volatile constituents
of kiwifruit (Actinidia chinensis Planch.) a rapid change in
composition of the concentrated extracts during storage was
observed. The purpose of this study was to identify the artifacts
formed during storage and to postulate mechanisms for their
formation.

Experimental

Sample Preparation. Kiwifruit of the major commercial variety,
Hayward, was obtained locally. The fruit was allowed to ripen at
room temperature at which point the soluble solids content of the
juice was 16-17%. After separation of the the skin, the pulp was
gently blended in a Waring blender taking care not to fracture the
seeds. The blended pulp was immediately subjected to vacuum steam
distillation.

Isolation of Volatiles. An aliquot of blended pulp (1.2 kg) was
diluted with distilled water (700 mL) in a 3-L three-neck flask and
vacuum distilled (25-30°C/1 mm Hg). Distillation continued for 2.5
to 3 h yielding approximately 500 mL of distillate which was
collected in two liquid nitrogen cooled traps. A total of 3.6 kg
of fruit pulp was distilled in three batches. The distillates were
combined and immediately frozen until use. The combined distillate
was extracted in 250 mL batches for 20 h with 60 mL trichlorofluoromethane (Freon 11, b p 23.8°C) using a continuous liquid-
liquid extractor. The trichlorofluoromethane was distilled through
a 120 x 1.3 cm glass distillation column, packed with Fenske
helices, prior to use. Each extract was carefully concentrated to
approximately 100 µL by distillation of solvent using a Vigreux
column (16 cm), and a maximum pot temperature of 30°C.

Gas Chromatography. A Hewlett-Packard 5880A gas chromatograph with
a FID, equipped with a 30 m x 0.32 mm i d DB-WAX column (d_f = 0.25
µm, bonded polyethylene glycol phase, J&W Scientific) was employed.
The column temperature was programmed as follows: 30°C (2 min isothermal), to 38°C at 1°C/min, then to 180°C at 2°C/min and held for
20 min. Hydrogen carrier gas was adjusted to an average linear
velocity of 49.7 cm/sec (30°C). The injector and detector were
maintained at 225°C. A modified injection splitter (J&W Scientific) was used at a split ratio of 1:30.

Gas Chromatography-Mass Spectrometry. A Finnigan MAT 4500 series
quadrupole gas chromatograph/mass spectrometer/data system equipped
with the same capillary column and using the same temperature
program described in the previous section was employed. Helium was
used as the carrier gas at an average linear velocity of 47.7
cm/sec (30°C). Injector temperature was 220°C, and the ion source
temperature was 180°C. The outlet end of the fused silica column
was inserted directly into the ion source block, which was
maintained at approximately 180°C.

Gas Chromatography-Fourier Transform Infrared Spectroscopy. An IBM
Instruments 9630 gas chromatograph was coupled with an IBM

Instruments IR-85 Fourier Transform infrared spectrometer, through
an IBM GC-IR interface. The interface consisted of a gold-coated
Pyrex light-pipe with potassium bromide windows. A scan rate of 6
scans/sec and a spectral resolution of 8 cm^{-1} were used for data
acquisition. Samples were introduced into the system via splitless
injections. A fused silica capillary column, 30 m x 0.32 mm i d
DB-WAX (d_f = 1.0 μm), was employed with the outlet end connected
directly to the GC-IR light-pipe entrance. Helium was used as the
carrier gas at an average linear velocity of 41.4 cm/sec (35°C).
No make-up gas was employed in the system. The column temperature
was programmed from 35°C to 180°C at 2°C/min. The GC-IR light-pipe
assembly was maintained at 170°C.

Results and Discussion

In our recent studies of kiwifruit (<u>Actinidia</u> <u>chinensis</u> Planch.)
volatiles (<u>10</u>), we noticed a rapid change in the composition of the
concentrated extracts during storage in the freezer at -10°C. The
artifacts formed during storage were analyzed and characterized by
capillary GC, GC/MS and GC/FTIR.

   To understand the sensitivity of the extracts to artifact
formation it is informative to review the volatiles in kiwifruit.
Quantitatively, peroxidation products of unsaturated fatty acids
(<u>11</u>, <u>12</u>), which include (E)-2-hexenal (77.87%), (E)-2-hexen-1-ol
(<u>6.80%</u>), 1-hexanol (3.40%), hexanal (1.78%), (Z)-2-hexenal (0.87%),
(E)-3-hexen-1-ol (0.32%) and (Z)-3-hexen-1-ol (0.17%), constitute
over 90% of the total volatiles. Other major constituents include
the esters, methyl butanoate (2.54%) and ethyl butanoate (3.52%).
The presence of large amounts of saturated and unsaturated
aldehydes in the extract is noteworthy since they are quite
susceptible to free-radical oxidation. We therefore expected that
at least some of the artifacts were the products of autoxidation.

   It has been suggested that autoxidation of saturated fatty
acids and aldehydes occurs through a free-radical mechanism (<u>13</u>,
<u>14</u>). Supporting evidence of a radical chain mechanism was provided
by Palamand and Dieckmann (<u>15</u>) who studied the autoxidation of
hexanal. The reaction involves peroxycarboxylic acid as an
intermediate (<u>16</u>) and probably proceeds via the mechanism shown in
Figure 1.

<u>Comparison of Fresh and Stored Samples</u>. Figure 2 shows GC/FID
chromatograms of a freshly prepared kiwifruit extract (top), an
extract which had been stored in the freezer for three months
(middle) and an extract which had been stored in the freezer for
four months (bottom). The artifacts which developed during storage
are numbered in the middle and bottom chromatograms. Table 1 lists
these artifacts. As can be seen in Figure 2, there is a dramatic
decrease in the levels of hexanal, (Z)-2-hexenal, (E)-2-hexenal,
(E)-3-hexenol, (Z)-3-hexenol and (E)-2-hexenol upon storage. There
is a corresponding increase in the level of artifacts which
generally elute at longer retention times than the native
constituents. Mass spectral identifications were verified by
comparison with Kovats retention indices of authentic reference
standards.

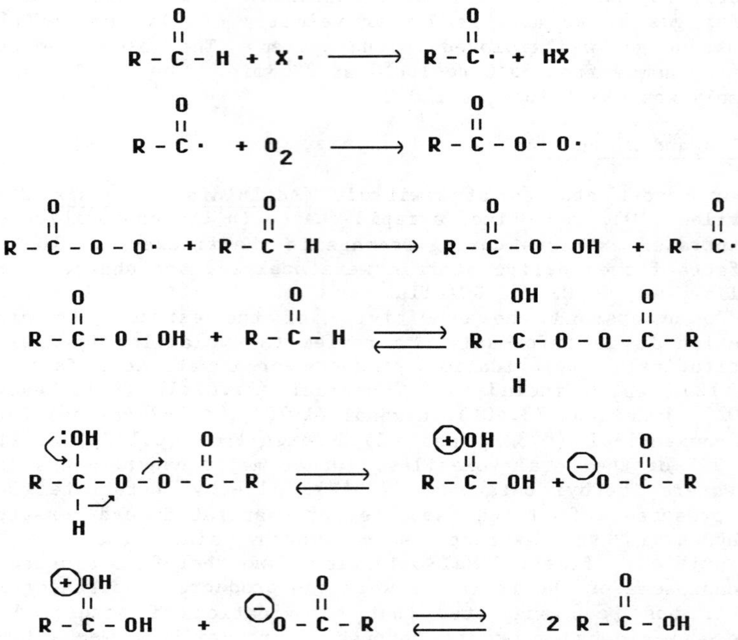

$$\underset{\substack{\text{O}\\ \|}}{\text{R}-\text{C}-\text{H}} + \text{X} \cdot \longrightarrow \underset{\substack{\text{O}\\ \|}}{\text{R}-\text{C}} \cdot + \text{HX}$$

$$\underset{\substack{\text{O}\\ \|}}{\text{R}-\text{C}} \cdot + \text{O}_2 \longrightarrow \underset{\substack{\text{O}\\ \|}}{\text{R}-\text{C}-\text{O}-\text{O}} \cdot$$

$$\underset{\substack{\text{O}\\ \|}}{\text{R}-\text{C}-\text{O}-\text{O}} \cdot + \underset{\substack{\text{O}\\ \|}}{\text{R}-\text{C}-\text{H}} \longrightarrow \underset{\substack{\text{O}\\ \|}}{\text{R}-\text{C}-\text{O}-\text{OH}} + \underset{\substack{\text{O}\\ \|}}{\text{R}-\text{C}} \cdot$$

$$\underset{\substack{\text{O}\\ \|}}{\text{R}-\text{C}-\text{O}-\text{OH}} + \underset{\substack{\text{O}\\ \|}}{\text{R}-\text{C}-\text{H}} \rightleftharpoons \underset{\substack{\text{OH}\\ |\\ \text{H}}}{\text{R}-\text{C}-\text{O}-\text{O}-\overset{\substack{\text{O}\\ \|}}{\text{C}}-\text{R}}$$

$$\underset{\substack{:\text{OH}\\ |\\ \text{H}}}{\text{R}-\text{C}-\text{O}-\text{O}-\overset{\substack{\text{O}\\ \|}}{\text{C}}-\text{R}} \rightleftharpoons \underset{\substack{{}^{\oplus}\text{OH}\\ \|}}{\text{R}-\text{C}-\text{OH}} + {}^{\ominus}\text{O}-\overset{\substack{\text{O}\\ \|}}{\text{C}}-\text{R}$$

$$\underset{\substack{{}^{\oplus}\text{OH}\\ \|}}{\text{R}-\text{C}-\text{OH}} + {}^{\ominus}\text{O}-\overset{\substack{\text{O}\\ \|}}{\text{C}}-\text{R} \rightleftharpoons 2 \underset{\substack{\text{O}\\ \|}}{\text{R}-\text{C}-\text{OH}}$$

Figure 1. Possible mechanism for the oxidation of aldehydes to carboxylic acids.

FTIR Studies. The identity of certain artifacts was confirmed by FTIR spectral data. The application of GC/FTIR in flavor research has been demonstrated by Schreier et al. (17). Examples of vapor phase IR spectra of selected artifacts are shown in Figure 3. There is a shift in the spectral bands of all molecules on a change of state. Therefore, IR spectra taken in the vapor phase will differ from those taken in the condensed phase. In general, stretching vibration bands move to higher wavenumbers in the vapor phase while deformation vibration bands move to lower wavenumbers (18).

Spectrum A in Figure 3 shows bands typical of an ester; the C=O stretching absorption has been shifted to about 1751 cm^{-1} from the 1728 cm^{-1} observed in the condensed phase while the C-O stretching vibration occurs at about 1173 cm^{-1}. The other three spectra represent carboxylic acids. Carboxylic acids constitute the majority of the artifacts formed. While carboxylic acids can exist either in the monomeric or dimeric form in the vapor state, higher temperatures favor the existance of monomers. It is evident from the sharp O-H stretching band between 3580 and 3587 cm^{-1} that the acids exist as monomers. The normal saturated acids display a strong C=O stretching absorption between 1778 and 1782 cm^{-1} with the C-O stretching band occurring between 1142 and 1153 cm^{-1}. The third spectrum (C) displays bands at 1753, 1663 and 980 cm^{-1}, indicative of an unsaturated acid with a conjugated trans configuration. Comparison with a standard of (E)-2-hexenoic acid revealed an almost identical spectrum. Another carboxylic acid (not shown) had absorption bands at 3587, 1763, 1645, 1138, 1111 and 818 cm^{-1} suggesting an unsaturated acid with a conjugated cis configuration. The spectrum was consistent with a standard of (Z)-2-hexenoic acid. The presence of the large amounts of carboxylic acids can lead to acid-catalyzed degradations and rearrangements of other constituents present in the extract.

Formation of 2-Ethyl-2(5H)-Furanone. The presence of artifacts with increased retention times suggests the formation of components of increased polarity and/or the formation of higher molecular weight constituents from condensation or addition reactions. The acids, aldehydes and alcohols present can undergo oxidation to form γ- and δ-lactones (14, 15). The formation of the lactone, 5-ethyl-2(5H)-furanone, probably occurs by the steps outlined in Figure 4. A plausible sequence would be reaction of 2-hexenoic acid to form a peroxy radical at the γ-position followed by production of the hydroperoxide. Cleavage of the O-O bond with the subsequent addition of H· could lead to 4-hydroxy-2-hexenoic acid. Intramolecular esterification would then produce the identified lactone.

Upon longer storage (see bottom chromatogram in Figure 2) a series of artifacts with Kovats indices ranging from 2100 to 2400 developed. Mass spectral data suggested that many of these artifacts are structurally related (similar mass ions), possibly longer chain δ-lactones. We are unable to elucidate their structures at present.

Conclusion

The formation of artifacts in aroma concentrates during storage is a potential problem in flavor research. Storage of concentrates

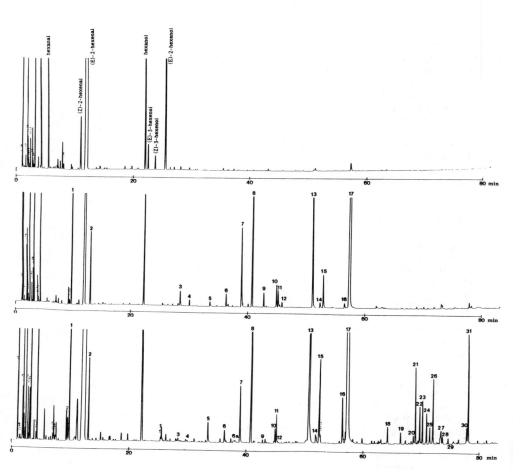

Figure 2. Capillary gas chromatograms of kiwifruit volatiles.
Top to bottom: a freshly prepared extract; an extract which
had been stored at -10°C for three months; an extract which
had been stored at -10°C for four months. The peak numbers
correspond to the numbers in Table 1.

TABLE 1. Artifacts formed during storage of kiwifruit extracts.

Peak no.[a]	compound[b]	Kovats index[g] DB-WAX	MW	mass spectral data, m/z (rel intensity)
1	4-methyl-3-hexanone?	1161	114	39(8), 41(8), 55(5), 57(100), 68(4), 71(3), 86(7), 114(12)
2	hexyl formate[c]	1212	130	41(33), 42(33), 43(34), 55(43), 56(100), 69(38), 84(13), 87(4)
3	unknown	1439		41(100), 43(34), 45(22), 47(14), 55(62), 56(58), 60(9), 73(33)
4	unknown	1464		
5	propanoic acid	1517	74	42(5), 45(54), 46(6), 55(13), 56(13), 57(29), 73(65), 74(100)
6	unknown	1563		39(13), 41(33), 43(44), 44(39), 45(14), 55(100), 57(34), 73(55)
6a	hexyl hexanoate	1592	200	43(74), 56(43), 61(12), 69(22), 71(18), 84(58), 99(68), 117(100)
7	butanoic acid[c]	1604	88	41(16), 42(17), 43(14), 45(13), 55(7), 60(100), 73(37), 88(3)
8	unknown	1637		39(13), 41(37), 43(50), 44(24), 45(10), 55(100), 57(23), 73(33)
9	unknown			41(100), 43(38), 45(30), 55(24), 57(84), 58(65), 59(73), 85(89)
10	5-ethyl-2(5H)-furanone[d]	1710	112	39(4), 41(2), 54(3), 55(19), 57(8), 83(100), 84(14), 112(7)
11	pentanoic acid[c]	1714	102	41(13), 42(8), 43(10), 45(10), 55(9), 60(100), 73(46), 87(3)
12	unknown			39(10), 41(36), 43(50), 55(100), 57(18), 71(8), 73(18), 85(3)
13	hexanoic acid[c]	1822	116	41(20), 43(16), 55(13), 57(11), 60(100), 73(58), 87(20), 99(2)
14	2-methyl-2-butenoic acid?	1846	100	39(39), 41(28), 45(26), 55(100), 60(15), 73(16), 82(18), 100(94)
15	(Z)-2-hexenoic acid[c,f]	1856	114	
16	(E)-3-hexenoic acid[e]	1924	114	41(98), 55(100), 60(62), 68(65), 69(53), 96(14), 99(14), 114(73)
17	(E)-2-hexenoic acid[c]	1944	114	41(33), 42(43), 55(35), 60(15), 68(34), 73(100), 99(39), 114(18)
18	unknown	2078		39(5), 41(9), 55(40), 68(5), 81(2), 85(2), 97(100), 98(7)
19	δ-lactone?	2127		57(54), 69(16), 83(14), 99(100), 113(13), 155(59), 171(5), 185(2)
20	monoterpendiol?	2176		43(59), 55(52), 62(31), 71(100), 73(29), 81(11), 109(10), 134(2)
21	δ-lactone?	2183		55(41), 69(26), 81(27), 99(100), 141(58), 171(24), 185(10), 213(2)
22	unknown	2200		43(24), 57(25), 61(100), 67(12), 85(16), 103(4), 121(14), 123(5)
23	δ-lactone?	2209		57(95), 69(21), 99(100), 113(17), 155(78), 171(19), 185(8), 213(2)
24	δ-lactone?	2226		57(100), 69(37), 81(84), 99(50), 171(46), 183(39), 185(20), 213(4)
25	δ-lactone?	2237		57(77), 69(15), 81(24), 99(100), 145(5), 171(13), 185(4), 213(1)
26	δ-lactone?	2250		55(39), 69(26), 81(28), 99(100), 141(58), 171(25), 185(11), 213(2)
27	δ-lactone?	2281		57(100), 69(47), 81(94), 99(79), 145(17), 171(55), 185(25), 213(2)
28	δ-lactone?	2287		57(63), 81(45), 97(53), 99(100), 145(10), 171(31), 185(14), 213(1)
29	δ-lactone?	2309		57(60), 69(25), 81(24), 97(32), 99(100), 171(14), 185(4), 213(1)
30	unknown	2383		43(59), 44(82), 55(98), 61(23), 73(100), 85(5), 91(6), 103(12)
31	δ-lactone?	2390		55(20), 57(65), 69(14), 81(18), 99(100), 171(12), 185(4), 213(1)

[a] The peak numbers correspond to the numbers in Figure 2. [b] Mass spectrum and Kovats index are consistent with that of reference compounds. [c] Identity also confirmed by GC/FTIR. [d] Tentatively identified by mass spectral data only. [e] Tentatively identified by retention data only. [f] Mass spectra of more than one component present. [g] Retention indices agree to within ± 2 units with reference compounds.

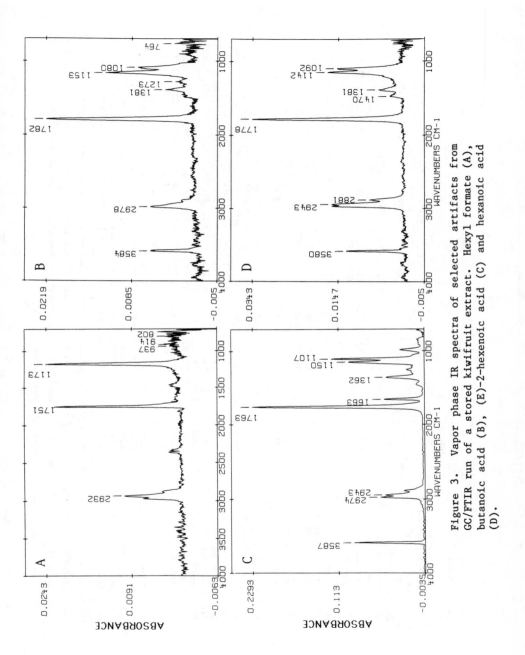

Figure 3. Vapor phase IR spectra of selected artifacts from GC/FTIR run of a stored kiwifruit extract. Hexyl formate (A), butanoic acid (B), (E)-2-hexenoic acid (C) and hexanoic acid (D).

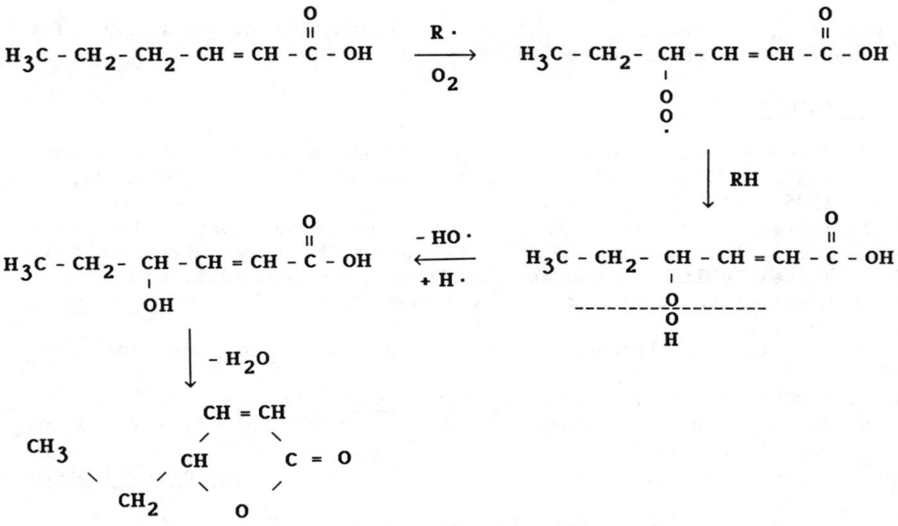

Figure 4. Possible mechanism for the formation of 5-ethyl-2(5H)-furanone from 2-hexenoic acid.

under inert atmosphere and at low temperatures slows degradative reactions. To inhibit free radical oxidations, researchers at the USDA Western Regional labs have routinely added trace amounts of antioxidants (0.001-0.01% of 1,3,5-trimethyl-2,4,6-tris(3,5-di-tert-butyl-4-hydroxybenzyl) benzene) to flavor extracts before concentration and storage (19). We perform our GC/MS analyses promptly (within two days) after the extracts have been concentrated and store the concentrated extracts at dry ice or liquid nitrogen temperatures, prior to GC/MS analyses.

Acknowledgment

The authors thank Dr. Robert A. Flath, USDA, Berkeley, for supplying the mass spectral data.

Literature Cited

1. Jennings, W.; Takeoka, G. In: "Methods in Enzymology"; Law, J.H.; Rilling, H.C., Eds.; Academic Press, Inc.: Orlando, FL 1985; Vol. 111; p. 149.
2. Maarse, H.; Belz, R. In: "Isolation, Separation, and Identification of Volatile Compounds in Aroma Research"; D. Reidel Publishing Company: Dordrecht, Holland 1981.
3. Koedam, A.; Scheffer, J.J.C.; Svendsen, A.B. Perf. Flavor. 1980, 5, 56.
4. Ohloff, G.; Flament, I.; Pickenhagen, W. Food Rev. Internat. 1985, 1, 99.
5. DeMets, M.; Verzele, M. J. Inst. Brew. 1968, 74, 74.
6. Badings, H.T.   Ph.D. Thesis, Agricultural University of Wageningen, 1970.
7. Kepner, R. E.; Ellison, B.O.; Maarse, H. J. Agric. Food Chem. 1975, 23, 343.
8. Paris, M. Riv. Ital. Essenze, Profumi, Piante Off., Aromi, Saponi, Cosmet., Aerosol 1975, 57, 83.
9. Lund, E.D.; Shaw, P.E.; Kirkland, C.L. J. Agric. Food Chem. 1981, 29, 490.
10. Takeoka, G.R.; Guntert, M.; Flath, R.A.; Wurz, R.M.; Jennings, W. (submitted to J. Agric. Food Chem.).
11. Schreier, P. In: "Chromatographic Studies of Biogenesis of Plant Volatiles"; Dr. Alfred Huethig Verlag: Heidelberg 1984.
12. Grosch, W. In: "Food Flavours Part A."; Morton, I.D.; MacLeod, A.J. Eds.; Elsevier: Amsterdam, The Netherlands 1982; p. 325.
13. Brodnitz, M.H. J. Agric. Food Chem. 1968, 16, 994.
14. Watanabe, K.; Sato, Y. Agr. Biol. Chem. 1970, 34, 464.
15. Palamand, S.R.; Dieckmann, R.H. J. Agric. Food Chem. 1974, 22, 503.
16. Horner, L. In: "Autoxidation and Antioxidants"; Lundberg, W.O., Ed.; John Wiley & Sons: New York 1961, Vol. I, p. 197.
17. Schreier, P.; Idstein, H.; Herres, W. In: "Analysis of Volatiles: Methods and Applications"; Schreier, P., Ed.; Walter de Gruyter & Co.: Berlin 1984, p. 293.
18. Welti, D. In: "Infrared Vapour Spectra"; Heyden & Son: London 1970.
19. Buttery, R.G., personal communication (1985).

RECEIVED January 3, 1986

Isolation of Nonvolatile Precursors of β-Damascenone from Grapes Using Charm Analysis

P. A. Braell[1], T. E. Acree[1], R. M. Butts[1], and P. G. Zhou[2]

[1]New York State Agricultural Experiment Station, Department of Food Science and Technology, Cornell University, Geneva, NY 14456
[2]Department of Soil and Agricultural Chemistry, Nanjing Agricultural University, Nanjing, People's Republic of China

A bioassay for the precursor compound to β-damascenone, *trans*-2,6,6-trimethyl-cyclohexa-1,3-dienyl-1-but-2-ene-1-one, found in grape skins was developed based on the generation of free β-damascenone and the sensory technique charm. Thin layer, column, and high pressure liquid chromatography were used to separate the precursor from 80 kg of grapes, *Vitis labruscana* , Bailey cv. Concord. There was a 22,000 fold enrichment in the precursor in 3.7 g of isolate.

Of the many odor-active compounds that have been isolated from natural products, β-damascenone is one of the most potent (**1**). Although this compound occurs at a level of less than 10 ng/g, it contributes strongly to the flavor of grape products (**2**). In fact, the contribution of β-damascenone to Concord grape (*Vitis labruscana* , Bailey) aroma may be equal to if not greater than that of methyl anthranilate, the compound generally considered characteristic of Concord flavor and the first odor-active compound identified in grapes (**3-4**). Intact grapes contain little or no free β-damascenone; rather they contain a non-volatile precursor that is converted to β-damascenone during post-harvest processing treatments (**3,5-6**). The purpose was to isolated a β-damascenone precursor preparation pure enough for chemical characterization.

If the precursor occurs at a concentration similar to that of β-damascenone, ca. 5 ng/g in thermally processed Concord grapes, its detection during isolation requires a

0097–6156/86/0317–0075$06.00/0

sensitive method. Since humans can smell β-damascenone at 2-20 pg/g in water
(1,3), a bioassay for β-damascenone precursor based on odor detection would have
the required sensitivity. In the bioassay reported here, the sensory technique called
charm analysis (7) was used to quantify the β-damascenone formed after acid
treatment of the precursor. This bioassay was used to optimize separation schemes
for the purification of β-damascenone precursors in Concord grapes.

MATERIALS AND METHODS

Grape extracts. Figure 1 outlines the procedure for preparation of Concord grape
extracts and their chromatography on C18 reversed-phase adsorbant (octadecyl
bonded silica). In a typical experiment 30 kg Concord grape skins were extracted
with 16 L methanol for 48 hrs after which the methanol was removed at 25 C and 0.1
Pa. Then, 250 mL aliquots of the aqueous residue were diluted 50 percent with
distilled water and passed onto a 2.4 cm x 10 cm column packed with 100g of 40
mm C18 reversed phase adsorbant. The column was rinsed with distilled water until
the eluate refractive index was less than 0.1 degrees Brix. Fractions were then eluted
from the column using 3 column volumes of 50% methanol/water followed by 3
column volumes 100% methanol. All fractions were analyzed for the presence of β-
damascenone precursor using the charm bioassay. The 50% methanol fractions were
pooled, extracted twice with equal volumes of Freon-113, stripped of methanol,
frozen at -40 C and freeze-dried at 10 C under 100 μPa.

Bioassay for β-damascenone. Figure 2 shows a flowchart of the bioassay for β-
damascenone. Sample extracts (0.5 mL) were placed in 5 mL .1 M citric acid and
heated 12 min at 90 C in 16 x 125 mm culture tubes covered with a vented metal cap.
Preliminary experiments showed little change in the β-damascenone formed after 8
min. Loss of free β-damascenone due to volatilization was prevented by cooling to
room temperature in ice water. Then 0.5 mL methanol plus 1.0 mL Freon-113 was
added with mixing. The Freon layer was removed, and a series of three-fold dilutions
were made from it. Thus each sample in the series was 1/3 the concentration of the
previous. These were stored until charm analysis.

Charm analysis. Charm is a formal procedure to quantify human response to odor
in gas chromatographic effluents, or for that matter, any technique that can deliver

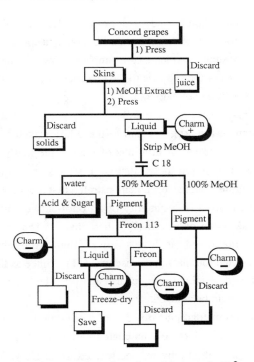

Figure 1. Initial methanol extraction and C18 reversed-phase chromatography of Concord grapes.

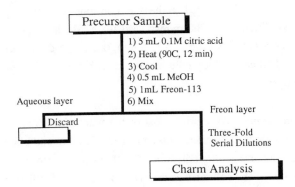

Figure 2. The bioassay for β-damascenone precursor isolated from Concord grapes.

vaporized chemicals into a humidified stream of air. Human subjects indicate the presence of β-damascenone by striking the mouse button on a computer which records the response. Figure 3 shows an idealized example of how charm data is generated from the on-off responses of the subjects to the presence of odor. The equation that produces the charm response chromatogram is

$$c = D^{n-1},$$

where c, the instantaneous charm, is equal to the dilution factor, D, raised to the power n-1 and n is the number of dilutions of the sample that produced an odor response at a particular retention index; that is, n is the sum of the unit responses shown at the top of Figure 3. It has been demonstrated that the integral of c, over the limits defined by the elution of a quantity of an odor-active compound as a gas chromatographic peak, is an odor-unit (7) called charm. The concentration of β-damascenone was quantified at levels lower than attainable using mass fragmentography (selected ion mass spectroscopy) (11). Since as little as 500 ftg β-damascenone could be determined using charm analysis, the β-damascenone in less than 1 gram of grapes could be assayed.

For each gas chromatographic run, 3 μl of a dilution was injected into a 0.32 mm x 25 m fused silica capillary column coated with 0.5 mm cross-linked methyl silicone (OV101). Chromatography was performed using He at 30 cm/sec. and a Hewlett Packard 5880 gas chromatograph held at 35 C for 3 min followed by a 10 C/min program to 225 C. As the separated compounds eluted they were mixed with 3 L/min humidified air, sniffed (8), and the response to β-damascenone, retention index of 1372, was recorded on an Apple Macintosh computer. The time data were converted to retention indices based on the retention times of carbon n-paraffins run under identical conditions (9). Charm responses were calculated using the programs described elsewhere (7). As little as 500 femtograms of β-damascenone standard injected in a fused silica capillary column coated with methyl silicone was detected using charm analysis.

Polyamide separations. To remove most of the plant pigments, the freeze-dried C18 50% methanol fractions were dissolved in 100 mL water and placed on a 2.4 cm x 30 cm column containing 100g polyamide CC (Universal Adsorbents Inc., Atlanta) as shown in Figure 4. The column was then rinsed with 5 L distilled water, followed by 5 L methanol. The β-damascenone charm was determined for both. The water fractions were pooled and saved.

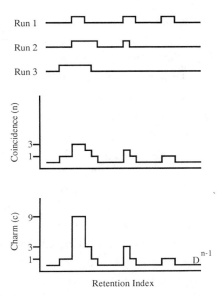

Figure 3. Construction of a charm response chromatogram from 3 serial dilutions of a model extract.

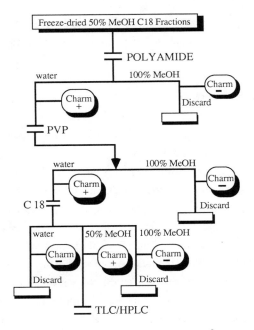

Figure 4. Polyamide, PVP, and secondary C18 chromatographic separations of β-damascenone precursor from extracts of Concord grapes.

PVP separation. The remaining pigment was removed from the polyamide water fractions by chromatography on a 2.4 cm x 15 cm column containing 50 g polyvinyl pyrollidone (Polyclar AT, GAF Corp., Wayne, N. J.) as shown in Figure 4. The PVP was rinsed first with 1 column volume of water followed by 1 column volume methanol. The β-damascenone charm of each fraction was then determined. The water rinses were concentrated on a 1 cm x 20 cm column containing 40 g of 40 mm C18 reversed phase adsorbant and eluted with 2 volumes of 50% methanol/water. The β-damascenone charm of the eluate was determined.

TLC and HPLC. To determine the best solvents for HPLC, 5μl of the C18 concentrated PVP fraction were separated on analytical TLC plates coated with either C18 reversed phase adsorbant or with silica. One cm sections of the plates were removed, placed in buffer, hydrolyzed, extracted and the β-damascenone charm determined. Standards containing 5 mg each of 1-O-n-octyl-β-D-glucopyranoside and α-methyl-D-mannoside were run simultaneously on the same plate and visualized by spraying with aniline-diphenylamine-acetone-phosphoric acid 80% (4 mL : 4 g : 200 mL : 30 mL) (**10**). HPLC separations were performed on a 0.46 mm x 25 cm column packed with 5 μm C18, eluted with 60/40 methanol:water at a flow of 0.8 mL/min. The refractive index of samples was recorded with a Knauer differential refractometer. Fractions were collected every minute and assayed for β-damascenone charm.

RESULTS AND DISCUSSION

Initial separations. The charm data showed that 97% of the β-damascenone precursor was located in the 'slip' skin of Concord grapes. Therefore, 80% of the grape mass was removed by pressing the grapes and extracting only the presscake or the skins. The methanol extracts of the grape skins not only contained the precursor but also the free sugars, organic acids, phenolics, etc. Williams, et al. (**6**), removed the free sugars and organic acids from terpenoid glycosides in grapes using C18 reversed-phase chromatography. Similarly, charm analysis of fractions from the C18 column showed that 97% of the β-damascenone precursor was contained in the 50% MeOH/water fractions as shown in Table 1, whereas the sugars and acids eluted with water. Extraction of the 50% MeOH eluate with Freon-113 removed the non-polar materials, including any free β-damascenone. Charm data showed that virtually all of the β-damascenone in the sample was generated from precursor.

Polyamide column chromatography separates phenolic glycosides (**12**) from phenolics and pigments. The precursor was entirely eluted from both polyamide and PVP columns with water while the majority of phenolics and pigments were eluted with methanol as shown in Table I. Concentrating the resulting PVP water rinses with C18 provided a 22,000 fold concentration of β-damascenone precursor which was used to develop a HPLC separation scheme.

Table I. Distribution of odor response in fractions obtained from C18, polyamide and PVP chromatographic separations of β-damascenone precursor from Concord grape extracts.

CHROMATOGRAPHY	FRACTION	% CHARM
C18	water	<1
	50% MeOH	97
	100% MeOH	3
Polyamide/PVP	water	100
	100% MeOH	<1

TLC . Analytical TLC separations on C18 and silica were used to determine the best HPLC solvent systems for chromatography of the β-damascenone precursor. Figure 5 shows the results from C18 TLC separation plotted as charm vs R_f. The best separation was obtained with a solvent containing 60/40 methanol : water.   In this system, the majority of the precursor had an R_f of 0.6 - 0.75, between the two glycoside standards. The charm distribution on the TLC plate indicated the presence of more than one β-damascenone precursor. Separation on analytical silica TLC plates, shown in Figure 6 as charm vs R_f, also gave evidence for more than one β-damascenone generating constituent. For silica separation, the best solvent was 6:3:1 ethyl acetate : 2-propanol : water. The majority of the precursor had an R_f of 0.4, much lower than either of the reference glycosides, while a second precursor was found between the references.

HPLC. An analytical HPLC separation on C18 was then performed using the solvent system developed by TLC. The results of this separation are shown in Figure 7 as charm vs elution time. The pattern of separation was the same as observed on TLC, with the majority of the β-damascenone precursor having an elution time

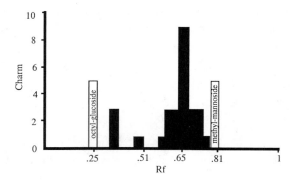

Figure 5.   Analytical C18 TLC separation of β -damascenone precursor.

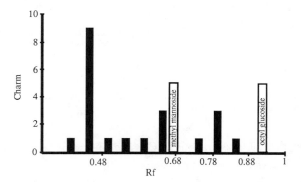

Figure 6.   Analytical silica TLC separation of β -damascenone precursor.

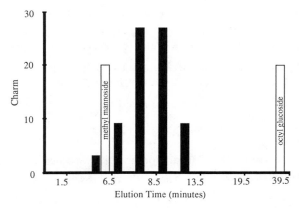

Figure 7.   Analytical HPLC separation of β-damascenone precursor on C18 packing.

between the glycoside references; closer to methyl mannoside than to octyl glucoside. The β-damascenone precursor was easily trapped from the HPLC and thus further purified.

Large Scale Preparation. Using the isolation scheme developed with the charm bioassay, a large scale preparative isolation of β-damascenone precursor from 80 kg Concord grapes was carried out thru the PVP procedure. The recoveries of precursor from the crude extract, 50% MeOH C18 eluate, polyamide and PVP steps were determined as total precursor charm based on the β-damascenone odor response. Also, the total precursor yield, in micrograms of β-damascenone equivalent weights, was calculated based on the minimum detectable quantity of β-damascenone (500 ftg). The results of these measurements, shown in Table II, indicated that overall 74% of the precursor was recovered thru the PVP procedure. The largest single loss of material (about 15%) occurred during the freeze-drying of the 50% MeOH C18 eluates. Subsequent improvements in the handling of the material during freeze-drying minimized this problem.

Table II. Charm-based recoveries from a preparative isolation of β-damascenone precursor from 80 kg Concord grapes.

FRACTION	TOTAL CHARM	% RECOVERY	YIELD μg
Crude extract	1,021,500	100	510
C18 50% MeOH	991,500	96	490
Polyamide	811,000	79	400
PVP	765,000	74	380

The overall yield estimated by charm analysis was 510 μg β-damascenone equivalents recovered from the 80 kg of Concord grapes. Of this quantity of precursor, 380 μg was recovered using the separation scheme. The precursor material was concentrated at least 22,000 fold based on the mass of the original sample and was separated from the sugars, organic acids, and phenolics in the grapes. To isolate several milligrams of the β-damascenone precursor we will apply the scheme to several hundred kg of grapes.

CONCLUSION

The β-damascenone precursor appears to be neutral, very polar, and have chromatographic properties similar to alkyl glycosides. It is more polar than octyl glucoside and similar in polarity to methyl mannoside. The evidence suggests that it is a damascone glycoside similar to the vetispirane and ionol glucosides in tobacco (**13-14**) or the terpene polyol glycosides in grapes (**15**). Furthermore, there is evidence from both silica and C18 chromatography that Concord grapes contain more than one compound with β-damascenone generating activity.

ACKNOWLEDGMENTS

The authors thank Welch Foods for supporting their research.

LITERATURE CITED

1. Ohloff, G. *Perf. Flav.* 1978, **3**, 11-22.
2. Acree, T. E.; Braell, P. A.; Butts, R. M. *J. Agric. Food Chem.* 1981, **29**, 688-690.
3. Braell, P. A. Masters Thesis, Cornell University, Ithaca, New York, 1984.
4. Power, F. B. ; Chestnut, V. K. *J. Am. Chem. Soc.* 1921, **43**, 1741-1742.
5. Masuda, M. ; Nishimura, K. *J. Food Sci.* 1980, **45**, 396-397 .
6. Williams, P. A.; Strauss, C. R.; Wilson, B. *J. Chrom.* 1982, **235**, 471-480.
7. Acree, T. E.; Barnard, J.; Cunningham, D. G. *Food Chem.* 1984, **14**, 273-286.
8. Acree, T. E.; Butts, R. M.; Nelson, R. R.; Lee, C. Y. *Anal. Chem.* 1976, **48**, 1821-1822.
9. Kovats, E. *Advan. Chromatogr.* 1965, **1**, 229-247.
10. Hansen, S. A. *J. Chrom.* 1975, **107**, 224-226.
11. Braell, P. A.; Acree, T. E.; Butts, R. M.; Barnard, J. Abstr. 184, 45th National Institute of Food Technologists Mtg., Atlanta, Ga., 1985.
12. Julkunen-Titto, R. *J. Agric. Food Chem.* 1985, **33**, 213-217 .
13. Anderson, R. C.; Gunn, D. M.; Murray-Rust, J.; Murray-Rust, P.; Roberts, J. S. *J. C. S. Chem. Comm.* 1977, 27-28.
14. Kodama, H.; Fujimori, T.; Kato, K. *Agric. Biol. Chem.* 1981, **45**, 941-944.
15. Williams, P. J., Strauss, C. R., Wilson, B. and Dimitriadis, E. In "Progress in Flavour Research 1984 : Proceedings of the 4th Weurman Flavour Research Symposium"; Adda, J. ed.; Elsevier Science Publishers, New York, 1985 ; 349-357.

RECEIVED March 6, 1986

Precursors of Papaya (*Carica papaya,* L.) Fruit Volatiles

P. Schreier and P. Winterhalter

Lehrstuhl für Lebensmittelchemie, Universität Würzburg, Am Hubland, D–8700 Würzburg, West Germany

Oxygenated terpenoids derived from linalool, a major constituent among papaya (C. papaya, L.) fruit volatiles,
were studied by capillary gas chromatography (HRGC) and
combined capillary gas chromatography–mass spectrometry
(HRGC–MS). Using a sample preparation technique suitable
for the separation and enrichment of polar compounds, the
two diastereoisomers of 6,7–epoxy–linalool, 2,6–dimethyl–
1,7–octadiene–3,6–diol, 2,6–dimethyl–3,7–octadiene–2,6–
diol, (E)– and (Z)–2,6–dimethyl–2,7–octadiene–1,6–diol,
and 2,6–dimethyl–7–octene–2,3,6–triol were identified for
the first time. Each of four diastereoisomeric epoxy–linalool oxides in their furanoid and pyranoid forms were
also detected for the first time as natural plant constituents. Biogenetic pathways of formation and metabolization of the oxygenated linalool derivatives are discussed.

Historically, most studies on plant volatiles were undertaken with
the aim of identifying the substances responsible for the characteristic aroma and flavor of plant materials. Since the 1970's studies on the pathways and control mechanisms of flavor formation
have been carried out (1), showing that fruits, vegetables and
spices contain volatiles originating predominantly from secondary
plant metabolism (2).
    Initially, researchers focused on studies of non–enzymic fla-
vor formation (3,4). More recently, a renaissance of investigations
of biogenesis of plant aromas is evident. The reasons for this new
trend are obvious. Worldwide, an increasing quantity of natural
flavors is required, and – as the natural sources are no longer sufficient – new ways of production are needed, e.g., plant cell or
tissue cultures, microorganisms or enzymes (5,6). At present, these
efforts are rather limited, since only a few experimentally demonstrated biogenetic pathways are known for plant volatiles (1,2).
    In this paper, results of our study on papaya (Carica papaya,
L.) fruit volatiles are presented, which extend our knowledge in
this field.

0097–6156/86/0317–0085$06.00/0

Experimental

Samples of papaya fruit (C. papaya, L.) were prepared by the scheme
shown in Figure 1, which involves direct extraction of the fruit
(7). In a typical experiment 0.7 kg of peeled papaya fruit (seeds
removed) were homogenized, extracted and concentrated to produce the
sample which was subjected to HRGC and HRGC-MS analysis.

Samples were separated on a Carlo Erba 4160 gas chromatograph
(HRGC) as well as on a Varian Aerograph 1400 interfaced by an open-
split system to a Finnigan MAT 44 mass spectrometer. The gas chromatographic columns were J & W fused silica DB-Wax, 30 m x 0.32 mm
id, df = 0.25 μm. On-column injections were used (0.4 μl). The columns were held at $50^{o}C$ for 3 min and then programmed at 4°/min to
$250^{o}C$. The flow rates of carrier gases (HRGC, N_2; HRGC-MS, He) were
2 ml/min. Linear HRGC retention indices were compared with those of
authentic reference samples.

HRGC-FTIR analysis was carried out with a Nicolet 20 SXB system
interfaced to a Dani 6500 gas chromatograph. A J & W fused silica
DB-5 column, 30 m x 0.32 mm id, df = 0.25 μm, was used. PTV injection (40°-200°C) was performed. The temperature program was 60°
to $250^{o}C$ at 5°/min. Light pipe and transfer line were held at $200^{o}C$;
He (2 ml/min) was employed as carrier gas. Vapor phase spectra were
recorded from 4000 - 700 cm^{-1} with a resolution of 8 cm^{-1}.

The preparation of authentic reference samples was carried
out as follows. Diastereoisomer (18) was obtained as colorless oil
by epoxidation of (-)R-linalool with m-chloroperbenzoic acid (8).
HRGC: R_t's 1781 and 1791, respectively. FTIR (vapor phase, cm^{-1}):
3659, 3086, 2983, 1842, 1645, 1459, 1373, 1230, 1073, 989, 915. EIMS
m/z (rel. int.): 155 (M-Me)$^{+}$(0.5), 137 (M-Me-H_2O)$^{+}$ (0.5), 97 (7), 85
(5), 79 (8), 71 (34), 68 (29), 59 (33), 55 (21), 43 (100). ^{1}H-NMR (60
Mhz, $CDCl_3$) was in close agreement with published data (9,10).

Compounds (19) and (20) were prepared by photooxygenation of
linalool, followed by $NaBH_4$-reduction (11,12). HRGC: R_t's 1927 and
2106, respectively.

Triol (21) was prepared according to Williams et al. (7) using
diastereoisomeric 6,7-epoxy-linalyl acetate. In our hands, (21) was
directly amenable to HRGC: R_t 2427. Derivatization afforded the a-
cetonide derivatives with R_t's 1828 and 1837,respectively (the natural (21) co-chromatographed with the isomer at R_t 1828). Synthesis
of (24) and (25) was accomplished by SeO_2 oxidation of linalool (13)
and afforded, after reductive work-up, the (E)- and (Z)-isomers in a
ratio of about 10:1. Spectral properties (^{1}H-NMR, MS, FTIR) corres-
ponded to those published (13,14,15). HRGC: R_t's 2294 and 2254.

Epoxidation of furanoid linalool oxides with m-chloroperbenzoic
acid afforded the diastereoisomeric (Z)- and (E)-epoxy-linalool oxides as colorless oils. ^{1}H-NMR spectra agreed with those published
(16). (E)-isomers (26) and (27): HRGC: R_t's 1867 and 1877, respectively. EIMS m/z (rel. int.):171 (M-Me)+(1), 153 (M-Me-H_2O)$^{+}$(1), 143
(6), 127 (4), 97 (5), 84 (31), 81 (27), 71 (20), 59 (55), 43 (100).
(Z)-isomers (28) and (29): HRGC: R_t's 1792 and 1797, respectively.
EIMS m/z (rel. int.): 171 (M-Me)$^{+}$ (0.5), 143 (M-epoxy)$^{+}$(6), 125
(M-epoxy-H_2O)$^{+}$(3), 110 (2), 97 (7), 84 (40), 81 (19), 71 (20), 59
(37), 43 (100). The epoxides of the pyranoid linalool oxides were
prepared analogously to the furanoid forms. (E)-isomers (30) and
(31): HRGC: R_t's 2141 and 2166, respectively. EIMS m/z (rel. int.):

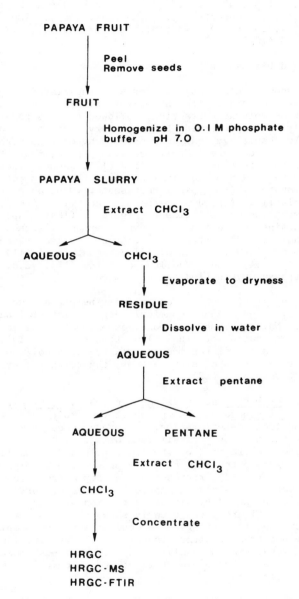

Figure 1.   Scheme of sample preparation at pH 7.0 suitable for separation and enrichment of polar compounds (7).

171 (0.5), 153 (0.2), 143 (14), 125 (5), 107 (3), 84 (21), 71 (20), 59 (69), 55 (25), 43 (100). (Z)-isomers (32) and (33): HRGC: R_t's 2116 and 2119, respectively. EIMS (rel. int.): 186 (M)$^+$(0.2), 171 (M-Me)(0.4), 153 (0.2), 143 (13), 125 (6), 84 (21), 71 (18), 59(71), 55 (28), 43 (100).

Results and Discussion

Figure 2 shows a classical way for studying flavor compounds involving separation and enrichment of volatiles by combined high-vacuum distillation-solvent extraction, followed by preseparation on silica gel. Analysis is performed by capillary gas chromatography (HRGC) and coupled capillary gas chromatography-mass spectrometry (HRGC-MS). If this procedure is followed with papaya, more than 130 volatiles can be detected and identified (17). Among these components benzyl isothiocyanate and linalool are the major constituents. The isothiocyanate is a typical representative of an enzymatically induced volatile, formed by myrosinase activity from glucosinolates after disruption of cell tissue (Figure 3) (18). Papaya fruit pulp also contains a series of terpene compounds as shown in Figure 4 (17). Linalool (13) is the major compound; also present are several hydrocarbons (2-9), geraniol (10), ho-trienol (11), α-terpineol (12), (E)- (14) and (Z)-linalool oxide (15), 2,6,6-trimethyl-2-vinyl-tetrahydropyran (16) and 1,8-cineole (17).

Degradation of Linalool. Papaya fruit pulp possesses a pH value of 5.6; thus, significant chemical degradation of linalool is not expected. Nevertheless, in model experiments, we investigated the possible degradation of linalool at this pH value (19). As shown in Figure 5, in spite of the modest acidity, linalool was partially decomposed to a series of hydrocarbons, the tetrahydropyran derivative (16) and α-terpineol (12).

Comparison of Distillation/Extraction (pH 5.6) vs. Direct Extraction (pH 7.0) of Papaya. Since the pH of the fruit pulp was shown to cause chemical deterioration, we prepared our samples at pH 7.0, following the procedure of Williams et al. (7). This procedure involves direct extraction of the fruit pulp and is outlined in Figure 1.In Figure 6 the HRGC analysis of the two samples are compared. The upper curve (a) represents the sample prepared by distillation/extraction as shown in Figure 2; the lower curve (b) represents the sample prepared by direct extraction at pH 7.0, as shown in Figure 1 (7). Figure 6 (b) clearly shows the presence of a series of polar components eluting at the end of the PEG capillary column which are not present after standard sample preparation (a).
    HRGC and HRGC-MS analysis of the compounds shown in Figure 6 (b) led to a number of identifications which will be covered in subsequent sections.

Terpene polyols. As indicated in Figure 7, three polyols were identified,i.e. 2,6-dimethyl-1,7-octadiene-3,6-diol (19), 2,6-dimethyl-3,7-octadiene-2,6-diol(20) and 2,6-dimethyl-7-octene-2,3,6-triol (21). The diendiols have been previously detected in Cinnamomum camphora (20), different grape cultivars (7,21,22) and recently in passionfruit (23), but not in papaya. During previous work (7), the iso-

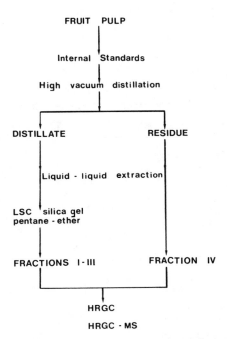

Figure 2.   Standard sample preparation procedure for the   study of papaya (C.papaya, L.) fruit volatiles (17).

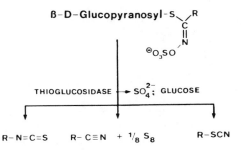

Figure 3.   Formation of   volatiles   from glucosinolates   by thio-glucosidase activity due to disruption of cell tissue (18).

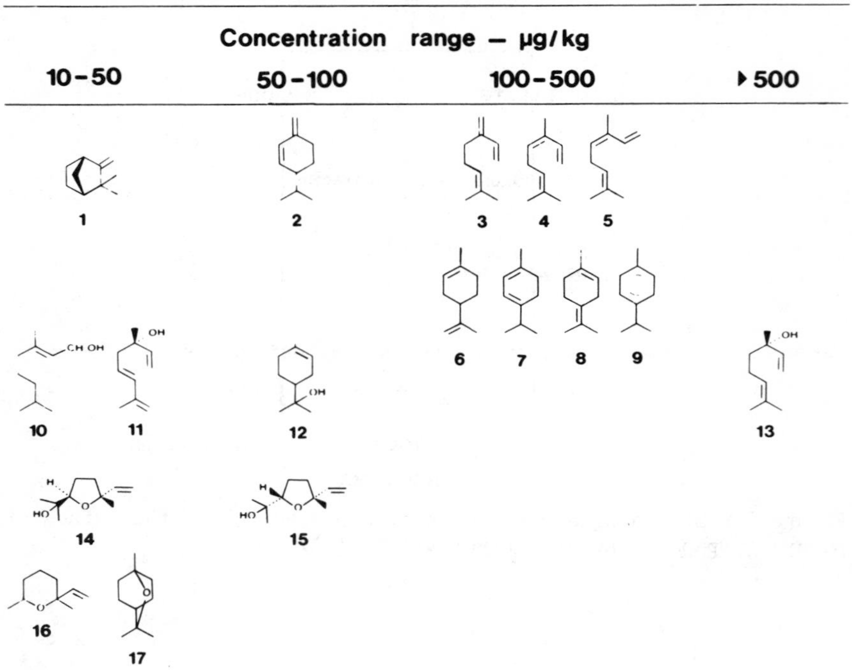

Figure 4.   Terpenoids   identified   by   HRGC   and HRGC–MS in papaya
(C.papaya, L.) fruit pulp (pH 5.6).

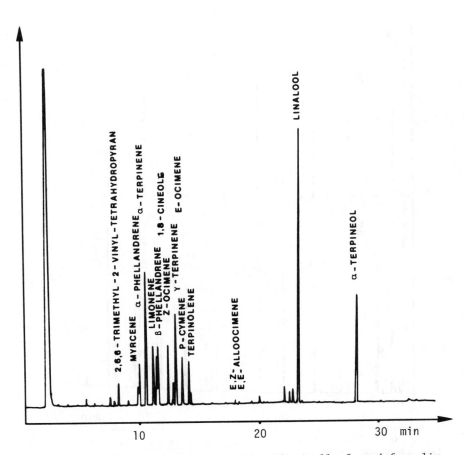

Figure 5.   HRGC analysis of volatiles chemically formed from linalool at pH 5.6 (J & W DB–Wax, 30 m x 0.32 mm id, df = 0.25 μm).

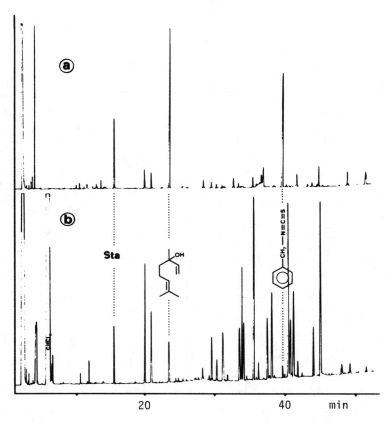

Figure 6.   HRGC comparison of papaya (C.papaya, L.) fruit volatiles obtained after high-vacuum distillation/extraction (pH 5.6) (a) and direct extraction (pH 7.0) (b) (J & W DB-Wax,   30 m x 0.32 mm id, df = 0.25 μm). Sta = external standard.

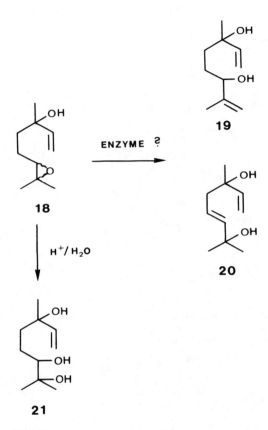

Figure 7.   6,7-Epoxy-linalool (18) as precursor of terpene polyols (19–21).

mers of 6,7-epoxy-linalool (18) had been proposed as the possible
precursors of the terpene polyols (19-21), but experimental evidence
for their occurrence in the natural plant material was lacking.
Using the above-mentioned sample preparation technique (Figure 1) we
identified the diastereoisomeric epoxy derivatives (18) in papaya
fruit pulp by HRGC and HRGC-MS.

Isomers of 6,7-Epoxy-linalool as Precursors of Linalool Oxides. Previously, the triol (21) had been proposed as a possible precursor of
the hydroxy ethers (14) and (15), the so-called linalool oxides(Figure 8). At an acidic pH (<3.5) and/or during heat treatment (e.g.,
steam distillation/extraction), the triol (21) had been found to be
decomposed to (14) and (15) (23,24). In these previous experiments
no formation of the corresponding pyranoid linalool oxides (22) and
(23) was observed. We evaluated the hypothesis that linalool oxides
are formed from triol (21) under natural conditions of papaya pulp
(i.e. pH 5.6) and could find no formation of linalool oxides. Even
in model experiments carried out at pH 3.5, only traces of linalool
oxides were detected after incubation of (21) for three days. As a
result of these experiments the isomers of 6,7-epoxy-linalool have
to be considered as the natural precursors of linalool oxides (14,
15) as recently suggested by Ohloff et al. (25). The latter's proposal was based on earlier findings obtained in a series of chemical
reactions (9).

Oxygenated Terpenoids. In addition to the polyols (19-21) and the
isomers of 6,7-epoxy-linalool (18) additional oxygenated terpenoids
were identified. Their structures, together with possible biogenetic
pathways, are represented in Figure 9. The isomers (Z)- (24) and
(E)-2,6-dimethyl-2,7-octadiene-1,6-diol (25) were found in a distribution of about 1:6 (Z:E). As to their biogenetic formation, a
direct ω-hydroxylation must be considered, as previously proposed
for several bacterial (26,27) and fungal (28) transformations of
linalool. In higher plants, the (E)-isomer (25) has been found in
different Nicotiana species (13,29). The ß-D-glucosides of (E)-
(25) and (Z)-2,6-dimethyl-2,7-octadiene-1,6-diol (24) have been
isolated from Betula alba leaves and the fruits of Chaenomeles
japonica (14). In contrast to the odorless polyols (19-21), compound (25) has been described as an useful perfume and flavor
substance (30).

Epoxy-Linalool Oxides. The presence of the four isomeric linalool
oxides (14,15,22,23) in papaya fruit pulp is known from earlier
investigations (17,31). In addition to these compounds, the four
diastereoisomeric epoxy derivatives of these hydroxy ethers in their
furanoid (26-29) and pyranoid forms (30-33) were identified in this
work. This is the first time that these epoxy-linalool oxides were
detected in a natural plant material. The natural occurrence of
eight epoxy-linalool oxides and their quantitative distribution in
the fruit pulp (e.g., furanoid forms, E:Z = 5:1) might prevent a
subsequent epoxidation of the corresponding linalool oxides. Therefore, in Figure 9, a hypothetical diepoxy-linalool derivative is
postulated as a possible biogenetic precursor of structures (26-33).
This hypothesis is supported by the findings of Osborne (32), who
detected the formation of the diepoxy derivatives of geraniol and
nerol after feeding these alcohols to Pisum sativum cell cultures.

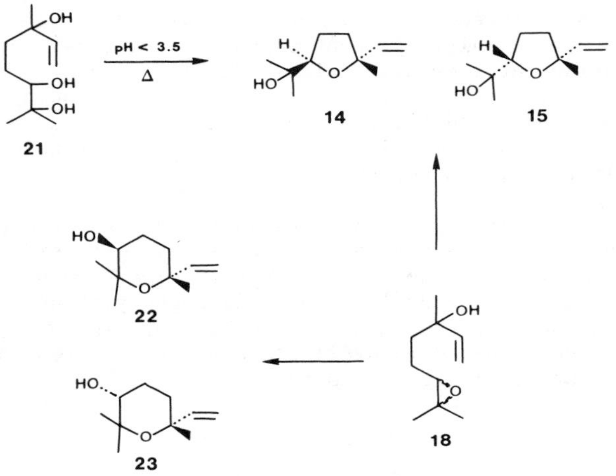

Figure 8.   Chemical formation of furanoid linalool oxides (14) and (15) from triol (21) and biogenetic formation of the four linalool oxides (14), (15), (22) and (23) from 6,7-epoxy-linalool (18).

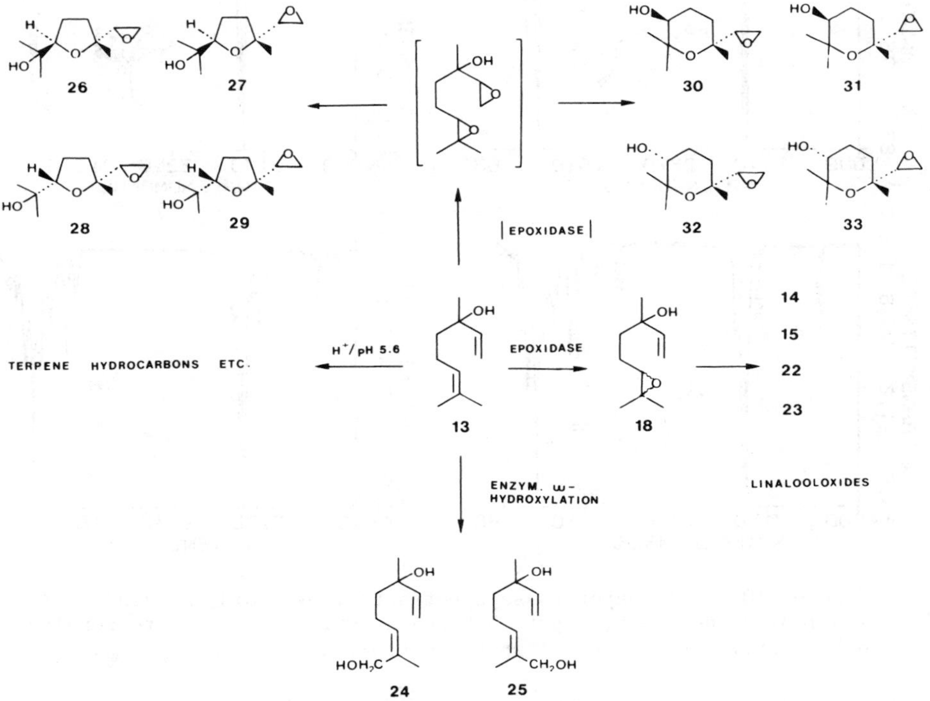

Figure 9.   Scheme summarizing chemical and biogenetic formation of linalool derivatives (cf. explanations in the text).

As to the stereochemistry of the newly identified epoxy-linalool oxides(26-33) only differentiation between (Z)- and (E)-isomers
can be provided at present. Using HRGC-FTIR technique (Figures 10
and 11) the occurrence of an intramolecular H-bonding in two diastereoisomers indicated (Z)-configuration. These bands were detected
at 3490 cm^{-1} and 3525 cm^{-1} in the pyranoid and furanoid (Z)-config-
urated derivatives, respectively.

Glycosides. Finally, it should be mentioned that in our studies on
papaya fruit volatiles and their precursors, glycosidic forms of
several volatiles were detected. At this time, we have identified
the ß-D-glucosides of benzyl alcohol and 2-phenylethanol. There are
also indications for the occurrence of glycosidic forms of terpenoids, but from the data available exact structures of the sugar
moieties cannot be elucidated as yet.

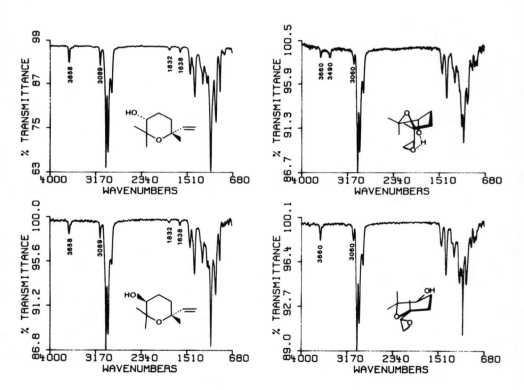

Figure 10. FTIR vapor phase spectra from pyranoid (Z)- and (E)-
linalool oxides(left, top to bottom) as well as from corresponding
epoxy derivatives (right, top to bottom).

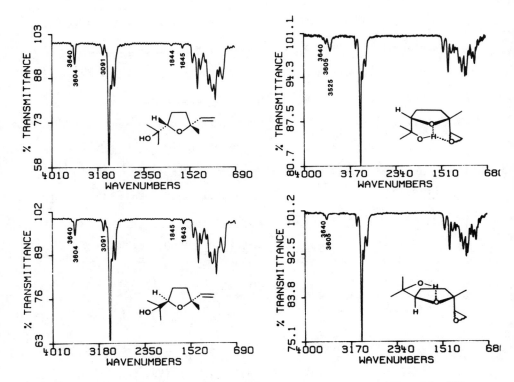

Figure 11.   FTIR vapor phase spectra from furanoid (Z)- and (E)-linalool oxides(left, top to bottom) as well as from corresponding epoxy derivatives (right, top to bottom).

Literature cited

1. Schreier, P. In "Chromatographic Studies of Biogenesis of Plant Volatiles"; Hüthig: Heidelberg, 1984.
2. Croteau, R. In "Isopentanoids in Plants: Biochemistry and Function"; Nes, W.D.; Fuller, G.; Tsai, L.S., Eds.; M. Dekker: New York, 1984, p.31-73.
3. Vernin, G. In " Chemistry of Heterocyclic Compounds in Flavours and Aromas"; Horwood: Chichester, 1982.
4. Waller, G.R.; Feather, M.S. "The Maillard Reaction in Foods and Nutrition", ACS Symposium Series No. 215, American Chemical Society: Washington, D.C., 1983.
5. Schreier, P. In "Topics in Flavour Research"; Berger, R.; Nitz, S.; Schreier, P., Eds. Eichhorn: Marzling, 1985, p.353-369.
6. Sariaslani, F.S.; Rosazza, J.P.N. Enzyme Microb. Technol., 1984, 6, 242-253.
7. Williams, P.J.; Strauss, C.R.; Wilson, B. Phytochemistry, 1980, 19, 1137-1139.
8. Winterhalter, P.; Katzenberger, D.; Schreier, P. Phytochemistry, in press.

9. Felix, D.; Melera, A.; Seibl, J.; Kovats, E.Sz. Helv. Chim. Acta
    1963, 46, 1513-1536.
10. Kametani, T.; Nemoto, H.; Fukumoto, K. Bioorg. Chem., 1978, 7,
    215-220.
11. Matsuura, T.; Butsugan, Y. J. Chem. Soc. Japan, 1968, 89, 513-
    516.
12. Kjosen, H.; Liaaen-Jensen, S. Acta Chem. Scand., 1973, 27, 2495-
    2502.
13. Behr, D.; Wahlberg, I.; Nishida, T.; Enzell, C.R. Acta Chem.
    Scand., 1978, 32B, 228-234.
14. Tschesche, R.; Ciper, F.; Breitmaier, E. Chem. Ber., 1977, 110,
    3111-3117.
15. Hirata, T.; Aoki, T.; Hirano, Y., Ito, T.; Suga, T. Bull. Chem.
    Soc. Japan, 1981, 54, 3527-3529.
16. Ohloff, G.; Schulte-Elte, K.H.; Willhalm, B. Helv. Chim. Acta,
    1964, 47, 602-626.
17. Idstein, H.; Schreier, P. Lebensm. Wiss. u. Technol., 1985, 18,
    164-169.
18. Kjaer, A. Progr. Chem. Org. Nat. Prod., 1960, 18, 122-176.
19. Morin, P.; Richard, H. In "Progress in Flavour Research 1984";
    Adda, J., Ed.; Elsevier: Amsterdam; 1985; p. 563-576.
20. Takaoka, D.; Hiroi, M. Phytochemistry, 1976, 15, 330.
21. Rapp, A.; Knipser, W. Vitis, 1979, 18, 229-233.
22. Rapp, A.; Knipser, W.; Engel, L. Vitis, 1980, 19, 226-229.
23. Engel, K.H.; Tressl, R. J. Agric. Food Chem., 1983, 31, 998-
    1002.
24. Williams, P.J.; Strauss, C.R.; Wilson, B. J. Agric. Food Chem.,
    1980, 28, 766-771.
25. Ohloff, G.; Flament, I.; Pickenhagen, W. Food Rev. Int., 1985,
    1, 99-148.
26. Devi, J.R.; Bhat, S.G.; Bhattacharyya, P.K. Indian J. Biochem.
    Biophys., 1977, 14, 359-363.
27. Madyastha, K.M. Proc. Indian Acad. Sci., 1984, 93, 677-686.
28. Bock, G.; Benda, I.; Schreier, P. J. Food Sci., 1986, in press.
29. Suga, T.; Hirata, T.; Hirano, Y.; Ito, T. Chem. Lett. , 1976,
    1245-1248.
30. Hasegawa, T. Jap. Pat. 58,140,032, 1982.
31. Flath, R.A.; Forrey, R.R. J. Agric. Food Chem., 1977, 25,103-
    109.
32. Osborne, M.J. In "Terpenoid Biosynthesis in Tissue Cultures of
    Jasmine"; Ph.D. Thesis, Univ. of London, 1979.

RECEIVED February 12, 1986

Volatile Compounds from Vegetative Tobacco and Wheat Obtained by Steam Distillation and Headspace Trapping

R. A. Andersen[1], Thomas R. Hamilton-Kemp[2], P. D. Fleming[1], and D. F. Hildebrand[3]

[1] Agricultural Research Service, U.S. Department of Agriculture, and Department of Agronomy, University of Kentucky, Lexington, KY 40546
[2] Department of Horticulture, University of Kentucky, Lexington, KY 40546
[3] Department of Agronomy, University of Kentucky, Lexington, KY 40546

Identification and quantitative analyses of 25 compounds in steam distillates of burley tobacco stalk were accomplished. Compounds included twelve C_6–C_9 compounds that were probable fatty acid oxidation products and 13 compounds $>C_9$ that varied in origin. The latter included oxidation products of fatty acids, a C_{10} prenyl pyrophosphate metabolite, and biodegradation products of carotenoids and chlorophyll. About 1/3 of the distillate mass was accounted for. Burley tobacco stalk headspace volatiles were also studied. When compared to the steam distillate, the headspace contained greater concentrations of sesquiterpenoids but lower concentrations of C_6 and C_9 aldehydes and alcohols. Volatiles in steam distillates of tobacco stalk were not quantitatively different in a fungal resistant and a fungal sensitive variety of tobacco. Yield comparisons were made of headspace volatiles from tobacco and wheat. Studies were also begun on the effects of fatty acid oxidation inhibitors on yields of tobacco and wheat steam distillates.

There have been relatively few reports on the occurrence and biogenesis of volatile compounds in vegetative tobacco. Investigators have reported on these constituents in air-cured burley tobacco leaf (1-4). Burton et al. (4) showed that the composition of some volatile components in burley leaf at harvest changed during air curing. Several investigators reported on volatiles in flue-cured tobacco leaf in terms of composition (3, 5, 6) and its changes during curing (6).

Volatile compounds emitted from the leaves and stems of agricultural crops during growth may affect plant host-parasite interactions that determine susceptibility or resistance to fungal infections and insect damage (7-9).

The purpose of this investigation was to characterize volatiles and study their biogenesis in stalk of burley tobacco and wheat during active growth. A comparision was made of volatile compounds obtained by steam distillation and headspace

0097-6156/86/0317-0099$06.00/0

trapping techniques that were similar to those described by
Buttery et al. (7).

Experimental

Reagents   Chemical compounds for comparisons of mass spectra and
GC retention times were obtained from commercial sources.
Pentadecanal was synthesized as before (10). Neophytadiene and
solanone were obtained from H. Burton, University of Kentucky.
Reagent-grade hexane was redistilled on a Vigreaux column. Water
was deionized, passed through activated charcoal and distilled.
Tenax GC adsorbent (60-80 mesh) was obtained from Alltech Assoc.
Inc.

Plant materials   Burley tobaccos (Nicotiana tabacum L. cv KY 14
and cv KY 17) were grown at various times in the soil floor of a
greenhouse. Recommended cultural and fertilization practices were
followed (11). A randomized complete block experimental design
(12) with four replications was used for comparison of volatile
yields between the two cultivars. Plants were harvested at 7, 13,
and 14 weeks after transplant, and leaves (including stems) were
removed from stalks. Stalks were used either fresh for headspace
volatile determinations, or they were immediately frozen at
-20°C for the isolation of volatiles by simultaneous steam
distillation-hexane extraction. Fresh leaves (including stems)
were used for headspace volatile determinations.
    Wheat (Triticum aestivum L. cv Arthur 71) was field-grown and
harvested when plants were 40-50 cm in height in May, or was grown
in flats under greenhouse conditions and harvested 2 weeks after
emergence. Field-grown whole plants were immediately frozen and
stored at -20°C prior to simultaneous steam distillation-hexane
extractions. Field-grown stems were used frozen for headspace
determination. Fresh greenhouse-grown plants were used for
headspace volatile determinations.

Isolation of volatile compounds by simultaneous steam
distillation-hexane extraction (SDE)   Frozen burley tobacco stalks
(1-4 kg) were cut perpendicular to the main axis into 0.5-cm
sections and placed in a 12-L flask with 4 L of distilled water.
The flask contents were steam distilled for 4 h in a continuous
extraction apparatus of the type described by Likens and Nickerson
(13) and modified in a manner similar to that described by Buttery
et al. (14). Forty ml of hexane was used as the extractant and
the additional condenser in the steam side arm was maintained at
-4°C. The apparatus was operated at a reduced pressure (200 mm
of Hg) with contents boiling at 65°C.
    After distillation, the hexane solution was separated from the
water layer and concentrated to a 1-ml volume in a microdistillation apparatus. When preparative GC separation and
analysis was carried out, the hexane concentrates from SDE
equivalent to 9 kg of tobacco stalk were combined and reduced to
an 80-μl volume in the same manner.

Isolation of volatile compounds by headspace trapping on Tenax   A
diagram of the glass-Teflon Tenax trap is shown in Figure 1. The

1.5-g Tenax plug (16 x 1 cm) was held in a glass tube by hexane-washed glass wool. The Tenax was activated before use by washing it with 200 ml hexane and then preheating it 2 h in the glass tube at 250°C in a stream of purified nitrogen.

Plant samples used in the apparatus were: fresh burley tobacco stalk (1 kg) cut perpendicular to the main axis into 8-cm sections; fresh burley tobacco whole leaf (100 g or 1 kg); and wheat leaves plus stems or stems alone (250 g). Purified air or purified nitrogen was used as the entrainment (sweep) gas at a flow rate of 500 ml/min for 20 h; a slight pressure drop was maintained at the system exit. Hexane (50 ml) was used to desorb Tenax-adsorbed components; the hexane solution was then concentrated to 1 ml in a microdistillation apparatus. This solution was used directly for capillary GC or GC-MS analysis.

Preparative GC fractionation of tobacco stalk SDE samples
Identification of volatiles in SDE equivalent to 9 kg stalk material was carried out by preliminary GC separation into fractions on temperature-programmed packed 20% methyl silicone as previously described (15). These fractions were stored at -20°C until analyzed by capillary GC or GC-MS.

Capillary GC and GC-MS analyses A Hewlett-Packard 5880A GC was used with a 60 m x 0.32 mm Supelcowax 10 or 30 m x 0.25 mm Carbowax 20M fused silica capillary column for capillary GC analyses. The following conditions were used: operation in splitless mode; He carrier linear velocity 31 cm/sec; column temperature held 1 min isothermal and then programmed from 60°C to 220°C at 3°/min and held at 220°C for 30 min. Kovats indices were calculated for separated components relative to hydrocarbon standards (16).

Yields of total volatile oils in SDE and in Tenax-trapped headspace eluates were estimated by GC results using octadecane as the internal or external standard.

Preparative GC fractions of tobacco stalk that were condensed in U-tubes were dissolved in acetone. An aliquot of the combined acetone solutions that contained pentadecane as internal standard was analyzed by GC-MS using a Hewlett-Packard Model 5985A instrument and as previously described (15), but with the capillary GC column conditions described in this subsection. In addition to electron impact MS, chemical ionization MS were obtained with methane. The identification of compounds was confirmed by matching EI- and CI-MS data and by co-chromatography of plant components with authentic compounds.

Fatty acid determinations The relative amounts of individual hydrolyzable fatty acids in 100 mg samples of freeze-dried tobacco stalk harvested 7 weeks after transplant and in wheat harvested at the 40- to 50-cm height plant stage were determined as fatty acid methyl esters by GC-FID (17).

Fatty acid oxidation inhibitor studies Each of the following solutions was used in place of 4L of distilled water during the isolation of volatiles by SDE from 1 kg tobacco (KY 14) stalk harvested 7 weeks after transplant: 10 mM $SnCl_2$ adjusted to pH

5.0 with dilute NaOH; 1 M NaCl; or 10 mM acetylsalicylic acid.   A
control run was made using 4 L of distilled water.
     Each of the following solutions was used in place of 4L of
distilled water during the isolation of volatiles by SDE from 250
g of wheat (whole plant):   100 μM phenidone; 100 μM nordihydroguaiaretic acid; or 100 μM acetylsalicylic acid.   A control run
was made using 4 L of distilled water.   The solution containing
the wheat was homogenized at high speed in a Waring blendor for 30
sec just before the distillation procedure was begun.

Results and Discussion

Tobacco stalk volatiles isolated by SDE   Total volatile oils
obtained by SDE of tobacco stalk harvested 7 weeks after transplant contained >80 detectable components on capillary GC columns
coated with Carbowax-type liquid phase.   Yields of total volatile
oils were about 1 μg/g of stalk on a wet weight basis.
     The reported stimulation of fungal spore germination by
volatile compounds identified from plant sources (8) and the
influence of host plant-emitted volatiles in the susceptibility to
fungal infection (9) led us to compare volatiles from stalks of
two genetic varieties of tobacco. These differed in resistance to black shank, a common disease of tobacco caused by
Phytophthera parasitica.   Resistant- and non-resistant tobacco
plant samples yielded no significant quantitative differences in
the 35 largest peaks.

Identification of volatile compounds in SDE of KY 14 tobacco
stalk   The identification and quantitation of 25 volatile
compounds in tobacco stalk which comprised about 30% of the total
volatile oil isolated was facilitated by a 2-dimensional GC
approach.   In this technique, there was a preliminary preparative
GC separation of total SDE volatiles from 9 kg stalk into about 25
fractions on a relatively non-polar column followed by capillary
GC and CG-MS analysis of 1 single or 2-3 combined fractions on a
strongly polar column.   Quantitative results of individual
components were based on tobacco stalk samples that were harvested
from a single plot and steam distilled in one distillation run.
     Identification and quantitative estimations of volatile
compounds in SDE samples of KY 14 tabacco stalk were carried out
(Table I).   Many 6- and 9-carbon volatile aldehydes and alcohols
in plant tissue are believed to form from the oxidation of
unsaturated fatty acids such as linoleic and linolenic acids
mediated by lipoxygenase, hydroperoxide cleavage enzyme systems,
cis 3:trans 2-enal isomerase enzymes and aldehyde dehydrogenases
(18-20).   Monoterpenes such as 1,8-cineole in plants may be
synthesized from neryl or geranyl pyrophosphate (21).   Diterpenes
may derive from geranylgeranyl pyrophosphate (21), but it is also
possible that neophytadiene from tobacco stalk may have formed
from a breakdown of chlorophyll during the isolation as proposed
in Figure 2.   Aliphatic aldehydes such as pentadecanal are
believed to form via α -oxidation of fatty acids.   The proposed
oxidation of palmitic acid to pentadecanal is shown in Figure 2.
Compounds such as 13-carbon solanone, β-ionone and geranylacetone
may form as carotenoid oxidation products.   A possible mechanism
of formation of β-ionone is given in Figure 2.

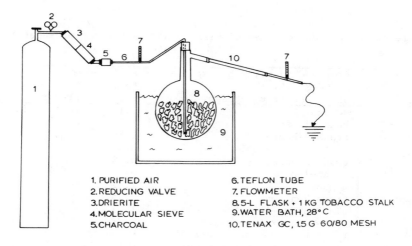

1. PURIFIED AIR
2. REDUCING VALVE
3. DRIERITE
4. MOLECULAR SIEVE
5. CHARCOAL
6. TEFLON TUBE
7. FLOWMETER
8. 5-L FLASK + 1 KG TOBACCO STALK
9. WATER BATH, 28°C
10. TENAX GC, 1.5 G 60/80 MESH

Figure 1.   Apparatus for collection of headspace components.

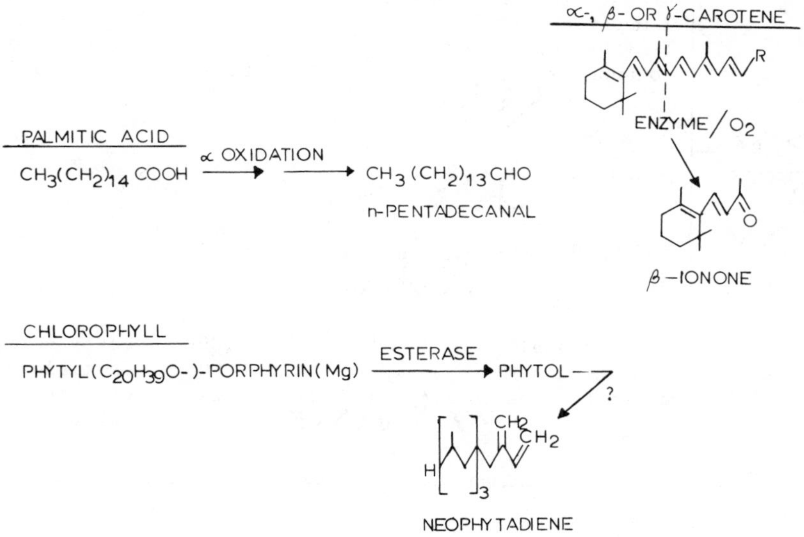

PALMITIC ACID

$CH_3(CH_2)_{14}COOH \xrightarrow{\alpha\,OXIDATION} CH_3(CH_2)_{13}CHO$

n-PENTADECANAL

α-, β- OR γ-CAROTENE

ENZYME/O₂

β—IONONE

CHLOROPHYLL

PHYTYL(C₂₀H₃₉O-)-PORPHYRIN(Mg) $\xrightarrow{ESTERASE}$ PHYTOL—?

NEOPHYTADIENE

Figure 2.   Possible pathways of formation for β-ionone, pentadecanal and neophytadiene.

Table I. Compounds in Steam Distillate-Hexane Extract of Vegetative
Tobacco Stalk

	Compound identity by capillary GC/GC-MS (EI + CI) and GC cochromatography	% of total oil	Possible precursor
C_6-C_9 Compounds	hexanal	2.18	fatty acid
	hexanoic acid	1.79	"
	trans 2-hexenal	0.86	"
	2-ethylhexan-1-ol	0.97	"
	trans 2-octen-1-ol	0.78	"
	octan-1-ol	0.25	"
	nonanal	2.32	"
	trans 2-nonenal	5.75	"
	cis 3-nonen-1-ol	1.51	"
	trans, cis 2,6-nonadienal	4.38	"
	trans cis 2,6-nonadien-1-ol	0.09	"
	cis, cis 3,6-nonadien-1-ol	0.63	"
$> C_9$ Compounds	1,8-cineole	0.10	C_{10} prenyl pyro-phosphate
	naphthalene	0.28	---
	n-decanal	1.56	fatty acid
	trans 2-decanal	0.25	"
	trans, trans 2,4-decadienal	0.34	"
	2-methylnaphthalene	0.35	---
	1-methylnaphthalene	0.02	---
	solanone	0.41	carotenoid
	β-ionone	1.64	"
	geranylacetone	0.16	"
	tetradecanal	0.09	fatty acid
	pentadecanal	2.00	palmitic acid
	neophytadiene	1.24	chlorophyll or phytol

Tobacco stalk headspace volatiles isolated by Tenax traps   Tobacco
stalk headspace volatiles eluted from Tenax contained at least 70
components on 60m x 0.32mm GC columns coated with Carbowax-type
liquid phase, which is comparable in number to that found in SDE
samples.   However, there were differences in the capillary GC
profiles.   Peaks corresponding to 6- and 9-carbon aldehydes and
alcohols in the headspace profiles were absent.   Presumably,
oxidative conversion of fatty acids to volatile aldehydes and
alcohols occurred to a lesser extent during collection of
headspace volatiles.   However, our results to date are not
complete enough to quantitatively compare the individual volatile
compounds obtained by these 2 methods.

Identifications were made for several compounds; these are
indicated on the capillary GC total ion chromatogram obtained
during an identification run in the GC-MS system (Figure 3).   In
addition, 2 sesquiterpenoids were present at scan numbers 784 and
836 (Figure 3), but their specific structure has not been
determined.   There is evidence for the presence of several
straight chain, saturated hydrocarbons including undecane in the
headspace.   The origin of the sesquiterpenoid caryophyllene is
thought to be farnesyl pyrophosphate (21), and the other
unidentified sesquiterpenoids are also presumed to have the same
precursor.   Speculation about the origin of 3-octanone(scan 344)
must include fatty acid oxidation and also the possibility that
oxidation of Tenax-absorbed components such as alcohols may occur
during the 20-h air entrainment period.   Other studies are planned
that will compare the results of headspace analyses of tobacco
stalk conducted with nitrogen versus air as entrainment gases.
The presence of methyl salicylate is of considerable interest to
us, especially since we have some evidence that mite resistance in
strawberry plants is associated with high levels of methyl
salicylate in volatile isolates obtained by SDE and headspace
trapping (22).

Comparisons of yields of total headspace volatiles from vegetative
tissues of tobacco and wheat   Quantitative estimates of headspace
total volatile yields are summarized in Table II.

Table II. Yields of Total Headspace Volatiles from Vegetative
Plant Materials

Plant material description	Weeks growth [a]	Sweep gas	Sample weight, kg	Yield, ng/g
tobacco (KY 14 burley)				
stalk	7	air	1	10
stalk	14	air	1	71
leaf	13	air	1	112
leaf	14	air	0.1	980
wheat (Arthur 71)				
whole plant	2	air	0.25	365
whole plant	2	N_2	0.25	444
stems	8 [b]	air	0.25	2597

(a)In greenhouse   (b)Wheat stem sample was field grown

These findings represent preliminary information obtained from a survey of several factors investigated for their effects on yields of headspace volatiles. The factors were plant age, plant part, plant genetics (genus, species, variety), growth conditions, sweep gas and weight of sample. Yields from tobacco leaf were greater than from tobacco stalk; yields from wheat plant were greater than from tobacco leaf. Nitrogen and air were both effective for entrainment of volatiles, and, in the case of wheat, the yields of total volatiles did not differ by more than 25%.

Fatty acids as precursors of volatiles in steam distillate-hexane extracts of tobacco and wheat   Relative abundances of specific fatty acids in hydrolysates of 2 varieties of burley tobacco stalk and 1 variety of wheat (whole plant including leaves and stems) are given in Table III.

Table III. Relative Abundances of Fatty Acids in Plant Samples

Fatty acid	#Carbons: #double bonds	Total hydrolyzable fatty acids, %		
		Tobacco (KY14)stalk	Tobacco (KY17)stalk	Wheat (Arthur71)
palmitic	16:0	31.6	29.9	8.5
palmitoleic	16:1	--	--	24.8
oleic	18:1	2.9	3.9	1.8
linoleic	18:2	39.6	40.8	14.4
linolenic	18:3	24.2	23.7	50.4

There were relatively small varietal differences in content between the 2 tobacco stalk materials, but there were large differences between the tobacco stalks and wheat.

Our previous results (15) on the composition of wheat SDE volatiles are shown in Table IV. These results support a direct precursor-product relationship between a fatty acid and a specific steam distilled volatile compound, because hexanal and trans 2-nonenal (which presumably originate from their linoleic acid precursor C18:2) account for a larger proportion of the total oil from steam distillates of tobacco stalk than is the case for wheat plants (cf. Table I and Table IV). On the other hand, trans 2-hexenal and the sum of nonadien-aldehydes and alcohols (which presumably originate from linolenic acid C18:3) account for a larger proportion of the oil of wheat than of tobacco stalk.

Several studies showed that chemical agents can inhibit lipoxygenase and cycloxygenase-mediated changes of fatty acid substrates. For example, Tappel et al. (23) and Takahama (24) determined that lipoxygenase-catalyzed oxidation of linoleic acid was inhibited by various antioxidants such as catechol, nordihydroguaiaretic acid and quercetin. Recently Josephson et al. (25) showed that formation of carbonyl-containing compounds and alcohols that characterize the fresh fish aroma of emerald shiners was inhibited when fish were sacrificed and immediately treated with inhibitors of cyclooxygenase and lipoxygenase, respectively. Also Blackwell and Flower (26) showed that phenidone was a potent inhibitor of lipoxygenase and cyclo-oxygenase in vitro.

Table IV.   Compounds in Steam Distillate-Hexane Extract of Wheat
Plants[a]

Compound identity by capillary GC/GS-MS (EI), GC cochromatography and IR in some cases	% of total oil [b]
hexanal	1.3
trans 2-hexenal	11.1
hexan-1-ol	0.6
trans 2-hexen-1-ol	1.7
heptanal	2.2 [c]
trans 2-heptenal	tr
heptan-1-ol	tr
trans 2-octenal	0.5
octan-1-ol	0.5
nonanal	9.9
trans 2-nonenal	2.2
trans, cis 2,6-nonadienal	7.7
nonan-1-ol	tr
trans 2-nonen-1-ol	1.2
cis 3-nonen-1-ol	2.5
trans, cis 2,6-nonadien-1-ol	5.6
cis, cis 3,6-nonadien-1-ol	5.0
decanal	0.5
trans 2-decenal	tr
β-ionone	3.6
benzaldehyde	tr
unidentified	43.9

(a) Reproduced after adaptation with permission from reference 15
(b) From frozen wheat plants 40-50 cm in height that were macerated prior to distillation
(c) tr indicates less than 0.5%

Studies were carried out with tobacco and wheat to determine the effects of known inhibitors of lipoxygenase and cyclo-oxygenase on the yields of volatile oils in SDE (Table V). Although there is considerable evidence that lipoxygenase is important in higher plant metabolism, it is not certain whether cyclooxygenase plays a similar role (18-20, 27). In the case of volatiles from tobacco stalk, small increases were observed which may have resulted from the higher ionic strength of the solutions.   We decided to try other specific agents using homogenized plant tissue, reasoning that homogenization would release more native enzymes and would allow maximum exposure of endogenous plant substrates to the inhibitors.   Wheat was used because it could be homogenized more easily than tobacco stalk.

Phenidone (1-phenyl-3-pyrazolidone) inhibited total volatile oil yield in wheat by 43%, which was the largest decrease we found (Table V).   Nordihydroguaiaretic acid, an aromatic polyphenol caused a 32% inhibition in total volatile oil yield. Acetylsalicylic acid, on the other hand, did not affect yields. Acetylsalicylic acid is a strong inhibitor of cyclooxygenase but a weak inhibitor of lipoxygenase (28).   On the basis of these

Table V.   Effect of Fatty Acid Oxidation Inhibitors on Yields
           of Total Volatiles in Steam Distillates

Plant sample description	Agent (conc.)	Reported as major in- hibitor of	Total volatiles, µg/g Treated	Control
tobacco, KY 14 burley [a]	SnCl$_2$ (10 mM)	L[c]	5.2	4.1
" "	NaCl (1 M)		4.6	4.1
" "	acetylsali- cylic acid (10 mM)	C[d]	4.9	4.1
wheat, Arthur 71[b]	phenidone (100 µM)	L,C	8.6	15.2
" "	nordihydro- guaiaretic acid (100 µM)	L	18.5	27.3
" "	acetylsali- cylic acid (100 µM)	C	14.6	14.5

(a) 0.5 cm stalk sections   (c) lipoxygenase
(b) whole plant, homogenized   (d) cyclooxygenase

results and the known target enzymes inhibited by these agents,
the lipoxygenase system appears to play a significant role in the
production of volatile compounds in wheat.   The effect of
phenidone on the yield of individual volatile compounds in SDE of
wheat is illustrated in Figure 4. Comparisons of phenidone-
treated and control yields at a given Kovats index indicate
relatively strong reductions of 6-carbon volatiles with
phenidone.   On the other hand, there were some increases of
9-carbon compounds with phenidone treatment.

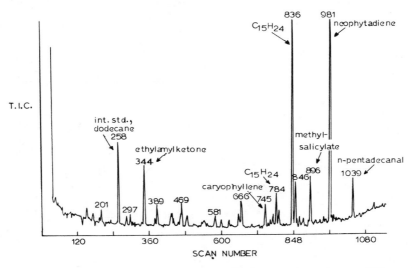

Figure 3.   Capillary GC-total ion current profile of Tenax-bound tobacco stalk headspace volatiles.

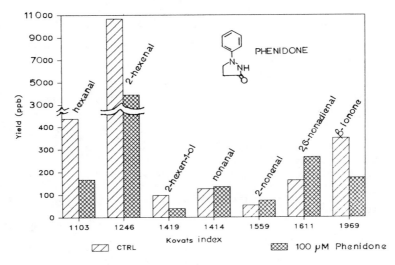

Figure 4.   Effect of phenidone on yields of steam distillate volatiles from wheat.

Acknowledgments

The authors thank Mr. John Loughrin for assistance in carrying out
analyses. We are indebted to the Tobacco Health Research
Institute and Mr. Charles Hughes for mass spectral analyses.

Literature Cited

1. Davis, D. L. Recent Advances in Tobacco Science, 1976, 2,
   80-111.
2. Demole, E.; Berthet, D. Helv. Chem. Acta. 1972, 55,
   1866-1882.
3. Rix, C. E.; Lloyd, R. A.; Miller, C. W. Tobacco Science
   1977, 21, 93-96.
4. Burton, H. R.; Bush, L. P.; Hamilton, J. L. Recent Advances
   in Tobacco Science, 1983, 9, 91-154.
5. Lloyd, R. A.; Miller, C. W.; Roberts, D. L.; Giles, G. A.;
   Dickerson, J. P.; Nelson, N. H.; Rix, C. E.; Ayers, P. H.
   Tobacco Science 1976, 20, 40-48.
6. Enzell, C. R.; Wahlberg, I. Recent Advances in Tobacco
   Science, 1980, 6, 64-122.
7. Buttery, R. G.; Ling, L. C.; Wellso, S. G. J. Agric. Food
   Chem. 1982, 30, 791-792.
8. French R. C. J. Agric. Food Chem. 1983, 31, 423-427.
9. Pharis, V. L.; Kemp, T. R.; Knavel, D. E. Scientia Horticulturae 1982, 17, 311-317.
10. Hamilton-Kemp, T. R.; Andersen, R. A. Phytochem. 1985 in
    press.
11. Andersen, R. A.; Lowe, R.; Vaughn, T. A. Phytochem. 1969,
    8, 2139-2147.
12. Cochran, W. G.; Cox, G. M. "Experimental Designs"; 2nd
    Edit.; John Wiley and Sons, Inc.: New York, 1957; Chapter
    4.
13. Likens, S. T.; Nickerson, G. B. Am. Soc. Brewing Chemists
    Proceedings 1964, 5-13.
14. Buttery, R. G.; Seifert, R. M.; Guadagni, D. G.; Black, D.
    R.; Ling, L. C. J. Agric. Food Chem. 1968, 16, 1009-1015.
15. Hamilton-Kemp, T. R.; Andersen, R. A. Phytochem. 1984,
    23, 1176-1177.
16. Perry, J. A. "Introduction to Analytical Gas Chroma-
    tography"; Marcel Dekker, Inc.: New York, 1981; p. 246.
17. Chaven, C.; Hymowitz, T.; Newell, C. A. J. Am. Oil Chem.
    Soc. 1982, 59, 23-25.
18. Galliard, T.; Phillips, D. R.; Reynolds, J. Biochimica et
    Biophysica Acta 1976, 441, 181-192.
19. Hatanaka, A.; Kajiwara, T.; Harada, T. Phytochem. 1975,
    14, 2589-2592.
20. Sekiya, J.; Kajiwara, T.; Hatanaka, A. Plant and Cell
    Physiol. 1984, 25, 269-280.
21. Loomis, W. D.; Croteau, R. In "The Biochemistry of Plants";
    Stumpf, P., Ed.; Academic Press: New York, 1980; 4, pp.
    364-410.
22. Hamilton-Kemp, T. R.; Rodriguez, J. G.; Andersen, R. A.,
    unpublished results.

23. Tappel, A. L.; Lundberg, W. O.; Boyer, P. D. Archiv. Biochem. Biophys. 1953, 42, 293-304.
24. Takahama, U. Phytochem. 1985, 24, 1443-1446.
25. Josephson, D. B.; Lindsay, R. C.; Stuiber, D. A. J. Agric. Food Chem. 1984, 32, 1347-1352.
26. Blackwell, G. J.; Flower, F. J. Prostaglandins 1978, 16, 417-425.
27. Douillard, R.; Bergeron, E. Physiol. Plant 1981, 51, 335-338.
28. Shafer, A. I.; Turner, N. A.; Handin, R. I. Biochim. Biophys. Acta 1982, 712, 535-541.

RECEIVED January 27, 1986

BIOSYNTHETIC PATHWAYS

10

Biogenesis of Aroma Compounds through Acyl Pathways

Roland Tressl and Wolfgang Albrecht

Technische Universität Berlin, Seestrasse 13, D-1000 Berlin 65, West Germany

Formation of aroma compounds from lipid
and amino acid metabolism is described.
Principle pathways have been studied with
tissue slices and ^{14}C-labeled precursors.
The enantiomeric composition of chiral
fruit aroma components such as alcohols,
3-hydroxyacid esters and lactones was
determined and possible pathways for
their biosynthesis were presented.

In the last decade the application of capillary GC/MS,
capillary GC/FTIR and FT-NMR made possible the identification of several hundred new aroma compounds in fruits
and plant materials. Most studies have been undertaken
with the aim of identifying the substances responsible
for the characteristic aroma or flavor. Our knowledge
about the pathways and control mechanisms of the biogenesis and accumulation of these substances is still
limited as recently documented by Schreier (1).
   In 1975 a scheme of principle pathways by which
fruit volatiles may be formed was presented and some of
the reaction sequences were demonstrated by labeling
experiments using (U-^{14}C)-precursors and fruit tissue
slices (2). Figure I illustrates the importance of acyl-
pathways in the formation of aroma and flavor compounds
while showing the central role of acetyl-CoA in the metabolism of fatty acids, terpenes, amino acids and carbohydrates. The growing unripe fruit synthesizes high-mole-
cular structures such as proteins, polysaccharides,
lipids and flavonoids by carbohydrate metabolism initiated by photosynthesis in the leaves. During ripening
catabolic reactions predominate and the production of
volatiles occurs during a short period and is influenced
by internal and external factors.

0097-6156/86/0317-0114$06.00/0

Lipid Metabolism

Many aroma compounds in fruits and plant materials are derived from lipid metabolism. Fatty acid biosynthesis and degradation and their connections with glycolysis, gluconeogenesis, TCA cycle, glyoxylate cycle and terpene metabolism have been described by Lynen (3) and Stumpf (4). During fatty acid biosynthesis in the cytoplasm acetyl-CoA is transformed into malonyl-CoA. The de novo synthesis of palmitic acid by palmitoyl-ACP synthetase involves the sequential addition of C_2-units by a series of reactions which have been well characterized. Palmitoyl-ACP is transformed into stearoyl-ACP and oleoyl-CoA in chloroplasts and plastides. During ß-oxidation in mitochondria and microsomes the fatty acids are bound to CoASH. The ß-oxidation pathway shows a similar reaction sequence compared to that of de novo synthesis. ß-Oxidation and de novo synthesis possess differences in activation, coenzymes, enzymes and the intermediates (S)-(+)-3-hydroxyacyl-S-CoA (ß-oxidation) and (R)-(-)-3-hydroxyacyl-ACP (de novo synthesis). The key enzyme for de novo synthesis (acetyl-CoA carboxylase) is inhibited by palmitoyl-S-CoA and plays an important role in fatty acid metabolism.

During the formation of aroma compounds in climacteric and postclimacteric fruits, catabolic reactions predominate in decreasing the amounts of poly- and oligosaccharides and acids. Figure II illustrates the formation of volatiles by Bartlett pears stored at 25 °C. The ripening fruit produces small amounts of ethylene (5) which is known as a ripening hormone inducing biochemical, physical and chemical changes (eg. the increase in enzyme activities, accumulation of fatty acids and amino acids and an increase in membrane permeability). Climacteric fruits such as pears, apples, bananas, mango, peach etc. show the characteristic climacteric rise in respiration (6) initiating the biosynthesis of volatiles. It can be seen that esters and **α**-farnesene were produced at rates subject to cyclic variations, indicating that the biogenesis of volatiles in fruit is a dynamic process.

In a series of labeling experiments, the biogenesis of esters and alcohols by fruit tissue slices was investigated (7,8). While (U-^{14}C)-acetate and (U-^{14}C)-butyrate were incorporated into the corresponding esters by postclimacteric banana tissue, (U-^{14}C)-octanoate was transformed in climacteric and postclimacteric banana tissues into caproic and butyric acid by ß-oxidation and into heptanoic acid by **α**-oxidation (9). In addition, 1-octanol, Z-4-hepten-2-ol, pentanol-2 and the corresponding esters were labeled. The biosynthesis of alcohols and esters is an analogous reaction to the formation of wax esters as outlined by Kolattukudy (10). Octanoyl-CoA is reduced by an acyl-CoA reductase (NADH dependent) to octanal which is further transformed into 1-octanol by an

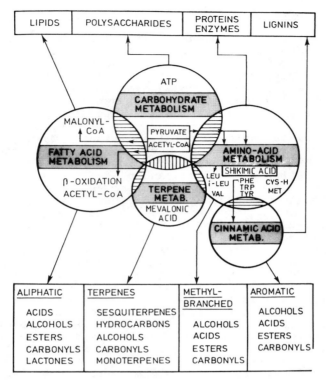

Figure I.   Biogenesis of fruit volatiles.

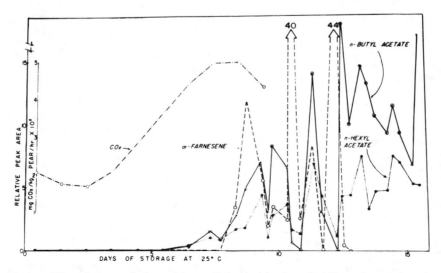

Figure II.   Rates of formation of individual volatiles from ripening Bartlett pear.

NADH dependent alcohol dehydrogenase. The formation of
esters is catalyzed by acyl-CoA : alcohol transacylases.
The biogenesis of (Z)-4-hepten-2-ol in banana cannot be
explained by ß-oxidation as for 2-heptanone and 2-hepta-
nol. Figure III presents a possible pathway for the conversion of caprylic acid to (Z)-4-hepten-2-ol. This unusual compound was also identified by Buttery et al. (11)
in corn husks. The enzymic conversion of (Z)-3-hexenoyl-CoA
to (E)-2-hexenoyl-CoA by an isomerase is known in the
ß-oxidation pathway of unsaturated fatty acids (12). The
elongation of (Z)-3-hexenoyl-CoA could also provide the
precursor for the formation of (Z)-4-hepten-2-one and (Z)-5-
octen-1-ol, which were recently identified in banana.
(Z)-4-hepten-2-one is reduced to (S)-(Z)-4-hepten-2-ol and
transformed into the corresponding acetates and butyrates.
In the analogous labeling experiment with strawberry
tissue, (8-^{14}C)-caprylic acid was transformed into
caproic and butyric acid, 2-heptanone and 2-pentanone
by ß-oxidation. Caprinic, dodecanoic and tetradecanoic
acids were also labeled in this experiment indicating de
novo synthesis. In addition, caprylic acid was reduced
to 1-octanol and transformed into methyl-, ethyl-, hexyl-
and octyl octanoates as well as into octyl acetate. Recently we developed gaschromatographic methods for the
investigation of chiral aroma compounds to gain insight
into their biosynthetic origins (13,14). 2-Pentanol,
2-heptanol and 2-nonanol and the corresponding esters
were identified in passion fruit. Figure IV presents the
enantiomeric composition of 2-pentanol and 2-heptanol in
yellow passion fruit and of 2-heptanol and 2-heptyl-
acetate in purple passion fruit. The free alcohols in the
yellow variety are mainly contained in the (S)-(+)-con-
figuration corresponding to the reduction of methyl ketones with alcohol-NADH-dehydrogenase of bakers yeast
(15). In contrast, purple fruit contains the alcohols in
the (R)-configuration. Additionally only the purple variety forms the corresponding esters. Figure V presents
a scheme to explain the possible formation of these
secondary alcohols. In the purple passion fruit 2-hep-
tanone may be reduced to (R)-(-)-2-heptanol by an enzyme
comparable in stereospecificity to dihydroxyacetone reductase (16). Only this variety formed (Z)-3-octenyl and
(Z)-3-decenyl esters of acetic, butyric, caproic and caprylic acids. (S)-(+)-3-hydroxyoctanoyl-CoA is presumably
dehydrated to (Z)-3-octenoyl-CoA, reduced to (Z)-3-octen-1-ol
and transformed into esters with acyl-CoA by alcohol acyl-
CoA:transacylases.
    Unsaturated esters may also be derived from the pro-
ducts of ß-oxidation of unsaturated fatty acids. Methyl-
and ethyl decadienoates are known as character impact
compounds of Bartlett pears and all presumptive intermediates of the pathway were identified by Jennings et
al. (17,18). Linoleyl-CoA is transformed into (E,Z)-2,4-
decadienoyl-CoA by ß-oxidation. The further degradation

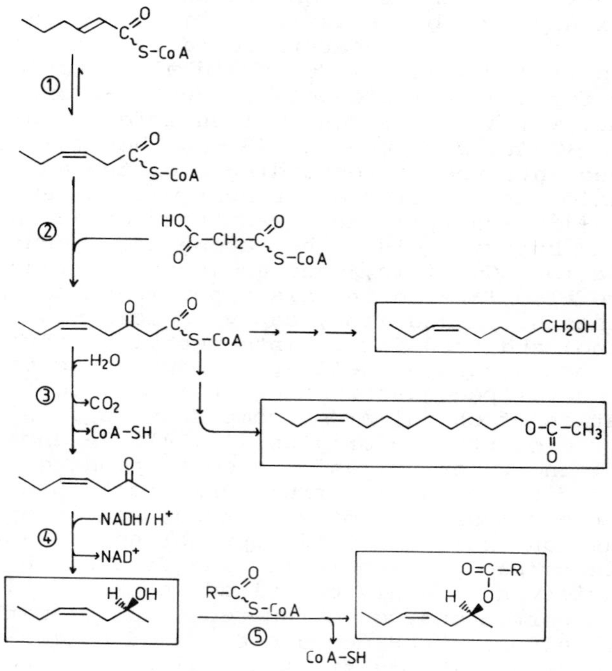

Figure III.   Scheme to explain the formation of (S)-(Z)-4-hepten-2-ol,
(S)-(Z)-4-hepten-2-yl acetate, and (Z)-5-octene-1-ol in banana.

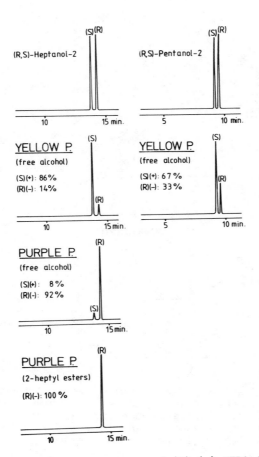

Figure IV.   Capillary GC-separation of (R)-(+)-MTPA-derivatives of
secondary alcohols and their esters, isolated by preparative GC from
yellow and purple passion fruits.

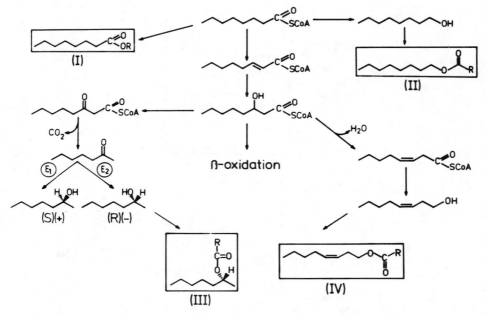

Figure V. Possible pathway for the biosynthesis of secondary alcohols, their esters, and other typical constituents of passion fruit. Enzyme (E1) is operative in yellow passion fruit. The antipodal reduction catalyzed by enzyme (E2) and the following esterification take place only in the purple variety.

of 2,4-decadienoates was recently investigated by Cuebas
and Schulz (19) with rat heart mitochondria. (E,Z)-2,4-
Decadienoyl-$\overline{\text{CoA}}$ is reduced by 2,4-dienoyl-CoA reductase
to (E)-3-decenoyl-CoA which is transformed into (E)-2-
decenoyl-CoA by an 2,3-enoyl-CoA isomerase and further
degraded by ß-oxidation. The proposed formation of (Z)-2-
octenoyl-CoA by Stoffel and Caesar (12) was not observed
during these experiments.

Boland et al. (20) demonstrated the oxidative degra-
dation of ω-6- and $\overline{\omega}$-3-unsaturated fatty acids by brown
algae into unsaturated hydrocarbons which act as pheromones. In deuterium labeling experiments pentadeca-6,9,
12-trienoic acid was transformed into (Z)-1,5,8,11-tetra-
decatetraene and 1,7,10,13-hexadecatetraene, respectively. The corresponding 3-hydroxy acids were the key
intermediates which were transformed into hydrocarbons
by this enzyme system. As outlined in Figure VI, 2,4-
decadienoate could be metabolized by an analogous reaction sequence forming 1,3,5-undecatriene which was
recently identified by Berger et al. (21) as an important
constituent in pineapple.

Chiral 3-hydroxyacid esters and 3-,5-acetoxyacid
esters have been identified in tropical fruits and other
products (14). These compounds are intermediates in
ß-oxidation and de novo synthesis of fatty acids, but
exhibit opposite configuration at the carbinol carbon.
The enantiomeric composition of 3-hydroxyacid esters in
passion fruit, mango and pineapple have been investigated.
Figure VII presents the separation of (R)- and (S)-3-
hydroxybutanoates. The compounds in yellow passion fruit
were mainly of the (S)-(+)-configuration as predicted for
intermediates of ß-oxidation. In the purple variety, and
in mango, the (R)-(-)-enantiomers predominate. These compounds may be found as an offshoot of de novo lipid synthesis or by hydration of (Z)-2-enoyl-CoA leading to
(R)-(-)-3-hydroxyacyl-CoA (12).

Figure VIII shows the enantiomeric composition of
various hydroxy- and acetoxyacid esters and of γ-hexa-
and δ-octalactone isolated from pineapple. Methyl
3-hydroxyhexanoate and methyl 3-acetoxyhexanoate are
mainly of the (S)-configuration corresponding to intermediates of ß-oxidation. The optical purity of the
5-acetoxy esters is lower than of the 3-acetoxy derivatives. The lactones were mainly of the (R)-configuration.
Figure IX presents a possible pathway to explain the
formation of these compounds. Methyl (S)-(+)-3-hydroxy-
hexanoate and methyl (S)-3-acetoxyhexanoate may be
derived from (S)-3-hydroxyhexanoyl-CoA by transacylation
with methanol and acetyl-CoA, respectively. The biosynthesis of 5-hydroxyacids is still unknown, but they may
be formed by elongation of 3-hydroxyacids with malonyl-
ACP. This hypothesis could explain their varying enantiomeric composition relative to the 3-hydroxyacids.
However, hydration of unsaturated acids and/or the
reduction of 5-oxoacids may be involved.

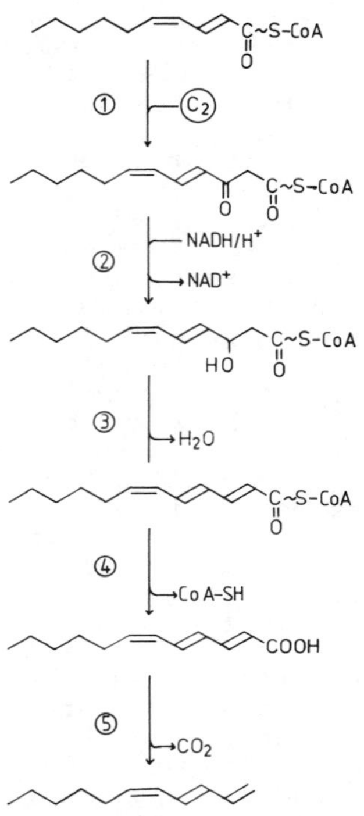

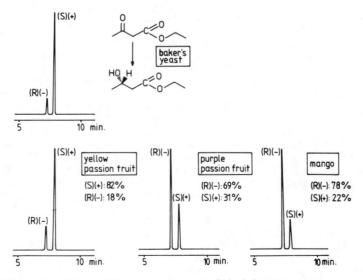

Figure VI. Possible pathway for the formation of (E,E,Z)-1,3,5-undecatriene.

Figure VII. Capillary GC-separation of (R)-(+)-MTPA-derivatives of ethyl 3-hydroxybutanoate obtained by yeast reduction and after isolation from passion fruit and mango.

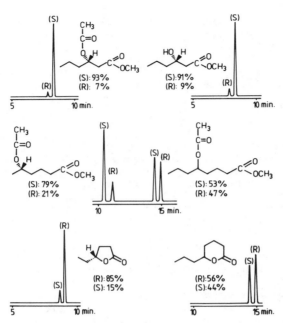

Figure VIII.   Capillary GC-separation of the (R)-(+)-PEIC-
derivatives of chiral (main) constituents isolated from pineapple.
3-Hydroxyhexanoate and 3-acetoxyhexanoate were separated as the (R)-
(+)-MTPA-derivatives.

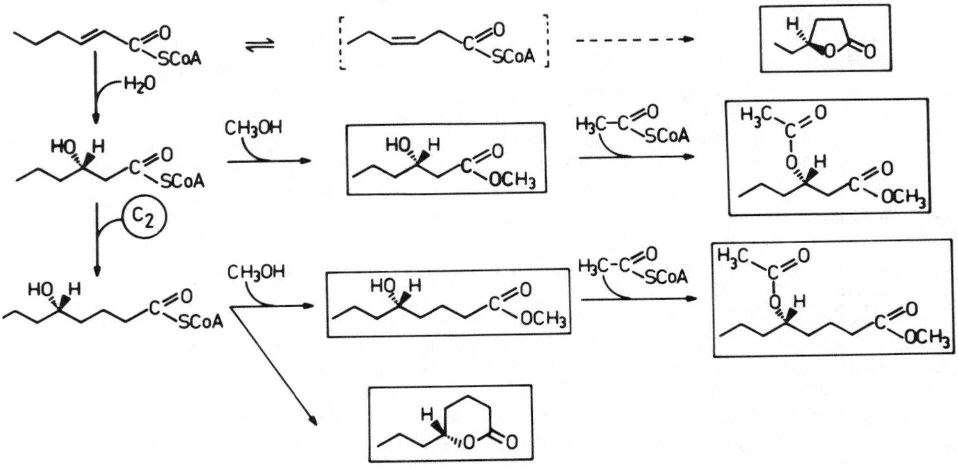

Figure IX.   Possible pathway for the formation of (S)-3-hydroxyacid
esters, (S)-3-acetoxyacid esters, and (S)-5-acetoxyacid esters in
pineapple.

In 1976 Winter et al. (22) identified 3-methylthio-
hexanol and a mixture of (Z)- and (E)-2-methyl-1,3-oxa-
thiane in yellow passion fruit. 3-Methylthiohexanol possesses a green fatty and sulfury note and imparts the
character of fresh fruit to the juice of passion fruits.
The oxathianes were described as key compounds for the
typical aroma of passion fruit and the (Z)-isomer had a
stronger flavor. Pickenhagen and Brönner-Schindler (23)
synthesized the enantiomers of (+)- and (-)-(Z)-2-methyl-
4-propyl-1,3-oxathiane and determined their thresholds
in water at 2 ppb and 4 ppb, respectively. The (2 S, 4 R)-
2-methyl-4-propyl-1,3-oxathiane possessed the typical
aroma of passion fruit. We have identified 8 sulfur containing constituents in yellow and purple passion fruit
(24) which fit the biosynthetic route shown in Figure X.
3-Mercaptohexanol and the corresponding esters especially contribute to the pleasant flavor of passion fruit.
The enantiomeric composition of the sulfur containing
compounds in yellow and purple passion fruit is still
unknown. The results of quantitative and qualitative investigation of sulfur containing components showed the
best correlation to the organoleptic properties of passionfruit concentrates.

γ- and δ-lactones are known as important flavor
components in fruits such as peach, apricot, strawberry, mango, coconut, milk products and in fermented
foods. The biosynthetic origins and enantiomeric composition of these compounds are not known. In 1983 we separated chiral δ-lactones by the formation of diastereoisomeric ortho esters with optically pure D-(-)-
2,3-butandiol and determined the enantiomeric compo-
sition of δ-octa-, δ-deca- and δ-dodecalactone isolated from coconut (25). This method was not applicable
for γ-lactones. Recently we solved this problem by
reduction of γ- and δ-lactones to the corresponding
1,4- and 1,5-diols and conversion to (R)-(+)-PEIC-deri-
vatives (PEIC = (R)-(+)-phenylethylisocyanate)(26). As far
as we know it is the first enantioresolution of these
γ- and δ-lactones by capillary GC. The configuration
of the lactones was determined by reduction of 4- and
5-oxoacid esters with bakers yeast and measurement of
the optical rotation (26). Mangos contain δ-lactones
from C_6 to C_{12} and γ-lactones from C_5 to C_{10}. The lactones were isolated by preparative GC, transformed into
(R)-(+)-PEIC-derivatives and then analyzed by capillary
GC as illustrated in Figure XI. The δ-lactones consisted mainly of the (R)-enantiomers comparable to the
results with lactones of coconut. γ-Hexa- and γ-octa-
lactone occur mainly in the (S)-configuration, whereas in
γ-decalactone the (R)-enantiomer predominates. These
unexpected results indicate that different biosynthetic
routes are operative, perhaps similar to the formation
of 3-hydroxyacid esters. Other biological systems bio-

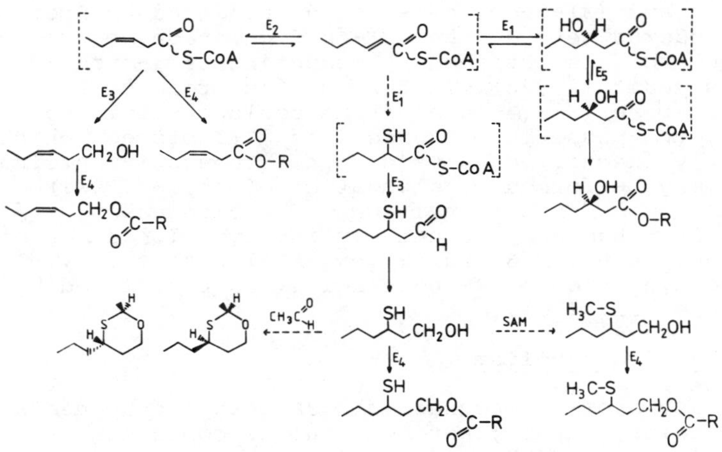

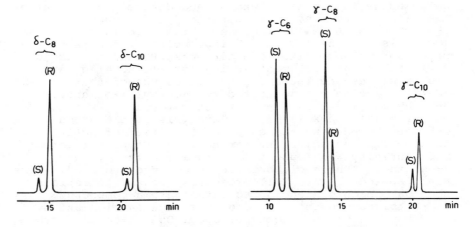

Figure X.   Possible pathway for the biosynthesis of (Z)-3-hexanoic acid esters, (Z)-3-hexanol, (Z)-3-hexenyl esters, 3-hydroxyhexanoic acid esters, and sulfur-containing components in passion fruit.

Figure XI.   Capillary GC-separation of (R)-(+)-PEIC-derivatives of δ- and γ-lactones isolated from mango.

synthesize γ-lactones in optically pure form: γ-deca-
and γ-dodecalactone in strawberries as well as γ-deca-
lactone and (Z)-6-dodecen-4-olide produced by Sporobolomyces odorus were of the (R)-configuration, corresponding
to the products obtained by reduction of 4-oxoacids with
bakers yeast (27,28). In the culture broth of Poria
xantha, pure (S)-γ-octalactone could be isolated. 4-Oxo-
acids, the possible precursors of γ-lactones, which were
recently identified as major constituents in hazelnut
(29), may be formed by elongation of acyl-ACP with
succinyl-CoA (or succinyl-ACP). The biogenesis of δ-lac-
tones in mango and coconut may involve elongation of
3-oxoacyl-ACP (or 3-hydroxyacyl-ACP) with malonyl-ACP,
or the hydration of 4-enoic acids, as postulated by
Dimick et al. (30).

Amino Acid Metabolism

Many aroma compounds such as methyl branched alcohols, acids, esters, ketones, sulfur containing and
aromatic compounds are derived from amino acid metabolism. The biosynthesis of amino acids in chloroplasts
and plastids and their degradation in mitochondria involve different reaction sequences. Wyman et al. (31)
and Myers et al. (32,33) demonstrated the transformation
of (U-^{14}C)-leucine to 3-methyl-1-butanol and 3-methyl-
butyl acetate by banana discs. We investigated the free
amino acids in ripening bananas and determined that after
the climacteric rise in respiration, the amounts of
L-leucine and L-valine increased from 5 to 15 mg per
100 g tissue. The level of L-isoleucine, which is only a
precursor of some minor constituents, remained constant
(34). Labeling experiments using postclimacteric banana
tissue slices and labeled amino acids showed that
(U-^{14}C)-leucine was converted to 3-methyl-1-butanol,
3-methylbutyl esters, 3-methylbutyrates and 2-ketoiso-
caproate (Figure XII) (35). Hydrolysis of the aroma
extract showed a nearly equal distribution of activity
in alcohols and acids. By far the largest proportion of
radioactivity was found in 3-methyl-1-butanol, but
2-heptanol, 1-hexanol, butyric acid and caproic acid were
also labeled. Figure XIII presents a scheme to explain
the conversion of leucine to banana aroma compounds.
    α-Ketoisocaproate is formed by transamination and by
decarboxylation to 3-methylbutanoyl-CoA which is further
transformed into 3-methylbutanoates or reduced to 3-me-
thylbutanol and incorporated into 3-methylbutyl esters.
The end products of the oxidative degradation of leucine
are acetoacetate and acetyl-CoA which are further degraded by the TCA-cycle or transformed into butyrates,
butanoic acid, caproic acid and 2-heptanol by well-known
transformations of lipid metabolism. In analogous experiments, L-valine was converted into 2-ketoisovaleric acid,
2-methylpropionic acid, 2-methyl-1-propanol, 2-methyl-

Banana disks	40 g (3 x 20 mm)
Incubation time	3 hr
Precursor	50 μCi of $(U-^{14}C)$-leucine
Radioactivity in the aroma extract, %	0,5

Distribution of radioactivity among volatile components, %

Ethyl acetate	0,5
3-Methylbutanal	
2-Methylbutyl acetate	1,1
Methyl 3-methylbutyrate	
Ethyl butyrate	0,1
n-Butyl acetate	2,4
Ethyl 3-methylbutyrate	
3-Methylbutyl acetate	16,0
3-Methyl-1-butanol	18,0
2-Methylpropyl 3-methylbutyrate	1,9
3-Methylbutyl 2-methylbutyrate	
n-Butyl butyrate	10,0
2-Pentanol butyrate	
3-Methylbutyl butyrate	10,0
3-Methylbutyl 3-methylbutyrate	9,0
Methyl 2-ketoisocaproate	4,0
n-Hexyl butyrate	
3-Methylbutyl caproate	25,0
2-Heptanol 3-methylbutyrate	
2-Heptenol 3-methylbutyrate	
3-Methylbutyric acid	2,0

Distribution of radioactivity among volatile components after saponification of the aroma extract

Alcohols (47 %)	%	Acids (53 %)	%
2-Methyl-1-propanol	0	Acetic acid	0,1
1-Butanol	0	Butyric acid	1,7
3-Methyl-1-butanol	76,0	3-Methylbutyric acid	35,0
2-Heptanol	23,0	Caproic acid	1,2
1-Hexanol	1,0	2-Ketoisocaproic acid	62,0

Figure XII. Conversion of $(U-^{14}C)$-L-leucine into volatile components by postclimacteric banana tissue slices.

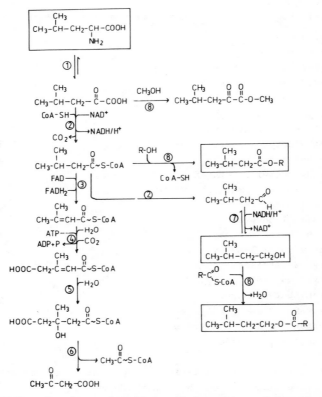

Figure XIII. Leucine catabolism and formation of 3-methyl butanoates and 3-methyl butylesters in banana.

propionates and 2-methylpropyl esters. The end product of valine catabolism is propionyl-CoA which is further degraded via 3-hydroxypropionyl-CoA to acetyl-CoA and CO_2. 2,3-Butandione, acetoine, 2,3-butandiol and the corresponding acetates and butyrates (which are formed as major constituents in ripening bananas) are derived from α -acetolactate, an intermediate of valine biosynthesis (Figure XIV). Therefore, (U-^{14}C)-valine was not converted into these constituents.

Figure XV presents possible biosynthetic routes to asparagusic acid from cysteine and valine, respectively, which were investigated using radiolabeled precursors (36). In labeling experiments (U-^{14}C)-cysteine and ($\overline{U}$-^{14}C)-acetate were not transformed into asparagusic acid as suggested by Metzner (37). The catabolism of valine forms intermediates which may act as precursors as outlined in the scheme. (U-^{14}C)-valine and (U-^{14}C)-isobutyrate were converted to some extent into 2-oxo-3-methylbutyrate, isobutyrate and methylacrylate. 3-Mercaptoisobutyric acid and 3-methylthioisobutyric acid were strongly labeled components. On the other hand asparagusic acid incorporated only a small amount of radioactivity. Isobutyric acid is more efficiently transformed into volatiles by asparagus discs than is valine. Valine, leucine and isoleucine act as precursors for methyl branched fatty acids, ketones and esters in hop oil, and they are incorporated into cohumulone, humulone and adhumulone, respectively, as demonstrated by Drawert et al. (38).

L-Phenylalanine,which is derived via the shikimic acid pathway,is an important precursor for aromatic aroma components. This amino acid can be transformed into phenylpyruvate by transamination and by subsequent decarboxylation to 2-phenylacetyl-CoA in an analogous reaction as discussed for leucine and valine. 2-Phenylacetyl-CoA is converted into esters of a variety of alcohols or reduced to 2-phenylethanol and transformed into 2-phenylethyl esters. The end products of phenylalanine catabolism are fumaric acid and acetoacetate which are further metabolized by the TCA-cycle. Phenylalanine ammonia lyase converts the amino acid into cinnamic acid, the key intermediate of phenylpropanoid metabolism. By a series of enzymes (cinnamate-4-hydroxylase, p-coumarate 3-hydroxylase, catechol O-methyltransferase and ferulate 5-hydroxylase) cinnamic acid is transformed into p-coumaric-, caffeic-, ferulic-, 5-hydroxyferulic- and sinapic acids,which act as precursors for flavor components and are important intermediates in the biosynthesis of flavonoides, lignins, etc. Reduction of cinnamic acids to aldehydes and alcohols by cinnamoyl-CoA : NADPH-oxidoreductase and cinnamoyl-alcohol-dehydrogenase form important flavor compounds such as cinnamic aldehyde, cinnamyl alcohol and esters. Further reduction of cinnamyl alcohols lead to propenyl- and allylphenols such as

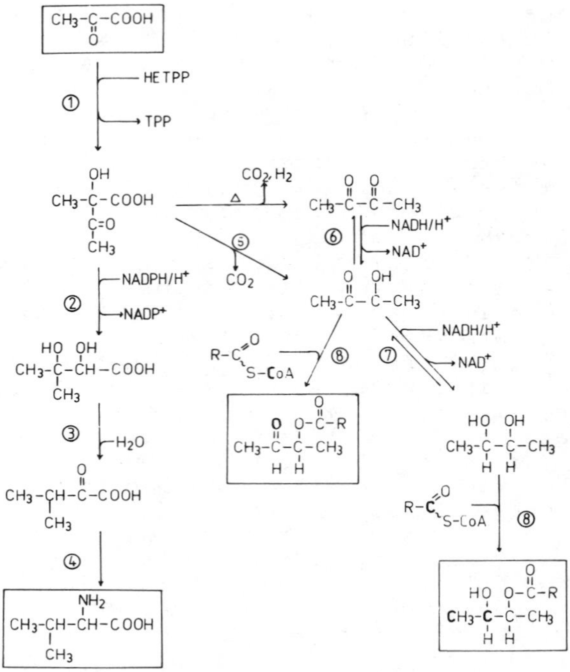

Figure XIV. Biosynthesis of L-valine and the related formation of acetoin, 2,3-butandiol, and esters in banana.

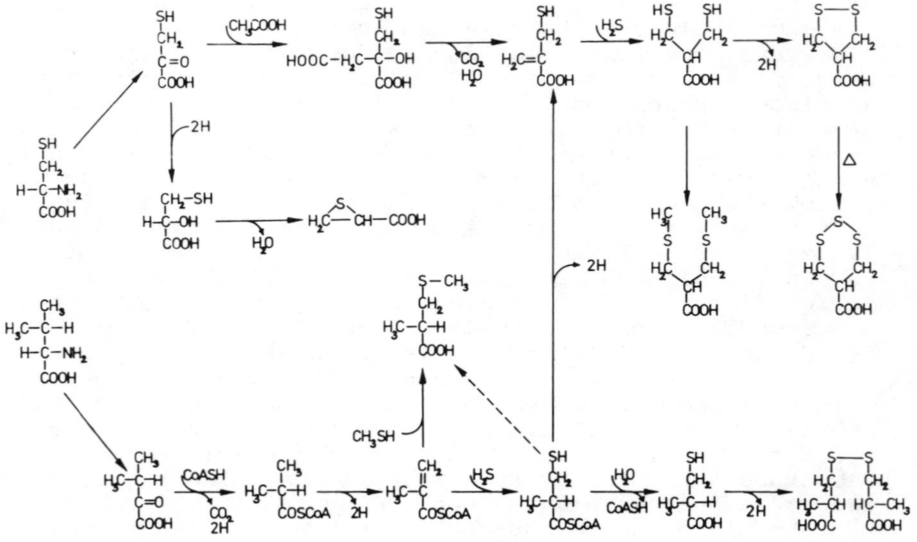

Figure XV. Possible biosynthetic pathways leading to asparagusic acid (VI).

anethole (39), eugenol, methyleugenol and elemicin. This
pathway is operative in banana as demonstrated by labeling experiments which (1-^{14}C)-caffeic acid which was
converted into eugenol, eugenol methyl ether and elimicin
by banana tissue slices in sucrose solution. Eugenol
accounted for 53 % of the radioactivity of volatiles
extracted from the sucrose solution and 5-methoxyeugenol
for about 10 %. In the extract of the banana slices
eugenol methyl ether and elimicin were more efficiently
labeled. The conversion of coniferylalcohol to eugenol
and methyleugenol was demonstrated by Klischies et al.
(40) using double labeled coniferylalcohol. No further
studies on the biogenesis of these important classes of
compounds have been performed, but they have been identified in a variety of plant materials (eg. eugenol in
tomatoe, nutmeg, cranberry, cocoa, current, peach, clove,
pepper, cinnamon, mace, mint, passion fruit and cloud-
berry). Eugenol methyl ether and 5-methoxy eugenol have
been identified in blackberry, nutmeg and mace.

Hydrogenation of the unsaturated side chain of cinnamic acids leading to dihydrocinnamic acids and to the
corresponding alcohols must also occur, but the enzymes
are still unknown. Certain microorganisms decarboxylate
cinnamic acids to the corresponding 4-vinylphenols which
have been found in beer, wine and other fermented products (41). A reaction sequence analogous to the ß-oxidation of fatty acids was proposed by Zenk (42) to
explain the conversion of cinnamic acids to benzoic
acids. The C_6 - C_1 acids can also be derived from the
shikimate pathway, via dehydroshikimic acid. Both pathways occur in higher plants and microorganisms. A third
pathway for the formation of aromatic compounds via
polyketides, as demonstrated for the origin of 6-methyl-
salicilic acid, is synthesized by a multienzyme complex
(Dimroth et al., 43). By analogy to cinnamic acids,
benzoic acids are also utilized as the precursors of
aroma compounds and a variety of other natural products.

Conclusions

More detailed studies on the biosynthesis of aroma
compounds through acyl pathways are essential for future
biotechnological applications. The reaction sequences
(enzymes and their regulation) leading to typical aroma
compounds are still unknown. The more important components, possessing cis- or trans-double bonds or chiral
hydroxy groups in distinct positions of the carbon chain
cannot be explained by the ß-oxidation sequence. The
sites and enzymes responsible for the biosynthesis of typical esters are not known. The characterized compounds
indicate discrete enzyme systems for chain elongation,
reduction of prochiral ketones and transacylation. More
studies with labeled precursors and isolated enzymes are
essential to establish these biosynthetic routes. As far as

we know callus and suspension cultures of fruits are not able to produce typical aroma compounds (44). These systems may, however be useful in investigating biochemical sequences of aroma compounds comparable to their biochemical application in alkaloid biosynthesis. It will also be necessary to call attention to conclusions drawn from work with microorganisms.

Literature Cited

1. Schreier, P."Chromatographic Studies of Biogenesis of Plant Volatiles"; Bertsch, W.; Jennings, W.G.; Kaiser, R.E., Eds.; Alfred Hüthig Verlag, Heidelberg 1984
2. Tressl, R.; Holzer, M.; Apetz, M. In "Aroma Research"; Maarse, H.; Groenen, P.J., Eds.; Int. Symp. Aroma Research, Zeist, 1975, pp. 41-62
3. Lynen, F. Angew. Chem. 1964, 77, 929-944
4. Stumpf, P.K. In "The Biochemistry of Plants"; Stumpf, P.K.; Conn, E.E., Eds.; Academic Press, New York, 1980, pp. 177-204
5. Baldwin, J.E. J. Chem. Soc. Chem. Commun. 1982, 1086
6. Rhodes, M.J.C.; Wooltorton, L.S.C. Phytochemistry 1967, 6, 1-12
7. Tressl, R.; Drawert, F. J. Agric. Food Chem. 1973, 21, 560-565
8. Tressl, R.; Drawert, F. Z. Naturforsch. 1971, 26b, 774-779
9. Stumpf, P.K. Annual Rev. Biochem. 1969, 38, 159
10. Kollatukudy, P.E. Annual Rev. Plant Physiology 1970, 21, 163-189
11. Buttery, R.G.; Ling, L.C.; Chan, B.C. J. Agric. Food Chem. 1978, 26, 866-869
12. Stoffel, W.; Caesar, H. Hoppe Seyler's Z. Physiol. Chem. 1965, 342, 76-83
13. Tressl, R.; Engel, K.-H. In "Progres in Flavour Research"; Adda, J., Ed.; Proc. 4. Weurmann Flavour Symposium, Dourdan 1984. Elsevier Science Publ., Amsterdam, 1985, pp. 441-455
14. Tressl, R.; Engel, K.-H.; Albrecht, W.; Bille-Abdullah, H. In "Characterization and Measurement of Flavor Compounds"; Bills, D.D.; Mussinan, C.J., Eds.; ACS Symposium Series No. 289, Washington, D.C., 1985, pp. 43-60
15. Neuberg, C. Adv. Carbohydr. Res. 1949, 10, 75-11)
16. Hochuli, E.; Taylor, K.E.; Dutler, H. Eur. J. Biochem. 1977, 75, 433-439
17. Jennings, W.G.; Creveling, R.K.; Heinz, D.E. J. Food Sci. 1964, 29, 730
18. Jennings, W.G.; Tressl, R. Chem. Mikrobiol. Technol. Lebensm. 1974, 3, 52-55
19. Cuebas, D.; Schulz, H. J. Biol. Chem. 1982, 257, 14140-14144
20. Boland, W.; Ney, P.; Jaenicke, L. In "Analysis of

Volatiles"; Schreier, P., Ed.; Walter de Gruyter, Berlin, 1984, p. 371
21. Berger, R.G., Kollmannsberger, H. In "Topics in Flavour Research"; Berger, R.G.; Nitz, S.; Schreier, P., Eds.; Proc. Int. Conf., Freising-Weihenstephan, 1985. H. Eichhorn, Malzling-Hangenham, 1985, pp.305-320
22. Winter, M.; Furrer, A.; Wilhalm, B.; Thommen, W. Helv. Chim. Acta 1976, 59, 1613-1620
23. Pickenhagen, W.; Brönner-Schindler, H. Helv. Chim. Acta 1984, 67, 947-952
24. Engel, K.-H.; Tressl, R.,in prep.
25. Tressl, R.; Engel, K.-H. In "Analysis of Volatiles"; Schreier, P., Ed.; Walter de Gruyter, Berlin, 1984, pp. 323-342
26. Engel, K.-H.; Tressl, R., in prep.
27. Naoshima, Y.; Ozawa, H.; Koudo, H.; Hayaishi, S. Agric. Biol. Chem. 1983, 47, 1431-1434
28. Muys, G.T.; Van der Ven, B.; De Jonge, A.P. Appl. Microbiol., 1963, 11, 389-393
29. Tressl, R.; Silberzahn, W., unpublished results.
30. Dimick, P.S.; Walker, N.J.; Patton, S. Biochem. J. 1969, 111, 395-399
31. Wyman, H.; Buckley, E.H.; McCharty, A.I.; Palmer, J. K. Annual Research Report, United Fruit Co., 1964
32. Myers, M.J.; Issenberg, P.; Wick, E.L. J. Food Sci. 1969, 34, 504
33. Myers, M.J.; Issenberg, P.; Wick, E.L. Phytochemistry 1970, 9, 1963
34. Drawert, F.; Rolle, K.; Heimann, W. Z. Lebensm. Unters. Forsch. 1971, 145, 7-15
35. Tressl, R.; Emberger, R.; Drawert, F.; Heimann, W. Z. Naturforsch. 1970, 25b, 704-707
36. Tressl, R.; Holzer, M.; Apetz, M. J. Agric. Food Chem. 1977, 25, 455-459
37. Metzner, H. "Biochemie der Pflanzen"; Enke Verlag, Stuttgart, 1973
38. Drawert, F. In "Aroma Research"; Maarse, H.; Groenen, P.J., Eds.; Int. Symp. Aroma Research, Zeist, 1975, pp. 13-39
39. Kaneko, K. Chemical and Pharmaceutical Bulletin 1960, 8, 611-614
40. Klieschies, M.; Stöckigt, J.; Zenk, M.H. J. Chem. Soc. Chem. Comm. 1975, 879-880
41. Wackerbauer, K.; Kossa, T.; Tressl, R. EBC-Proc. Amsterdam, 1977, p. 495
42. Zenk, M.H. In "Biosynthesis of aromatic compounds"; Billek, G., Ed.; Pergamon, Oxford, 1966, p. 45
43. Dimroth, P.; Walter, H.; Lynen, F. Eur. J. Biochem. 1970, 13, 98-110
44. Drawert, F.; Berger, R.G. Lebensm.-Wiss. u. -Technol. 1983, 16, 209-214

RECEIVED March 10, 1986

11

Biosynthesis of Cyclic Monoterpenes

Rodney Croteau

Institute of Biological Chemistry, Washington State University, Pullman, WA 99164-6340

Cyclization of an allylic pyrophosphate is a key
step in the biosynthesis of most monoterpenes.
Early hypotheses concerning the nature of the
acyclic precursor and the cyclization process are
first described, and chemical models for the
cyclization presented. Following a review of
several representative cyclase enzymes and the
reactions that they catalyze, a series of
stereochemical and mechanistic experiments with
partially purified cyclases are reported. The
results of these studies have allowed a detailed
description of events at the active site and the
formulation of a unified stereochemical scheme for
the multistep isomerization-cyclization reaction by
which the universal precursor geranyl pyrophosphate
is transformed to cyclic monoterpenes.

The monoterpenes constitute a large group of natural
products synthesized and accumulated in distinct glandular
structures by some 50 families of higher plants (1). The vast
majority of these compounds are cyclic and represent a
relatively small number of skeletal themes multiplied by a very
large range of simple derivatives, positional isomers and
stereochemical variants. (2). Research on the biochemistry of
the monoterpenes can be divided into three general areas;
cyclization reactions, secondary transformations of the parent
cyclic compounds, and catabolism. These and related topics have
been described in a number of comprehensive reviews (3-5) and
the field is periodically surveyed (6). The focus of this
chapter is recent enzyme-level studies directed toward the
development of a general model for monoterpene cyclization. The
topic has been briefly described elsewhere (7-9).

0097-6156/86/0317-0134$06.75/0
© 1986 American Chemical Society

Historical Perspective

Such early findings that isoprene (methylbutadiene) was produced by pyrolysis of turpentine (primarily pinenes) and that heating of isoprene produced dipentene (racemic limonene) led Wallach (10), through a series of structural investigations of nearly a century ago, to formulate the "isoprene rule" – that a certain category of natural products (terpenoids) can be regarded as being constructed of isoprene units, commonly joined in a head-to-tail fashion (Figure 1). Many years later, Ruzicka was to extend and reformulate the "isoprene rule" in mechanistic terms as the "biogenetic isoprene rule" (11). As applied to the monoterpenes, the rule posits intramolecular electrophilic attack of C1 of the neryl cation on the distal double bond to yield a monocyclic (α-terpinyl) intermediate, which by a series of subsequent internal electrophilic additions, hydride shifts, and Wagner–Meerwein rearrangements can be seen to give rise to the cationic equivalents of most known skeletal types (Figure 1). The latter species, by deprotonation or capture by a nucleophile, could yield many of the common monoterpenes, and by subsequent, often oxidative, modification of these cyclic progenitors most other monoterpenes could be generated. This profound contribution by Ruzicka set the foundation for nearly all biogenetic investigations to follow, including in vivo studies using basic precursors, such as [2-^{14}C]acetate and [2-^{14}C]mevalonate, with which labeling patterns consistent with the "biogenetic isoprene rule" were demonstrated (12).

   With the growing appreciation of the central role of allylic pyrophosphates in isoprenoid metabolism, and, specifically, the finding that geranyl pyrophosphate (Figure 2) was the first C_{10} intermediate to arise in the general isoprenoid pathway (13), attempts were made to apply the biogenetic isoprene rule to monoterpene biosynthesis in more explicit mechanistic terms. These attempts brought considerable confusion and some controversy to the field. It had long been known that geraniol and its derivatives cannot cyclize directly (because of the trans-double bond at C2) whereas the cis-isomer, nerol, and the tertiary isomer, linalool, readily cyclize (see Figure 2) (14,15). Thus, to circumvent the topological barrier to the direct cyclization of geranyl pyrophosphate, a number of theories, often conflicting and sometimes poorly supported, were proposed for the origin of neryl pyrophosphate and linalyl pyrophosphate by direct condensation of C5 units or by various isomerizations of geranyl pyrophosphate (Figure 2) (see references 4 and 16 for discussion of these now largely historical ideas). With the availability of partially purified cell-free enzyme systems in the late 1970s, it became apparent that geranyl pyrophosphate was efficiently converted to cyclic monoterpenes without loss of hydrogen from C1 of the trans-precursor and without isomerization to free neryl or free linalyl pyrophosphate, or to any other detectable intermediate (3,4,17). Furthermore, geranyl pyrophosphate was, in most cases, cyclized more efficiently than neryl pyrophosphate and with efficiencies comparable to that of (±)-linalyl pyrophos-

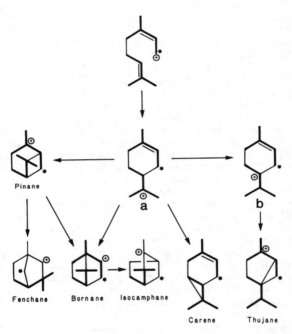

Figure 1.   Postulated ionic mechanism for the formation of
monoterpenes via the α-terpinyl cation (a) and the
terpinen-4-yl cation (b). Regular (head-to-tail) structures
are divided into isoprene units and the labeling patterns from
Cl-labeled acyclic precursor are illustrated.

phate (9,18,19). It was thus clear that monoterpene cyclases (synthases) were capable of catalyzing a multistep process whereby the enzyme carried out the isomerization of geranyl pyrophosphate to a bound intermediate capable of cyclizing, as well as the cyclization reaction itself. The recognition of geranyl pyrophosphate as the universal precursor of cyclic monoterpenes provided a critical insight which, while not immediately clarifying the detailed mechanism of cyclization, did result in considerable simplification of the problem by eliminating alternatives no longer tenable. All recent proposals for the biosynthesis of cyclic monoterpenes have been based on the concept of a tightly coupled isomerization-cyclization reaction (5,7-9).

Enzymology

The crucial cyclization reactions by which the parent monoterpene carbon skeletons are generated are catalyzed by enzymes collectively known as cyclases. Multiple cyclases, each producing a different skeletal arrangement from the same acyclic precursor, often occur in higher plants while single cyclases which synthesize a limited variety of skeletal types are also known (18-20). Individual cyclases, each generating a simple derivative or positional isomer of the same skeletal type, have been described, as have distinct cyclases catalyzing the synthesis of enantiomeric products (4,19). The number of monoterpene cyclases in nature is presently uncertain, yet studies on the known examples from a limited number of plant species (3) suggest the total is at least fifty.

Cyclases, with apparently rare exception (21), are operationally soluble enzymes possessing molecular weights in the 50,000-100,000 range. Based on relatively few examples (19,22), they appear to be rather hydrophobic and to possess relatively low pI values. The only co-factor required is a divalent metal ion, Mg^{2+} or Mn^{2+} generally being preferred (K_m in the 0.5 to 5 mM range). Most monoterpene cyclases can utilize geranyl pyrophosphate, neryl pyrophosphate, and linalyl pyrophosphate (Figure 2) as acyclic precursors, without detectable interconversion among these substrates or preliminary conversion to other free intermediates. Michaelis constants for all three substrates with all cyclases studied are in the low μM range. Since geranyl pyrophosphate is efficiently cyclized without formation of free intermediates, it is clear that monoterpene cyclases are capable of catalyzing both the required isomerization and cyclization reactions, the overall process being essentially irreversible in all cases. It is assumed, but not yet proven in all cases, that the isomerization and the cyclization (of all the various precursors) take place by the same general mechanism at the same active site of the enzyme. The cyclases, as a class, are thus notable for the complexity and length of the reaction sequence catalyzed while maintaining complete regio- and stereochemical control of the product. In general properties, the monoterpene cyclases (the enzyme type is more properly termed an isomerase-cyclase) resemble the few

sesquiterpene (23-25) and diterpene (26-28) cyclases that have
been examined, as well as prenyl transferases which catalyze
similar electrophilic reactions (29).

Many cyclases have been partially purified (i.e., freed of
competing activities such as phosphatase) permitting accurate
determination of kinetic constants and the demonstration, in
most cases, that geranyl pyrophosphate is a more efficient
substrate than neryl pyrophosphate based on comparison of
respective V/Km values. Preparation to homogeneity has been
hindered by the somewhat unsavory nature of the enzyme sources
which contain high levels of resins, phenolics and competing
activities (9). Operational limitations are also imposed by the
facts that the cyclases, like most enzymes involved in the
biosynthesis of natural products, do not occur in very high
intracellular concentration and the reactions that they catalyze
are rather slow (9,24). Selective extraction of monoterpene
cyclases from leaf epidermal oil glands and affinity
chromatography techniques have partially overcome these
preparative difficulties (9).

The focus of this chapter is the cyclization of the
universal precursor, geranyl pyrophosphate, to bicyclic
monoterpenes of the bornane, pinane and fenchane type, where the
absolute configuration of the relevant products allows
stereochemical correlation and certain mechanistic deductions to
be made with little ambiguity. The geranyl pyrophosphate:bornyl
pyrophosphate cyclases catalyze the first committed step in the
biosynthesis of camphor (via hydrolysis of the pyrophosphate
ester (Figure 2) to borneol and oxidation to the ketone)
(18,30-33). A (+)-bornyl pyrophosphate cyclase has been
isolated from common sage (Salvia officinalis; Lamiaceae) (18),
whereas an enantiomer-producing (-)-bornyl pyrophosphate cyclase
has been partially purified from extracts of common tansy
(Tanacetum vulgare; Asteraceae) (34). These cyclizations are
unique, thus far, in the retention of the pyrophosphate moiety
of the substrate in the product. In looking ahead to a
subsequent section, this feature has allowed examination of the
role of the pyrophosphate group in cyclization reactions.

Two pinene cyclases have been isolated from sage (19,35).
Electrophoretically pure pinene cyclase I converts geranyl
pyrophosphate to (+)-α-pinene and to lesser quantities of
(+)-camphene and (+)-limonene, whereas pinene cyclase II, of
lower molecular weight, converts the acyclic precursor to
(-)-β-pinene and to lesser quantities of (-)-α-pinene,
(-)-camphene and (-)-limonene. Both purified enzymes also
utilize neryl and linalyl pyrophosphate as alternate substrates
for olefin synthesis. The availability of enzyme systems
catalyzing formation of enantiomeric products from a common,
achiral substrate has provided an unusual opportunity to examine
the stereochemistry of cyclization.

(-)-endo-Fenchol cyclase has been obtained from fennel
(Foeniculum vulgare; Apiaceae) (36). Fenchol cyclase, unlike
the aforementioned cyclases, prefers Mn^{2+} to Mg^{2+} as co-factor.
Detailed investigation has confirmed that the initial cyclic
product is endo-fenchol, not the corresponding pyrophosphate

ester (36), and that this product arises by rearrangement of a
pinyl intermediate (37).

Model Reactions

The solvolyses of geranyl, neryl and linalyl derivatives have
provided useful models for monoterpene biosynthesis (38-43).
Results obtained using diverse systems and conditions have
generally shown that neryl and linalyl systems yield monocyclic
products in good yield and at higher rates than geranyl systems,
which afford primarily acyclic products. Geranyl derivatives do
give rise to monocyclic products via preliminary conversion to
the tertiary, linalyl, intermediate (44-46), and this mode of
cyclization is favored under conditions where nucleophilic
trapping is slow relative to re-ionization of the tertiary
allylic system, and where stabilization of intermediate cationic
species is favored by ion-pairing (i.e., in non-nucleophilic
media with the presence of a large counter ion) (45-49). The
importance of ion-pairing in both chemical (50,51) and enzymatic
(16,52) transformations of allylic pyrophosphates has been
repeatedly emphasized.

Solvolysis of a number of (-)-linalyl esters leads to
(+)-α-terpineol in high enantiomeric excess (53,54) and the
cyclization can be formulated as either a syn-exo or anti-endo
process (Figure 3) (The alternate syn-endo and anti-exo
cyclizing conformations are precluded by the absolute
configuration of the product which indicates the face of the
C6-C7 double bond attacked). Arigoni and co-workers, in a
definitive study of the fate of the hydrogens at C1 of linalyl
p-nitrobenzoate, deduced that the anti-endo conformation of the
linalyl system was preferred during cyclization (55). Although
a concerted cyclization with π-participation by the distal
double bond had seemed an attractive explanation for the net
stereochemistry observed in this allylic displacement (54),
Poulter and King (56,57) have shown the cyclization of the
isomeric neryl system to α-terpineol to be both stepwise and
stereospecific, providing an elegant demonstration of the
preference for ionic over concerted pathways for this reaction
type.

Several groups have examined the influence of divalent
cations, such as Mg^{2+} and Mn^{2+}, in catalyzing the solvolysis of
allylic pyrophosphates such as geranyl pyrophosphate (58-60).
The results strongly suggest that the role of the metal ion in
enzymatic transformations of allylic pyrophosphates is to
neutralize the negative charge of the pyrophosphate moiety and
thus assist in the ionization of the substrate to produce the
allylic cation.

Consideration of chemical models, now largely in hindsight,
allows broad outlines of a cyclization scheme to be delineated.
Reaction of geranyl pyrophosphate is initiated by ionization
which is assisted by low pH and divalent metal ion. Conversion
of the geranyl to the linalyl system precedes cyclization to the
monocyclic intermediate by the established stereochemical
course. The overall process occurs stepwise via a series of

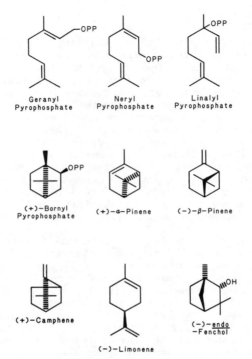

Geranyl
Pyrophosphate

Neryl
Pyrophosphate

Linalyl
Pyrophosphate

(+)-Bornyl
Pyrophosphate

(+)-α-Pinene

(-)-β-Pinene

(+)-Camphene

(-)-Limonene

(-)-endo
-Fenchol

Figure 2. Structures of acyclic precursors and some cyclic
monoterpene products.

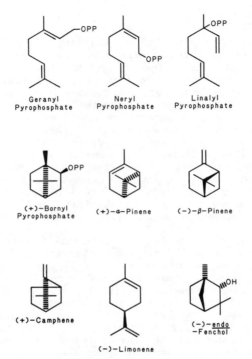

(-)-3R-LPP

(+)-4R-α-Terpineol

(-)-3R-LPP

Anti
Endo

Syn
Exo

Figure 3. Linalyl derivatives cyclize preferentially from the
anti-endo conformation.

carbocation-pyrophosphate anion paired intermediates where topology is maintained between the initial ionization and the termination steps.

Two elements of the cyclization have yet to be addressed: the isomerization of geranyl pyrophosphate to linalyl pyrophosphate (or the equivalent ion-pair) and the construction of bicyclic skeleta. Studies on the biosynthesis of linalool (61), and on the analogous nerolidyl system in the sesquiterpene series (52), have shown this allylic transposition to occur by a net suprafacial process, as expected. On the other hand, the chemical conversion of acyclic or monocyclic precursors to bicyclic monoterpenes, under relevant cationic cyclization conditions, has been rarely observed (47,62-65) and, thermodynamic considerations notwithstanding (66), bicyclizations remain poorly modeled.

Cyclization Scheme

With the preceding reviews of the enzymology of monoterpene cyclization and of model studies relevant to the cyclization process, it is possible to formulate a unified stereochemical scheme for the enzymatic cyclization of geranyl pyrophosphate (Figure 4). The proposal which follows is consistent with the implications of parallel advances in related fields, most notably the contributions of Cane (8,16,24,25,52), Arigoni (67) and Coates (68,69) on the stereochemistry of sesquiterpene and diterpene cyclizations, and of Poulter and Rilling (29,70) on the stepwise, ionic mechanism of prenyl transferase, a reaction type of which several monoterpene, sesquiterpene and diterpene cyclizations are, in a sense, the intramolecular equivalents.

It is generally agreed that the cyclase catalyzes the initial ionization of the pyrophosphate moiety to generate an allylic cation-pyrophosphate anion pair, with the assistance of the divalent metal ion and in a manner completely analogous to the action of prenyl transferase. This step is followed by stereospecific syn-isomerization to either a 3R- or a 3S-linalyl intermediate and rotation about the C2-C3 single bond. (Collapse of the initially formed ion pair to enzyme-bound linalyl pyrophosphate is proposed here since the free energy barrier for isomerization of an allylic cation is relatively high, even when the system is a tertiary-primary resonance hybrid (71).) Subsequent ionization and cyclization of the cisoid, anti-endo conformer affords the corresponding monocyclic 4R- or 4S-α-terpinyl ion, respectively. Following the initial generation of the common α-terpinyl intermediate, the further course of the reaction may involve additional cyclizations via the remaining double bond, hydride shifts, and/or rearrangements before termination of the cationic reaction by deprotonation to an olefin or capture by a nucleophile. For example, deprotonation of the α-terpinyl cation affords limonene (72), whereas further cyclization to the most highly substituted position of the cyclohexene double bond and capture of the resulting cation by the paired pyrophosphate anion generates the bornyl pyrophosphates. Internal addition to the least

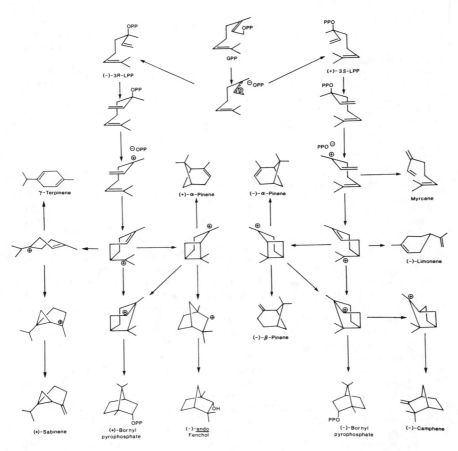

Figure 4. A unified stereochemical model for monoterpene cyclization.

substituted position followed by deprotonation yields the pinenes, while Wagner-Meerwein rearrangement of the (+)-pinyl skeleton and capture of the cation by H_2O provides (-)-endo-fenchol. A 1,2-hydride shift in the original α-terpinyl cation yields the terpinen-4-yl intermediate, which, by deprotonation produces γ-terpinene and related olefins (22), or by internal electrophilic attack on the cyclohexene double bond generates the cyclopropane ring of (+)-sabinene and related products (73). The overall process can be viewed as a series of steps: ionization; pyrophosphate migration; bond rotation; ionization; cyclization; termination, diverging enantiospecifically in the isomerization sequence, and involving numerous regiochemical variants in the cyclization to the various parent skeleta. The highly reactive electrophilic intermediates are presumed to remain paired with the pyrophosphate anion throughout the course of the reaction, even where charge separations may exceed 3Å.

As can be seen, the scheme includes the necessary coupled isomerization component of the multi-step reaction, and reformulates earlier concepts of cyclization in explicit stereochemical terms which take into account the conformational constraints imposed by the required 2p orbital alignment for cyclization of an eight carbon chain containing two trisubstituted π-systems and the apparent imperative for allylic displacement of the tertiary pyrophosphate moiety in an anti-sense. The scheme accounts for the cyclization of geranyl pyrophosphate without free intermediates and can obviously rationalize the cyclization of neryl and linalyl pyrophosphates as alternate substrates – all routes merging at the cisoid conformer of linalyl pyrophosphate (or the ion-paired equivalent). Although linalyl pyrophosphate is the first explicitly chiral intermediate in the cyclization scheme, it should be emphasized that the eventual configuration is pre-determined by the helical conformation of geranyl pyrophosphate achieved on initial binding to the enzyme (for a more detailed discussion of conformational considerations in mono- and sesquiterpene cyclization, see Cane (8)). The scheme applies equally well to monocyclic and bicyclic monoterpenes of either enantiomeric series, but with an obvious degree of uncertainty to symmetrical products, such as 1,8-cineole and γ-terpinene, in which absolute stereochemical inferences based on absolute product configurations are not possible. The following sections present a description of current efforts to test and probe further the implications of this stereochemical model.

Stereochemistry

As noted in an earlier section, the labeling patterns of several monoterpenes derived in vivo from basic precursors such as [2-^{14}C]mevalonic acid are consistent with the basic cyclization scheme (Figure 1). More recently, the labeling patterns of antipodal bornane and pinene monoterpenes from [1-^{3}H]geranyl pyrophosphate have been determined ((+)- and (-)-bornyl

pyrophosphate are labeled at C3, and (+)- and-(-)-pinenes at C7)
(31,34,35). These results have indicated that the antipodes are
derived via enantiomeric cyclizations involving antipodal
linalyl and α-terpinyl intermediates (Figure 4), rather than by
way of a hydride shift or other rearrangement from a common
cyclic progenitor. Cane (8) has recently re-emphasized that for
many monoterpenes the corresponding conformation of the
presumptive intermediates can usually be deduced based on the
assumption of least motion during the course of the cyclization.
Thus, there exists a direct correspondence between the observed
relative and absolute configuration of the terpenoid product and
the inferred configuration and conformation of the linalyl
intermediate.

Although it is not yet certain whether linalyl
pyrophosphate is the true enzyme-bound intermediate of
cyclization processes or simply an efficient substrate analog,
it is abundantly clear that cyclases are capable of ionizing and
subsequently cyclizing linalyl pyrophosphate which, at minimum,
must mimic the corresponding bound intermediate actually formed
at the active site. This feature allows the configuration of
the cyclizing intermediate to be determined, in principal, in
two ways: by measuring which enantiomer of a racemic mixture is
depleted by the cyclase or by testing directly, and
independently, each linalyl pyrophosphate enantiomer as a
cyclase substrate. Both approaches were taken with
(-)-endo-fenchol cyclase and it was unequivocally demonstrated
that 3R-linalyl pyrophosphate was the preferred enantiomer for
the cyclization, as predicted (74). The optically pure linalyl
pyrophosphates were also employed to probe the cyclizations to
(+)- and (-)-bornyl pyrophosphate and to (+)- and (-)-pinene.
Consistent with stereochemical considerations, 3R-linalyl
pyrophosphate proved to be the preferred substrate for
cyclization to (+)-bornyl pyrophosphate and (+)-pinene, whereas
3S-linalyl pyrophosphate was preferred for the corresponding
enantiomeric cyclizations to (-)-bornyl pyrophosphate and
(-)-pinene (75).

Curiously, certain cyclases, notably (+)-bornyl
pyrophosphate cyclase and (-)-endo-fenchol cyclase, are capable
of cyclizing, at relatively slow rates, the 3S-linalyl
pyrophosphate enantiomer to the respective antipodal products,
(-)-bornyl pyrophosphate and (+)-endo-fenchol (74,75). Since
both (+)-bornyl pyrophosphate cyclase and (-)-endo-fenchol
cyclase produce the designated products in optically pure form
from geranyl, neryl and 3R-linalyl pyrophosphate, the antipodal
cyclizations of the 3S-linalyl enantiomer are clearly abnormal
and indicate the inability to completely discriminate between
the similar overall hydrophobic/hydrophilic profiles presented
by the linalyl enantiomers in their approach from solution. The
anomalous cyclization of the 3S-enantiomer by fenchol cyclase is
accompanied by some loss of normal regiochemical control, since
aberrant terminations at the acyclic, monocyclic and bicyclic
stages of the cationic cyclization cascade are also observed
(74). The absolute configurations of these abnormal co-products
have yet to be examined.

The pinene cyclases convert the anomalous linalyl enantiomer to abnormal levels of acyclic (e.g. myrcene) and monocyclic (e.g. limonene) terpenes, these aberrant products perhaps arising via ionization of the tertiary substrate in the transoid or other partially extended (exo) form (see below). In any event, for all "normal" cyclizations examined thus far, the configuration of the cyclizing linalyl intermediate has been confirmed to be that which would be expected on the basis of an anti-endo conformation. Scattered attempts at intercalating the cyclization cascade with analogs of proposed cyclic intermediates (e.g. α-terpinyl and 2-pinyl pyrophosphate) have been unsuccessful (20,35,36).

A critical prediction of the anti-cyclization, in which the leaving group departs from the allylic system on the side opposite to that which the incoming nucleophilic group of the substrate becomes attached, is that configuration at C1 of the geranyl substrate will be retained. Thus, following the syn-allylic transposition, transoid to cisoid rotation about the newly generated C2-C3 single bond of the linalyl system brings the face of C1 from which the pyrophosphate moiety has departed into juxtaposition with the neighboring si-face of the C6-C7 double bond from which C1-C6 ring closure occurs (Figure 5). Neryl pyrophosphosphate, on the other hand, should exhibit inversion of configuration at C1 since the cyclization is either direct, or involves isomerization to the linalyl intermediate and cyclization without the attendant C2-C3 rotation. These predictions have been confirmed directly by using 1R-[1-^{3}H]- and 1S-[1-^{3}H]-geranyl and neryl pyrophosphates as substrates for the (+)- and (-)-bornyl pyrophosphate cyclases (76). The products so isolated were converted to the corresponding tritiated camphors, and the label located by stereoselective exchange of the exo-α-hydrogens (Figure 5). A similar approach has been applied to the (+)-and (-)-pinenes (which can be stereoselectively converted to (+)- and (-)-camphor, respectively) and to (-)-endo-fenchol (via fenchene to fenchocamphorone), in each case comparing tritium exchange rates to those of the product generated from racemic C1-labeled precursor (75). These results, in addition to confirming a nearly universal preference for anti-stereochemistry in mono-, sesqui- and diterpene cyclizations (8,16,77-79), establish the conformation of the cyclizing tertiary intermediate as anti-endo and rule out all other possible conformers. When taken together with the previously described configurational preferences, the summation of these studies unequivocally establishes a consistent overall stereochemistry for monoterpene cyclizations in the bicyclic series. It should be noted that complete stereochemical analysis has yet to be applied to monocyclics in general or to anomalous cyclization products in particular, and it is conceivable that such cyclizations involve syn-processes and/or extended (exo) conformations.

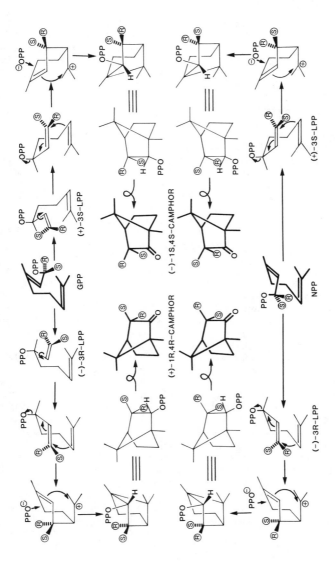

Figure 5. Stereochemistry at C1 of geranyl pyrophosphate and neryl pyrophosphate in the cyclization, via linalyl pyrophosphate, to bornyl pyrophosphate. "Reproduced with permission from Ref. 76. Copyright 1985, Journal of Biological Chemistry".

Mechanism

The ability of monoterpene cyclases to utilize linalyl pyrophosphate as an acyclic precursor has permitted determination of configuration of the cyclizing intermediate and has additionally allowed separate focus on the cyclization component of the reaction sequence by simply bypassing the normally tightly coupled isomerization step. For bornane, pinane and fenchane monoterpenes, the K_m values for linalyl pyrophosphate are lower, in most cases, than those for geranyl pyrophosphate while the relative velocities for cyclization are higher in all instances, resulting in catalytic efficiencies (V/K_m) up to ten-times higher for the appropriate linalyl pyrophosphate enantiomer than for the achiral primary allylic isomer. These results tend to suggest that the isomerization of geranyl to linalyl pyrophosphate is the slow step of the reaction sequence (compared to the cyclization of the more reactive linalyl intermediate), however the limiting component of the coupled process is not yet clear in any instance.

Because the coupled isomerization-cyclization of geranyl pyrophosphate proceeds without detectable free intermediates, no ready means has been available to examine the isomerization to linalyl pyrophosphate in isolation. In an attempt to dissect the normally cryptic isomerization step from the tightly coupled reaction sequence, a non-cyclizable analog of geranyl pyrophosphate, 6,7-dihydrogeranyl pyrophosphate, was prepared. Although the C6-C7 double bond determines the partitioning to cyclic products, it cannot, for topological reasons, assist in the generation of the initial allylic cation. Thus, saturation of the isopropylidene function prevents cyclization, with minimal effect on binding behavior or on the initial pyrophosphate ionization-migration process. The analog inhibited the cyclization of geranyl pyrophosphate to (+)-bornyl pyrophosphate and (+)-α-pinene, and was itself catalytically active, yielding (at rates comparable to the normal cyclization) acyclic olefins and alcohols as products (Figure 6) (80). The enzymatic products resembled the solvolysis products of 6,7-dihydrolinalyl pyrophosphate, yet comprised a far higher proportion of olefins suggesting that enzymatic product formation occurred in an environment relatively inaccessible to water, presumably the active site of the cyclase. 6,7-Dihydrolinalyl pyrophosphate was itself sought as an enzymatic reaction product from the 6,7-dihydrogeranyl substrate, but was observed only in trace levels. It therefore appeared that the analog underwent the normal pyrophosphate ionization-migration step, giving rise to 6,7-dihydrolinalyl pyrophosphate which was re-ionized and, because the subsequent cyclizations were precluded, the resulting cation (and its allylic isomers) either deprotonated or captured by water. These results provided strong, albeit indirect, evidence for the cryptic isomerization component of the normally coupled reaction sequence (80).

The availability of the bornyl pyrophosphate cyclases has provided a unique opportunity to directly examine the function

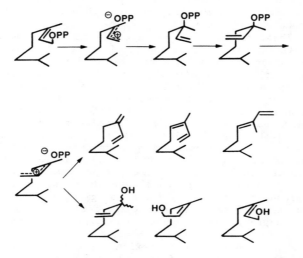

Figure 6. Conversion of 6,7-dihydrogeranyl pyrophosphate, via 6,7-dihydrolinalyl pyrophosphate, to dihydro olefins and alcohols by (+)-bornyl pyrophosphate and (+)-pinene cyclases.

of the pyrophosphate moiety in the coupled isomerization-cyclization reaction leading to monoterpenes. Initial studies on the mechanism of the pyrophosphate migration in the conversion of geranyl pyrophosphate to (+)- and (-)-bornyl pyrophosphate established that the two ends of the pyrophosphate moiety of the substrate retained their identities in the cyclization to both products, and also indicated that there was no appreciable exchange with exogenous inorganic pyrophosphate in the reaction. Thus, separate incubations of $[1-{}^3H_2;\alpha-{}^{32}P]$ and $[1-{}^3H_2;\beta-{}^{32}P]$geranyl pyrophosphates with partially purified preparations of each enantiomer-generating cyclase gave $[{}^3H:{}^{32}P]$bornyl pyrophosphates (of unchanged isotope ratio) which were selectively hydrolyzed to the corresponding bornyl phosphates. Measurement of ${}^3H:{}^{32}P$ ratios of these monophosphate esters indicated that label from only the α-phosphate of the substrate was retained in the derived product (81).

With the absence of tumbling or end-to-end interchange of the pyrophosphate established for both cyclizations, it became critical to examine the fate of the C-O-P bridge oxygen of the precursor in these transformations. To this end, collaborative studies with Professor D.E. Cane were undertaken in which $[8,9-{}^{14}C;1-{}^{18}O]$geranyl pyrophosphate was prepared and converted to (+)- and (-)-bornyl pyrophosphate by large-scale incubations. Analysis of the products by mass spectrometry of the derived benzoates demonstrated an ${}^{18}O$ enrichment identical to that of the original substrate, indicating that the isomerization-cyclization of $[1-{}^{18}O]$geranyl pyrophosphate involves essentially no positional oxygen isotope exchange (i.e., the original pyrophosphate ester oxygen of the precursor is the exclusive source of the pyrophosphate ester oxygen of the product) (81).

These results, although implying very tight coupling of the pyrophosphate and terpenoid reaction partners within the enzyme active site, could not distinguish between an initial [1,3]-sigmatropic rearrangement or a tight ion pair in which rotational equilibrium about the Pα-OPβ bond is not achieved, a [3,3]-sigmatropic rearrangement involving the initial attachment of a non-bridge oxygen to C3 of the linalyl system and return of the formerly bridged ${}^{18}O$ atom to C2 of the bornyl system, or processes involving bonding of the β-phosphate group at the tertiary center. To examine these various possibilities with respect to the formation and subsequent cyclization of the tertiary intermediate, both α- and β-${}^{32}P$-labeled and 3-${}^{18}O$-labeled linalyl pyrophosphates were prepared. Analysis of products derived from the ${}^3H;{}^{32}P$-labeled substrates, as in the previous experiments, indicated the two ends of the pyrophosphate retained their identities in this cyclization, excluding direct involvement of the β-phosphate of geranyl pyrophosphate in the allylic transposition (82). Similarly, preparative-scale enzymatic conversions of (±)-1E-$[1-{}^3H;3-{}^{18}O]$linalyl pyrophosphate to the enantiomeric bornyl pyrophosphates, followed by mass spectrometric analysis of the derived benzoates, yielded an ${}^{18}O$ enrichment of the carbinol oxygen atom of the benzoate esters virtually identical to that of the precursor (82). The alternative [3,3]-sigmatropic rearrangement was therefore eliminated.

The summary of the results clearly indicates that it is solely the pyrophosphate ester oxygen of geranyl pyrophosphate which is involved in all the critical bonding processes in the coupled isomerization-cyclization leading to formation of both (+)- and (-)-bornyl pyrophosphate (Figure 7), and thus, that the pyrophosphate moiety remains closely associated with its terpenyl partner throughout the course of the reaction. These findings strongly support tight ion-pairing in the transformation. The observed absence of P_α-P_β interchange and complete lack of positional ^{18}O-isotope exchange in the case of bornyl pyrophosphate cyclase is particularly notable since the reaction seemingly involves a formal 1,3- followed by a 1,2-migration of the pyrophosphate, with the intervening generation of the transient α-terpinyl cation-pyrophosphate anion pair in which the charge separation is at least 3Å.

Results obtained with the (+)-and (-)-bornyl pyrophosphate cyclases add to a growing body of evidence supporting the general involvement of ion-pair intermediates in the enzymatic transformation of allylic pyrophosphates (16,29,52,83,84) and imply this common feature for monoterpene cyclases. However, most monoterpene cyclases terminate the reaction by deprotonation of a carbocation to afford an olefin, or carbocation capture by water rather than by the pyrophosphate anion. Thus, for example, (-)-endo-fenchol derived from [1-^{18}O]geranyl pyrophosphate in enzyme preparations from F. vulgare, bears no detectable ^{18}O label (85). This result, implying water as the source of oxygen in the cyclic product, is nevertheless fully consistent with the unified cyclization model (Figure 4), in that the stereochemistry of (-)-endo-fenchol formation is incompatible with internal return of the pyrophosphate (and subsequent P-O bond hydrolysis) (85).

A rather different approach to examining the question of ion-pairing was taken when Professor A.C. Oehlschlager made available the sulfonium ion analogs of the linalyl and α-terpinyl cationic intermediates of the cyclization reaction (Figure 8). Both analogs were effective inhibitors of the cyclizations of geranyl pyrophosphate to (+)-bornyl pyrophosphate and (+)-α-pinene, with K_i values in the micromolar range (86). In the presence of inorganic pyrophosphate, however, the K_i values dropped to sub-micromolar levels (into the range of K_m values for the substrate). Similarly, the K_i values for inorganic pyrophosphate, itself a modest inhibitor, were decreased many fold by the presence of either sulfonium analog. That the combination of sulfonium analog and pyrophosphate provided synergistic inhibition of the electrophilic cyclizations suggested that the cyclases bind the paired species more tightly than either partner alone and therefore implicate ion pairing in the transformation of the allylic pyrophosphate substrate. Other anions were not effective in this role, and other diverse trialkylsulfonium salts were ineffective inhibitors of cyclization, thus

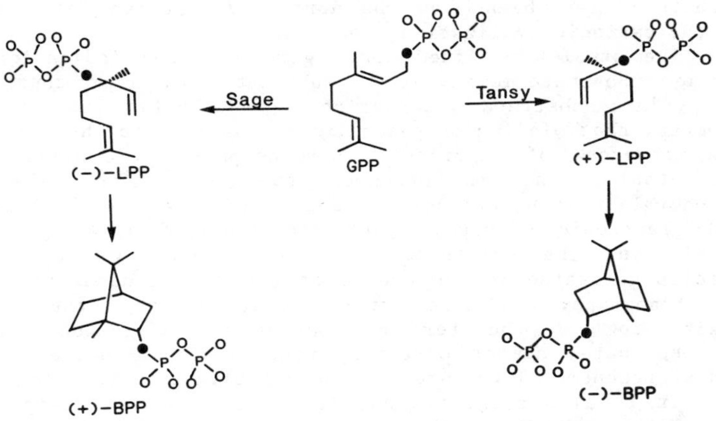

Figure 7. Conversion of [1-^{18}O]geranyl pyrophosphate and [3-^{18}O]linalyl pyrophosphate to (+)- and (-)-bornyl pyrophosphate. The darkened oxygen atom indicates ^{18}O.

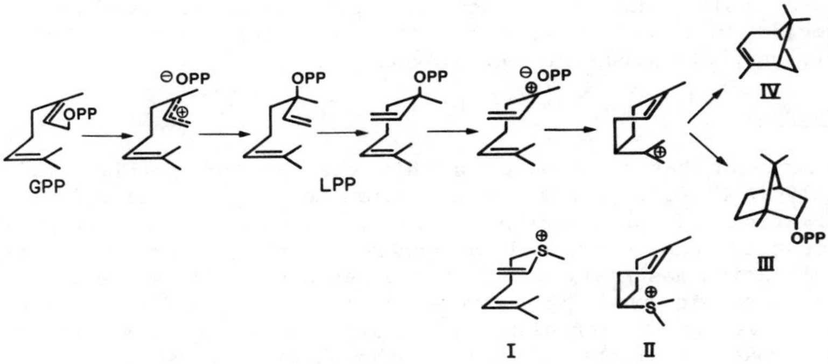

Figure 8. Sulfonium ion analogs of the linalyl (I) and α-terpinyl (II) carbonium ion intermediates in the cyclization of geranyl pyrophosphate to (+)-bornyl pyrophosphate (III) and (+)-α-pinene (IV).

indicating that inhibition by the terpenoid-like analogs in the presence of pyrophosphate was due to both electronic and structural resemblance to the normal, ion-paired intermediates of the cationic cyclization reactions.

The studies outlined above give a strong indication that the pyrophosphate moiety of the substrate is a major contributor to cyclase-substrate interactions. It should also be noted, however, that binding of the sulfonium analogs to the cyclase in the absence of inorganic pyrophosphate is still quite respectable, and is presumed to result from the same non-covalent interactions, probably a combination of structural and electronic effects, which are involved in aligning and stabilizing the substrate as well as the various cationic species generated in the course of the normal catalytic cycle. The topography of these interactions, and thus the relative positioning of the terpenyl partner with respect to the pyrophosphate, almost certainly underlies the inherent regio- and stereochemical features of these cyclization reactions.

It is also possible that the pyrophosphate moiety of the substrate functions as the base in the terminating deprotonation step of some cyclizations. The assistance of the pyrophosphate in deprotonation has been implicated in the prenyl transferase reaction (70), and the process has been recently modeled (87). Should such assistance apply in monoterpene cyclization, a spatial correlation must exist between the position of the pyrophosphate and the proton removed from the proximal face of the transient cation. Such an experimental observation would provide indirect evidence for this additional function of the pyrophosphate group in olefin synthesis.

Prospect

No attempt has been made in this chapter to provide complete analysis of each possible cyclization mode or to rationalize the formation of all parent skeletal forms. Yet, from the investigations described a coherent scheme for monoterpene cyclization has emerged and it now seems safe to assume that all cyclohexanoid types are generated by simple variations on the same general mechanism involving only a few possible conformations of the allylic pyrophosphate precursor. It is not known, however, in what way the cyclase enforces the required conformation on the reacting substrate, or what factors determine formation of distinct products arising from ostensibly identical substrate conformations. A more precise formulation of substrate-cyclase interactions, within the context of this still evolving model, must take into account the substantial changes in charge distribution, hybridization, configuration, and bonding which comprise the cyclization process and its component initiation, migration, rearrangement, , and termination events. Indeed, half of the carbon atoms of geranyl pyrophosphate undergo such alterations in the cyclization to bornyl pyrophosphate. It is not yet understood how the cyclase lowers the activation energy for any of these fundamental bond transformations or enforces the selection of a single regio- and stereochemical reaction channel by the precise folding of the acyclic geranyl substrate and the positioning of the counter ion. At the same time, it is clear that the active site of the cyclase cannot rigidly complement only the initial substrate conformation, but rather must be sufficiently flexible to accommodate the complete morphogenesis of substrate to product. Future challenges of research in this area will be to elucidate the means by which the enzyme accelerates the rate of cyclization of geranyl pyrophosphate while steering a single reaction course through the many possibilities available, and in the process survives the transient generation of highly reactive electrophilic species at the active site. Such studies, with pure cyclases, will require more suitable means of dissecting the multistep reaction into its component functional parts and of probing the many intermediate structures generated in the construction of this interesting group of natural products.

Acknowledgments

The work from the author's laboratory described herein was supported in part by grants from the National Science Foundation, the National Institutes of Health and the Department of Energy. It is with pleasure that I acknowledge the collaborative efforts of Professor D.E. Cane of Brown University and of Professor A.C. Oehlschlager of Simon Fraser University, and the contributions of my dedicated co-workers whose names are cited in the references.

Legend of Abbreviations

In the Figures, GPP indicates geranyl pyrophosphate, NPP indicates neryl pyrophosphate, LPP indicates linalyl pyrophosphate, and BPP indicates bornyl pyrophosphate.

Literature Cited

1. Croteau, R.; Johnson, M.A. In "Biology and Chemistry of Plant Trichomes"; Rodriguez, E.; Healey, P.L.; Mehta, I., Eds.; Plenum: New York, 1984; p. 133.
2. Devon, T.K.; Scott, A.I. "Handbook of Naturally Occurring Compounds"; Academic: New York, 1982; Vol. II. Terpenes, p.3.
3. Croteau, R. In "Biosynthesis of Isoprenoid Compounds"; Porter, J.W.; Spurgeon, S.L., Eds.; Wiley: New York, 1981; Vol. 1, p. 225.
4. Croteau, R. In "Isopentenoids in Plants"; Nes, W.D.; Fuller, G.; Tsai, L.-S., Eds.; Dekker: New York, 1984; p. 31.
5. Croteau, R. In "Herbs, Spices and Medicinal Plants"; Craker, L.E.; Simon, J.E., Eds.; Oryx: Phoenix, 1985; Vol. 1, Chap.3.
6. Hanson, J.R. <u>Nat. Prod. Reports</u> 1984, <u>1</u>, 443.

7. Croteau, R. In "Model Building in Plant Physiology/Biochemistry"; Newman, D.W.; Wilson, K.G., Eds.; CRC: Boca Raton, 1985, in press.

8. Cane, D.E. Accts. Chem. Res. 1985, 18, 220.

9. Croteau, R.; Cane, D.E. Methods Enzymol. 1985, 110, 352.

10. Wallach, O. "Terpene und Campher"; Vit: Leipzig, 1914; 2nd ed.

11. Ruzicka, L.; Eschenmoser, A.; Heusser, H. Experientia 1953, 9, 357.

12. Charlwood, B.V.; Banthorpe, D.V. Prog. Phytochem. 1978, 5, 65.

13. Lynen, F.; Agranoff, B.W.; Eggerer, H.; Henning, U.; Moeslein, E.M. Angew. Chem. 1959, 71, 667.

14. Stephan, K. J. Prakt. Chem. 1898, 58, 109.

15. Zeitschel, O. Ber. Dtsch. Chem. Ges. 1906, 39, 1780.

16. Cane, D.E. Tetrahedron 1980, 36, 1109.

17. Croteau, R.; Felton, M. Arch. Biochem. Biophys. 1981, 207, 460.

18. Croteau, R.; Karp, F. Arch. Biochem. Biophys. 1979, 198, 512.

19. Gambliel, H.; Croteau, R. J. Biol. Chem. 1984, 259, 740.

20. Croteau, R.; Karp, F. Arch. Biochem. Biophys. 1977, 179, 257.

21. Bernard-Dagan, C.; Pauly, C.; Marpeau, A.; Gleizes, M.; Carde, J.; Baradat, P. Physiol. Veg. 1982, 20, 775.

22. Poulose, A.J.; Croteau, R. Arch. Biochem. Biophys. 1978, 191, 400.

23. Croteau, R.; Gundy, A. Arch. Biochem. Biophys. 1984, 233, 838.

24. Cane, D.E. In "Enzyme Chemistry. Impact and Applications"; Suckling, C.J., Ed.; Chapman and Hall: London, 1984; p. 196.

25. Cane, D.E. In "Biosynthesis of Isoprenoid Compounds"; Porter, J.W.; Spurgeon, S.L., Eds.; Wiley: New York, 1981; Vol. 1, p. 283.

26. West, C.A. In "Biosynthesis of Isoprenoid Compounds"; Porter, J.W.; Spurgeon, S. L., Eds.; Wiley: New York, 1981; Vol. 1, p. 375.

27. Duncan, J.D.; West, C.A. Plant Physiol. 1981, 68, 1128.

28. Moesta, P.; West, C.A. Arch. Biochem. Biophys. 1985, 238, 325.

29. Poulter, C.D.; Rilling, H.C. In "Biosynthesis of Isoprenoid Compounds"; Porter, J.W.; Spurgeon, S.L., Eds.; Wiley: New York, 1981; Vol. 1, p. 161.

30. Croteau, R.; Karp, F. Biochem. Biophys. Res. Commun. 1976, 72, 440.

31. Croteau, R.; Karp, F. Arch. Biochem. Biophys. 1977, 184, 77.

32. Croteau, R.; Karp, F. Arch. Biochem. BIophys. 1979, 198, 523.

33. Croteau, R.; Hopper, C.L.: Felton, M. Arch. Biochem. Biophys. 1978, 188, 182.

34. Croteau, R.; Shaskus, J. Arch. Biochem. Biophys. 1985, 236, 535.

35. Gambliel, H.; Croteau, R. J. Biol. Chem. 1982, 257, 2335.
36. Croteau, R.; Felton, M.; Ronald, R.C. Arch. Biochem. Biophys. 1980, 200, 534.
37. Croteau, R.; Felton, M.; Ronald, R.C. Arch. Biochem. Biophys. 1980, 200, 524.
38. Cramer, F.; Rittersdorf, W. Tetrahedron 1967, 23, 3015.
39. Rittersdorf, W.; Cramer, F. Tetrahedron 1967, 23, 3023.
40. Bunton, C.A.; Hachey, D.L.; Leresche, J. J. Org. Chem. 1972, 37, 4036.
41. Stevens, K.L.; Jurd, L.; Manners, G. Tetrahedron 1972, 28, 1939.
42. Valenzuela, P.; Cori, O. Tetrahedron Lett. 1967, 3089.
43. Astin, K.B.; Whiting, M.C. J. Chem. Soc. Perkin Trans. II 1976, 1160.
44. Baxter, R.L.; Laurie, W.A.; MacHale, D. Tetrahedron 1978, 34, 2195.
45. Haley, R.C.; Miller, J.A.; Wood, H.C.S. J. Chem. Soc.(C) 1969, 264.
46. Bunton, C.A.; Cori, O.; Hachey, D.; Leresche, J.-P. J. Org. Chem. 1979, 44, 3238.
47. McCormick, J.P.; Barton, D.L. Tetrahedron 1978, 34, 325.
48. Kitagawa, Y.; Hashimoto, S.; Iemura, S.; Yamamoto, H.; Nozaki, H. J. Am. Chem. Soc. 1976, 98, 5030.
49. Bunton, C.A.; Leresche, J.-P.; Hachey, D. Tetrahedron Lett. 1972, 2431.
50. Bordwell, F.G. Accts. Chem. Res. 1970, 3, 281.
51. Sneen, R.A. Accts. Chem. Res. 1973, 6, 46.
52. Cane, D.E.; Iyengar, R.; Shiao, M.-S. J. Am. Chem. Soc. 1981, 103, 914.
53. Rittersdorf, W.; Cramer, F. Tetrahedron 1968, 24, 43.
54. Winstein, S.; Valkanas, G.; Wilcox, C.F., Jr. J. Am. Chem. Soc. 1972, 94, 2286.
55. Gotfredsen, S.; Obrecht, J.P.; Arigoni, D. Chimia 1977, 31, 62.
56. Poulter, C.D.; King, C.-H.R. J. Am. Chem. Soc. 1982, 104, 1420.
57. Poulter, C.D.; King, C.-H.R. J. Am. Chem. Soc. 1982, 104, 1422.
58. Brems, D.N.; Rilling, H.C. J. Am. Chem. Soc. 1977, 99, 8351.
59. Vial, M.V.; Rojas, C.; Portilla, G.; Chayet, L., Perez, L.M.; Cori, O.; Bunton, C.A. Tetrahedron 1981, 37, 2351.
60. Chayet, L.; Rojas, M.C.; Cori, O.; Bunton, C.A.; McKenzie, D.C. Bioorg. Chem. 1984, 12, 329.
61. Gotfredsen, S.E. Ph.D. Thesis, ETH, Zurich, No. 6243, 1978, as cited in references 16 and 52.
62. Sorensen, T.S. Accts. Chem. Res. 1976, 9, 257.
63. Coates, R.M. In "Progress in the Chemistry of Organic Natural Products"; Herz. W.; Grisebach, H.; Kirby, G.W., Eds.; Springer: Wien, 1976; Vol. 33, p. 73.
64. Luft, R. J. Org. Chem. 1979, 44, 523.
65. Money, T. Prog. Org. Chem. 1973, 8, 29.
66. Gascoigne, R.M. J. Chem. Soc. 1958, 876.
67. Arigoni, D. Pure Appl. Chem. 1975, 41, 219.

68. Guilford, W.J.; Coates, R.M. J. Am. Chem. Soc. 1982, 104, 3506.

69. Coates, R.M.; Cavender, P.L. J. Am. Chem. Soc. 1980, 102, 6358.

70. Poulter, C.D.; Rilling, H.C. Accts. Chem. Res. 1978, 11, 307.

71. Allinger, N.A.; Siefert, J.M. J. Am. Chem. Soc. 1975, 97, 752.

72. Kjonaas, R.; Croteau, R. Arch Biochem. Biophys. 1983, 220, 79.

73. Karp, F.; Croteau, R. Arch. Biochem. Biophys. 1982, 216, 616.

74. Satterwhite, D.M.; Wheeler, C.J.; Croteau, R. J. Biol. Chem. 1985, 260, 13901.

75. Croteau, R., unpublished data.

76. Croteau, R.; Felton, N.M.; Wheeler, C.J. J. Biol. Chem. 1985, 260, 5956.

77. Overton, K.H. Chem. Soc. Rev. 1979, 8, 447.

78. Dregler, K.A.; Coates, R.M. J. Chem. Soc. Chem. Commun. 1980, 856.

79. Anastasis, P.; Freer, I.; Gilmore, C.; Mackie, H.; Overton, K.; Picken, D.; Swanson, S. Can. J. Chem. 1984, 62, 2079.

80. Wheeler, C.J.; Croteau, R. Arch. Biochem. Biophys. 1986, in press.

81. Cane, D.E.; Saito, A.; Croteau, R.; Shaskus, J.; Felton, M. J. Am. Chem. Soc. 1982, 104, 5831.

82. Croteau, R.; Shaskus, J.J.; Renstrøm, B.; Felton, N.M.; Cane, D.E.; Saito, A.; Chang, C. Biochemistry 1985, 24, 7077.

83. Mash, E.A.; Gurria, G.M.; Poulter, C.D. J. Am. Chem. Soc. 1981, 103, 3927.

84. Davisson, V.J.; Neal, T.R.; Poulter, C.D. J. Am. Chem. Soc. 1985, 107, 5277.

85. Croteau, R.; Shaskus, J.; Cane, D.E.; Saito, A.; Chang, C. J. Am. Chem. Soc. 1984, 106, 1142.

86. Croteau, R.; Wheeler, C.J.; Aksela, R.; Oehlschlager, A.C. J. Biol. Chem. 1986, in press.

87. Jacob, L.; Julia, M.; Pfieffer, B.; Rolando, C. Tetrahedron Lett. 1983, 24, 4327.

RECEIVED February 3, 1986

Carotenoids
A Source of Flavor and Aroma

W. W. Weeks

Department of Crop Science, North Carolina State University, Raleigh, NC 27695-7620

Over four hundred carotenoids have been found in nature of which α-, β-, γ-carotene and lycopene are encountered most frequently. β-Carotene, a direct precusor to Vitamin A, is the most widely known tetraterpene in plants. The polyene chain of the carotenoids is readily oxidized, giving rise to cyclic and acyclic compounds often having an oxygen-containing functional group on a trimethylcyclohexane ring, or an oxygen containing functional group on the allylic side chain. In plants used for food, improved nutrition and palatibility are often associated with high carotene levels. Higher concentrations of carotenes in flue-cured tobacco have resulted in a wider variety of carotenoid derivatives some in increased concentration, in the tobacco leaf and in the smoke.

The yellow carotenoid pigments are tetraterpenes of which over 400 occur in higher plants (1), algae and bacteria, and in some animals that depend on plants for their existence (2). The carotenoid color from plants is a precursor for pigmentation in marine animals, egg yolks and fat globules and serves as a source for Vitamin A for mammals. Carotenoids are synthesized from products of acyclic $C_{40}H_{56}$ polyene lycopenes by hydrogenation, dehydrogenation, cyclization, methyl-migration, and chain elongation (2). They are companion pigments to chlorophyll in plants and their sensitivity to light and oxygen is demonstrated by xanthophylls and other pigments containing hydroxyl, carboxyl, carbonyl and esters as terminal groups. These oxygenated pigments are associated with yellow colors so very prominent in the fall. Figure 1 illustrates examples of acyclic, monocyclic, bicyclic, and oxygenated tetraterpenes. The four most common tetraterpenes in nature are lycopene, α- β-, and γ-carotene. β-Carotene is the most widely recognized member of the group because of its association with Vitamin A.

Carotenoid levels in mature cells of leaves and fruits remain relatively constant until the onset of senescence. In the case of seed plants, senescence occurs following flowering; in fruits it

0097–6156/86/0317–0157$06.00/0
© 1986 American Chemical Society

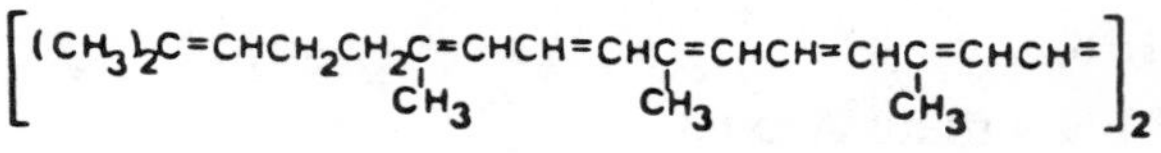

Lycopene (Acyclic)

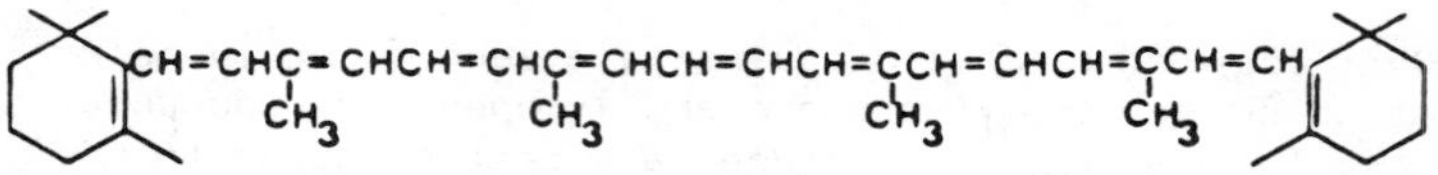

β-carotene (Bicyclic)

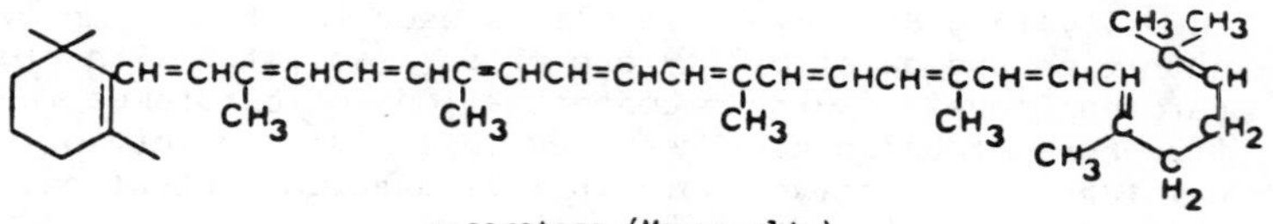

γ-carotene (Monocyclic)

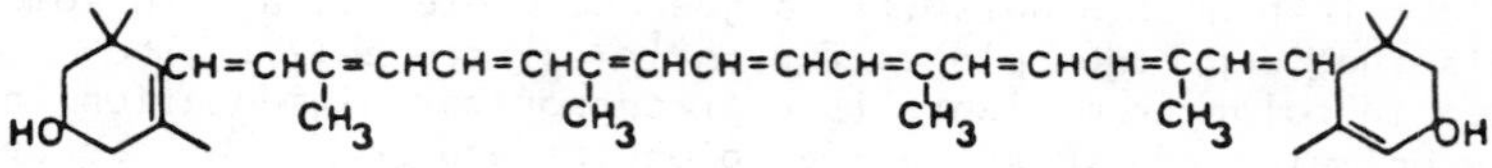

Lutein

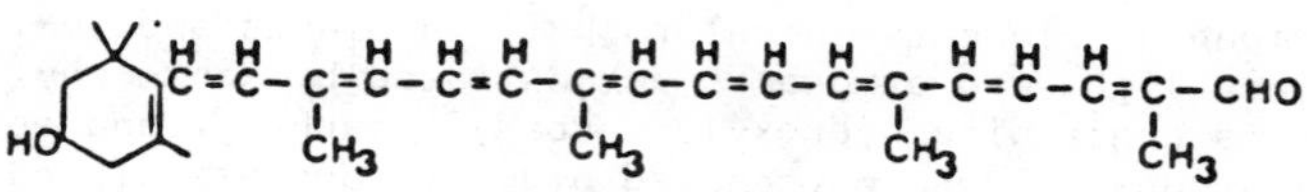

β-citraurin

Figure 1.   Examples of acyclic, monocyclic, bicyclic, and oxygenated pigments.

begins at the outset of ripening following the disappearance of
chloroplasts and appearance of chromaplasts (4). Yellow color
following degradation of chlorophyll is indicative of the presence
and effects of carotene degrading enzymes. Two types of enzymes,
one from chloroplasts and the second from mitochondria, are responsible for oxygenation and degradation of carotenoids. These systems
are designated as lipoxidases and peroxidases and both require molecular oxygen and cofactors for activity. Lipoxygenases catalyze the
conversion of polyunsaturated lipids to both desirable and undesirable flavor components in plants (3).

A variety of degradative oxygenations produce shorter chain terpenoid compounds from tetraterpenes in plants. These degradations
involve cleavage of (C9,C10), (C8,C9), (C7,C8), and (C6,C7) bonds of
the polyene chain to produce cyclic compounds containing 13, 11, 10,
and 9 carbon atoms respectively (Fig. 2) (4,5). The volatility of
these compounds has been of interest to investigators for a long
time.

Carotenoid Derivatives of Commercial Importance

Ionones: The discovery of ionones initiated one of the most fasci-
nating eras in organic chemistry. In 1893, Tiemann and Kruger, isolated a compound from orris roots with a pleasing aroma very similar
to violets. The study of this $C_{14}H_{22}O$ ketone later identified as
irone, led to the identification and synthesis of ionone, and revolu-
tionized the perfume industry. Following the application for a
patent describing synthesis of this compound, it became well known
and the price of violet fragrance dropped very sharply to less than
half its original cost (6).

Modern instrumentation has helped with the identification of
many ionones, dehydroionones, ionols and epoxy ionones isolated from
naturally occuring substances. The strong floral odor which upon dilution resembles violets occompanies α-ionone and (±) cis-α-ionone.
Either optically active isomer of ionone in the pure form provides a
stronger and more pleasing odor than the racemic mixture (6).

In 1970, Demole, et al. (7) isolated from rose oil a compound
having a structure similar to the ionones which was named demasce-
none. This compound was also found in geranium oil. Although the
structure was similar to ionone the odor differed. The odor has been
described as "sweet" by Leffingwell (8) and it is used frequently
for flavoring. The characteristic sweet and "ripe-like" odor of
pears, blackberries, raspberries, cranberries, and cooked apples has
been attributed to the presence of isomers of damascenone and damascone. Numerous efforts to provide economical synthesis of damascenones and damascone have been attemped, but a successful commercial
synthesis has not been achieved. Figure 3 illustrates ionones,
damascenones and damascones commonly found as flavorants and odorants.

Compounds other than ionone, and without ionone-like structures,
derived from carotenoids are also of commerical importance. Safranal,
$C_{10}H_{14}O$ is isolated from the styles of the autumn blooming crocus
and is accompanied by a $C_{20}H_{24}O_4$ yellow pigment. The fresh safranal
oil is very unstable, but it has a very strong and pleasant odor. It
is used as an additive to violet-like perfumes and the results are
very satisfying. Addition of tinctures of safranal to bitter and

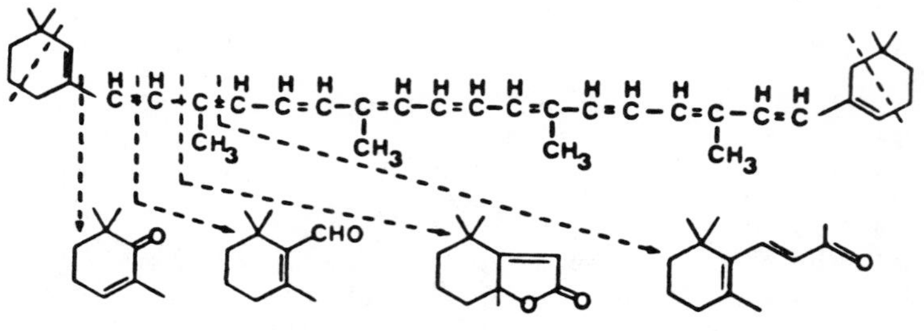

Oxidation of α-carotene to smaller compounds (5).

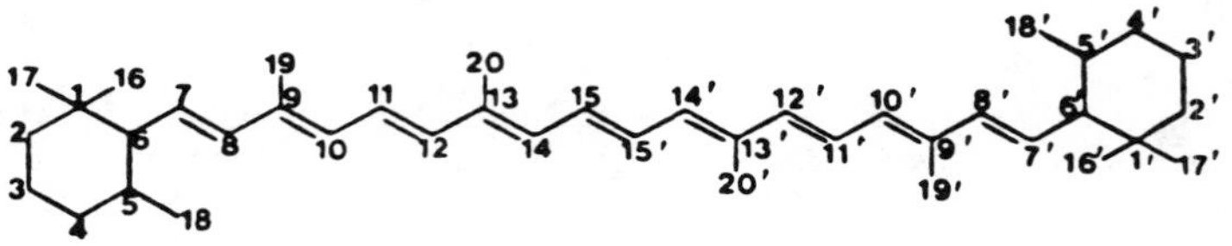

Numbering system for a carotene compound.

Figure 2. The oxidation of α-carotene to produce smaller volatile compounds and the numbering system for tetraterpene.

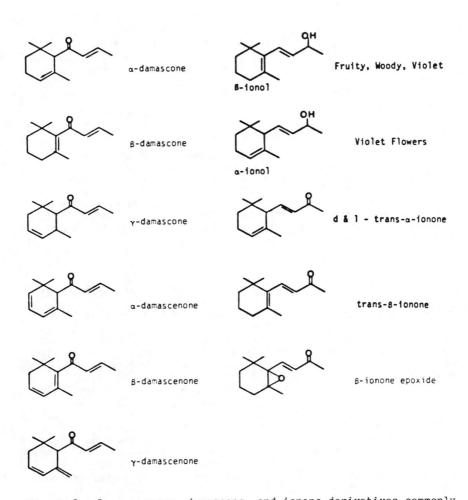

Figure 3.   Damascenones, damascone, and ionone derivatives commonly found in nature as flavorants and perfumes.

sweet orange flavors, and to apricot, plum, and cherry wines is common practice to obtain added flavor and body (9).
    Oxidation of tetraterpenes yields many compounds, and the cyclic compounds with a trimethylcyclohexane ring are easily associated with degradation of monocyclic and bicyclic carotenes, but allylic compounds are now as conspicuous in their biogenetic origin. Many such compounds are derived from lycopene, phytoene and phytofluene, three very common acyclic carotenes. These allylic compounds are often identified only as terpene aldehydes and ketones and not as carotene degradation products (10).

Carotenoid Derivatives From Tea

Tea is one of the most widely consumed beverages available and is marketed in every country in the world. Under ordinary brewing conditions 25 to 30% of the weight of the dry leaf is extracted into solution with only 0.1% being released as volatile components. The volatiles include several compounds derived from carotenes found in the green tea leaves. During drying and processing these carotenes give rise to smaller oxygenated compounds responsible for tea flavor and aroma. Sanderson et al. (10) suggested the compounds illustrated in Figure 4 as being largely responsible for tea flavor, particularly black tea.

Carotenes and Carotenoid Derivatives From Tobacco and Tobacco Smoke

Tobacco, unlike most other commodities, is not produced as a food crop, but it is used for manufacture of smoking materials and other products. The essential oils in tobacco are important for impact and balance in smoking (11). Smoking pleasure is derived from a balance of nicotine and volatile components. Tobacco chemists and flavorists are certain that carotenoid derivatives contribute to smoke flavor and aroma (5). Over a hundred compounds related to carotenoids have been isolated from tobacco and tobacco smoke.
    Eighteen carotene compounds have been characterized in green tobacco and of these, β-carotene and lutein are the primary pigments. Tobacco differs from other plants in that during senescence the oxygenated pigments are degraded before α- and β-carotenes, and degradation in the plant preceding harvesting occurs enzymatically. During curing, chemical changes occur very quickly in the tobacco leaf. Chlorophyll degrades within 36 to 48 hours and yellowing occurs, particularly in the flue-cured leaf. During the rest of curing oxidation of carotenes occurs producing the oxygenated constituents having flavoring effects. One of the unique characteristics of many of the carotenoid derivatives is an oxygen containing functional group on the trimethylcyclohexane ring (11).
    The effect of carotenoids upon smoke flavor and aroma has resulted in new endeavors in research to increase levels of carotenoid derivatives. Plant breeders, attempting to improve tobacco quality, have developed cultivars with higher carotenoid levels. Beatson and Wernsman increased carotenoid content 40% by backcrossing and selection in only five generations. Statistical analyses of these data showed positive correlations between carotenoids and nicotine in these flue-cured tobaccos ($r=.81$) (12). This phenomenon is practical for improving tobacco quality because

low nicotine tobaccos ripen and cure poorly and the smoke from these tobaccos is flat, hot and lacking in flavor.

Leaf length, width, density, maturity and biochemistry are controlled genetically and these properties in turn influence the flavor and aroma of tobacco and tobacco smoke. An example of this phenomenon are tobaccos containing genomes that produce 3% nicotine with higher levels of carotenoid derivatives. They ripen and mature normally and, therefore, handle easily and produce a higher quality leaf. Tobaccos with the recessive genome produce 0.25% nicotine and do not ripen well. The cured tobaccos from such genotypes are of poor quality and are low in oils and body. Experienced buyers readily recognize and preferentially purchase tobaccos that exhibit good body and high oil content to insure a product that is both flavorful and satisfying.

Capillary gas chromatography has provided the capability for the reproducible separation, identification, and quantification of volatile constituents from relatively small samples of cured tobacco from genetic, management and curing experiments. The chromatogram in Figure 5 exhibits a typical tobacco profile of the essential oils from ten grams of flue-cured tobacco. Compounds derived from carotenoid degradation are important constituents in the volatile profile, qualitatively and quantitatively. Peaks 7, 14, 15, 18, 23, 28, 33, 49, and 50 have been identified by gas chromatography and mass spectrometry as carotenoid derivatives; the structural formulae of these compounds are provided in Figure 6.

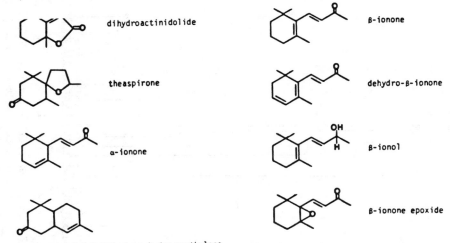

dihydroactinidolide

β-ionone

theaspirone

dehydro-β-ionone

α-ionone

β-ionol

1,1,6 trimethyl-3 keto-hexa hydro napthelene

β-ionone epoxide

Figure 4.   Carotenoid derivatives found in black tea; identified as flavor compounds.

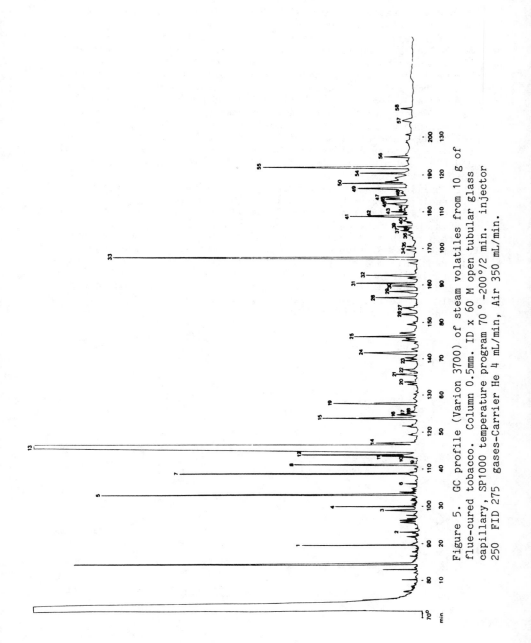

Figure 5.   GC profile (Varion 3700) of steam volatiles from 10 g of
flue-cured tobacco. Column 0.5mm. ID x 60 M open tubular glass
capillary, SP1000 temperature program 70° -200°/2 min. injector
250 FID 275 gases-Carrier He 4 mL/min, Air 350 mL/min.

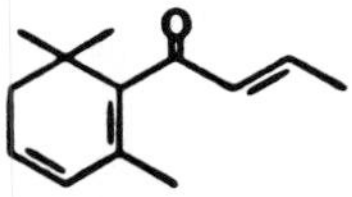

Pk. 7 Damascenone

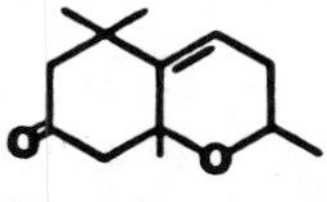

Pk. 14 1,3,7,7-tetramethyl-9-oxo-2-
        oxabicyclo-(4,4,0)-dec-5-ene

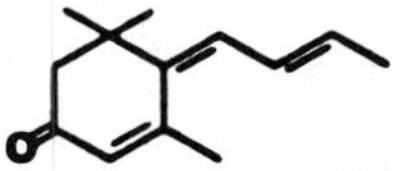

Pk. 15 4,6,8-megastigmatrien-3-one

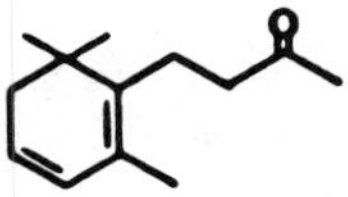

Pk. 18 4(2,6,6-trimethyl-1-3
        cyclohexadenyl)buta-2-one

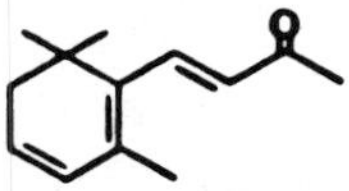

Pk. 23 Dehydro-β-ionone

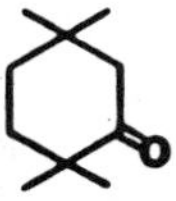

Pk. 28 2,2,5,5-tetramethyl
        cyclohexanone

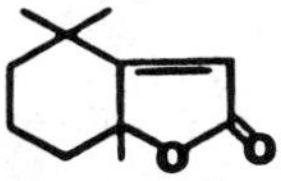

Pk. 33 Dihydroactinidolide

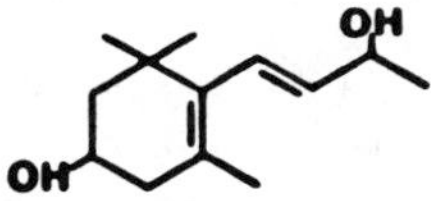

Pk. 49 3-hydroxy-β-ionol

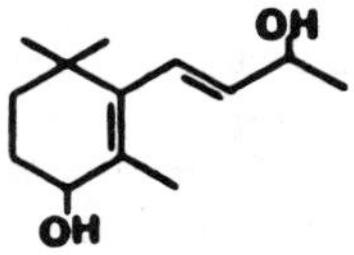

Pk. 50   4-hydroxy-β-ionol

Figure 6. Compounds identified as carotenoid derivatives in flue-
cured tobacco.

Acknowledgment

This chapter is Paper No. 10088 of the Journal Series of North
Carolina Agricultural Research Service, Raleigh, NC 27695. The
use of trade names in this publication does not imply endorsement
by the North Carolina Research Service of the products named, nor
criticism of similar ones not mentioned.

Literature Cited

1.   Bauernfeind, J. C. "Carotenes as Colorants and Vitamin A Precur-
     sors; Academic Press, New York and London. 1981, p. *xiii*.
2.   Goodwin, T. W. "The Biochemistry of the Carotenoids";
     Vol. 1. Plants 2nd edition, Chapman and Hall, London and
     New York, 1980, 1, 77-90.
3.   Goodwin, T. W.; Mercer E. I. "Introduction to Plant Biochem-
     istry"; 2nd. edit Pergaman Press, Oxford, New York, Toronto,
     Paus and Frankfort, 1983, 304-306.
4.   Enzell, C. R. Recent Advances in Tobacco Science 1976,
     2, 32-66.
5.   Ohloff, G. Flavour Industry. 1972, 3, 501-508.
6.   Theimer, E. T. "Fragrance Chemistry The Science of the Sense
     of Smell"; Academic Press, New York and London, 1972, 301-308.
7.   Apt, C. "Flavor: Its Chemical, Behavioral and Commercial
     Aspects"; Proceedings of the Arthur D. Little Inc. Flavor
     Symposium, 1977, 12-29.
8.   Leffingwell, J. H.; Young H.; Bernasek E. "Tobacco Flavoring
     for Smoking Products"; R. J. Reynolds Tobacco Company, Winston-
     Salem, N.C., 1972.
9.   Guenther, E. "The Essential Oils"; Vol. 2, D. Van Nostrand Co.,
     Inc. Toronto, New York and London, 1949, 348-349.
10. Sanderson, G.; Graham H. N. Journal of Agriculture and Food
     Chemistry, 1973, 21, 576=584.
11. Kaneko, H. Koryo. 1980, 128, 23-33.
12. Beatson, R.A.; Wernsman E. A.; Long R. C. Crop Science, 1984,
     24, 67-71.

RECEIVED February 18, 1986

Fatty Acid Hydroperoxide Lyase in Plant Tissues
Volatile Aldehyde Formation from Linoleic and Linolenic Acid

Akikazu Hatanaka, Tadahiko Kajiwara, and Jiro Sekiya

Department of Agricultural Chemistry, Yamaguchi University, Yamaguchi 753, Japan

Fatty acid hydroperoxide lyase is one of the enzymes responsible for volatile C_6- and C_9-aldehyde formation from linoleic and linolenic acid. This enzyme cleaves 9- and/or 13-hydroperoxides derived from linoleic and linolenic acid. The enzyme is distributed in a wide range of plant species in membrane bound forms; both chloroplastic and non-chloroplastic. Three types of hydroperoxide lyases have been reported; 9-hydroperoxide-specific, 13-hydroperoxide-specific and non-specific. Other properties of the hydroperoxide lyase including substrate specificity and reaction mechanism are discussed in this review.

Since 2-hexenal was isolated from plant tissue by Curtius et al ([1]) and 3-hexenal by von Romburgh ([2]), many higher plants have been reported to have the ability to produce volatile C_6-compounds such as hexanal (I), (3Z)-hexenal (II), (2E)-hexenal (leaf aldehyde, III) and (3Z)-hexenol (leaf alcohol). These C_6-compounds are largely responsible for the characteristic odor of green leaves of vegetables and trees and are also constituents of aroma from various fruits. Drawert et al ([3]) reported the enzymic formation of I and III from linoleic (IV) and linolenic acids (V), respectively, in fruit tissues. The reactions involved in the major biosynthetic pathway for these C_6-compounds consist of four sequential enzymic steps and a non-enzymic one; acyl hydrolysis of lipids, peroxidation of IV and V, cleavage of the fatty acid hydroperoxides, reduction of aldehydes to alcohols and isomerization of II to III (Fig. 1)([4-6]). The four enzymes and a factor involved in this pathway are lipolytic acyl hydrolase (LAH), lipoxygenase, fatty acid hydroperoxide lyase (hydroperoxide lyase), alcohol dehydrogenase (ADH) and isomerization factor (IF) as shown in Fig. 1. Volatile C_9-aldehydes and the corresponding alcohols, which are responsible for flavor of some cucurbitaceae fruits ([7],[8]), are formed by a similar pathway (Fig. 1) ([4],[9],[10]).

Among these enzymes, lipoxygenase and hydroperoxide lyase seem to be of the most importance, since product specificity of lipoxyge-

0097-6156/86/0317-0167$06.00/0

nase and substrate specificity of hydroperoxide lyase determine the
pattern of volatile aldehydes formed.   Lipoxygenase from many plant
sources catalyzes formation of 13-hydroperoxy-(9Z,11E)-octadecadi-
enoic acid (VI) and 13-hydroperoxy-(9Z,11E,15Z)-octadecatrienoic
acid (VII) from IV and V, respectively, while lipoxygenase from
sources such as potato tuber, corn seed, cucumber fruit and rice
embryo catalyzes formation of 9-hydroperoxy-(10E,12Z)-octadecadi-
enoic acid (VIII) and 9-hydroperoxy-(10E,12Z,15Z)-octadecatrienoic
acid (IX)(11,12).   From VII, II is formed by the action of hydroperoxide lyase along with 12-oxo-(9Z)-dodecenoic acid (X) as the co-
product.   This C12-oxo-acid (X) is presumed to be a precursor of
traumatin, a wound hormone (13).   (3Z,6Z)-Nonadienal and 9-oxo-
nonanoic acid are formed from IX by hydroperoxide lyase specific for
the 9-hydroperoxides (9).   We review here recent studies on hydroperoxide lyase.   Extensive general reviews on plant lipoxygenase
have been published elsewhere (11,12).

Enzyme Assay

The name hydroperoxide lyase was tentatively given to the enzymes
cleaving VI and VII between carbon-12 and -13 of the chains or VIII
and IX between carbon-9 and -10.   There are several methods to
detect the enzyme activity.   The first is based on the high molar
extinction coefficient of the hydroperoxides due to the conjugated
diene (i.e. 27,000 at 234 nm).   Enzyme activity is assayed by a
decrease in absorbance at 234 nm using the hydroperoxide as a
substrate (14).   This method, however, is not specific for hydroperoxide lyase activity, since the activity of fatty acid hydroperoxide isomerase (15) and fatty acid hydroperoxide cyclase (16) also
result in decrease of absorbance at 234 nm.   The second type of
assay method is based on analyses of the aldehyde reaction products.
After steam-distillation of the reaction mixture, the aldehydes
formed are determined by GLC analysis (17).   This method, however,
is time consuming and is not suitable for many samples.   Additionally, during steam-distillation, the labile aldehyde, (II), is
isomerized to III.   To prevent isomerization, headspace gas
analysis may be used for the determination of hydroperoxide lyase
activity (18,19).   A reaction mixture in a sealed vessel is
incubated, and the volatiles released into the headspace are analyzed by GLC.   In another method, aldehydes formed are extracted
and converted to 2,4-dinitrophenylhydrazones.   The resulting hydra-
zone are determined directly by absorbance (20) or after separation
by HPLC (21).   One of these methods, or a combination is employed
for enzyme assay.   Routinely the hydroperoxides (VI and VIII) are
used as substrates.   The volatile products are I and (3Z)-nonenal,
respectively.

Occurrence

Since hydroperoxide lyase was first identified in non-green tissues,
watermelon seedling (14) and cucumber fruit (9,22), the occurrence
of the enzyme has been reported in various non-green and green plant
tissues.   Non-green tissues in which hydroperoxide lyase occurs
include etiolated seedlings of watermelon (14), cucumber (17), and
alfalfa (17), soybean seed (21,23), fruits of cucumber (9,22,24),

tomato (5,25), pear(20), and apple (25), and cultured tobacco cells
(26). Hydroperoxide lyases occur in leaves of kidney bean (27),
tea (28) and others (29), peel tissue of cucumber fruit (30), and
cultured tobacco green cells (26). Hydroperoxide lyase activity in
green tissues may consist of at least two enzymes, of chloroplastic
and non-chloroplastic origin, while non-green tissue contain only
the non-chloroplastic type (29,30). Cultured tobacco cells can
convert reversibly between green and non-green cells depending on
illumination (26). Green cells showed twice the hydroperoxide
lyase activity as non-green cells. One of the hydroperoxide lyases
in green cells is associated with chloroplasts. Hydroperoxide
lyase activity was altered reversibly according to the light regime
of the cells (Fig. 2 and 3).

Subcellular Localization

Hydroperoxide lyase was found first in the 12,000 g supernatant of
watermelon seedling extracts (14). On the other hand, Galliard
et al reported that cucumber fruit hydroperoxide lyase was associated with a particulate fraction and was solubilized with Triton
X-100 (9,22). Galliard et al also used Triton X-100 to extract
hydroperoxide lyase effectively from cucumber fruit (24), tomato
fruit (32) and kidney bean leaf (27). Wardale et al attempted to
determine the subcellular localization of hydroperoxide lyase in
cucumber fruit by sucrose density gradient centrifugation (30).
The enzyme activity from the fresh tissue of cucumber fruit was
located in three fractions: plasma and Golgi membranes and endo-
plasmic reticulum. In the peel tissue of cucumber fruit,
chloroplasts contained a large amount of hydroperoxide lyase
activity (30). These results suggest that hydroperoxide lyase
exists in multiple locations in plants; non-chloroplastic particles
and chloroplasts. In other non-green tissues such as cucumber
seedlings (18), and fruits of pear (20), tomato (25,32) and apple
(25), hydroperoxide lyase also has been reported to be membrane
bound. Thus, it appears that the major subcellular location of
hydroperoxide lyase activity in non-green tissues is the microsomal
fraction. By contrast, green tissues appear to have the major
hydroperoxide lyase activity associated with chloroplasts (28,30).
Chloroplast thylakoid membranes appear to be the major site of location, since activity is not removed by repeated washing of chloroplast fragments with hypotonic buffer solution (28). The results
obtained with cultured tobacco cells, converting reversibly between
green and non-green cells according to illumination, indicate that
chloroplast hydroperoxide lyase activity is induced and then reduced
during development and disappearance of chloroplasts (26). Thus
hydroperoxide lyase appears to be located in at least two types of
particles; chloroplasts and non-chloroplastic particles (microsomes).

Purification

Watermelon hydroperoxide lyase was purified 42-fold by $(NH_4)_2SO_4$
fractionation and gel filtration on Sephadex G-200 without any particular solubilization treatment (14). From non-green tissues such
as fruits of cucumber (24), pear (20) and tomato (25), hydroperoxide
lyase was solubilized initially with Triton X-100 and then extracted.

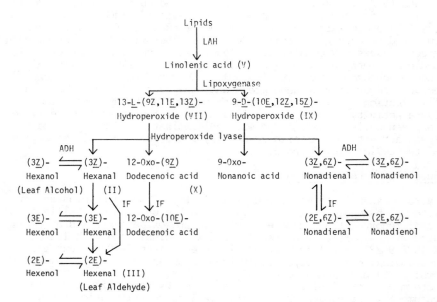

Fig. 1. Biosynthetic pathway of C6- and C9-compounds from C18-fatty
acids. LAH, Lipolytic Acyl Hydrolase; ADH, Alcohol Dehydrogenase; IF, Isomerization Factor.

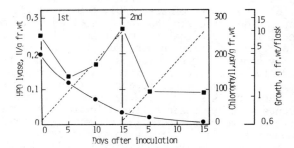

Fig. 2. Effect of transfer from light to dark on chlorophyll content
and hydroperoxide lyase activity of tobacco cells. The green
cells cultured in the light were inoculated in a fresh medium
and cultured in the dark (1st panel). The cells were again
inoculated and cultured in the dark (2nd panel). Cell growth
(---); Chlorophyll content (●); and Hydroperoxide lyase
activity (■) are indicated. Reproduced with permission
from reference 26. Copyright 1984 Pergamon Press.

After concentration, the solubilized enzyme was applied to a gel filtration column. The enzymes from pear and tomato were purified further (1.5-3 fold) by isoelectric focusing (20,25). Chloroplastic hydroperoxide lyase was solubilized from tea leaves with Tween 20 and partially purified 8.5-fold with 34% recovery by hydroxyapatite column chromatography (33). Although attempts at purification of hydroperoxide lyase have been made, the enzyme has not been purified to homogeneity thus far.

Schreier and Lorenz found that hydroperoxide lyase activities of tomato fruit were eluted at the molecular weight of 200,000 and 2,000,000 (25) on gel filtration. Gel filtration of the tea chloroplast enzyme on Sephadex G-200 gave two species of activity, one of 150,000-180,000 and the other eluting in the void volume (33). Inclusion of Tween 20 in the elution buffer reduced the proportion of activity in the void volume suggesting disaggregation of an aggregated form.

Substrate Specificity

Major products of the lipoxygenase reaction are 9- and 13-hydroperoxides of IV and V, VI, VII, VIII and IX. Hydroperoxide lyase utilizes the 9- and/or 13-hydroperoxides (VI, VII, VIII and IX). Based on substrate specificity, hydroperoxide lyase is classified into three types. The first type is 9-hydroperoxide-specific. This enzyme cleaves exclusively the 9-hydroperoxides (VIII and IX) to give C_9-aldehydes and a C_9-oxo acid. Pear fruit hydroperoxide lyase (20) belongs to this type. Hydroperoxide lyase of the second group is 13-hydroperoxide-specific. This enzyme cleaves exclusively the 13-hydroperoxides (VI and VII) to give C_6-aldehydes and C_{12}-oxo acid. Watermelon seedling (14), tea leaf (28), cultured tobacco cells (26), tomato fruit (26), alfalfa seedling (17) and soybean seed (21) contain this type of the enzyme. The third type of hydroperoxide lyase is non-specific. Kidney bean leaf (27), and cucumber fruit (9,22) and seedling (17) contain this type of the enzyme which can cleave both the 9- and 13-hydroperoxides to give the corresponding products. The hydroperoxide cleaving enzyme found in mushroom is another type of the enzyme; it is specific for 10-hydroperoxide (XI) and the volatile product in this instance is an alcohol, 1-octen-3-ol (31).

To investigate substrate specificity in greater detail, other hydroperoxides were examined with the tea leaf enzyme (Table I)(28). Hydroperoxides VI and VII were the best substrates for hydroperoxide lyase solubilized from tea leaf chloroplast. When 13-hydroperoxy-(6Z,9Z,11E)-octadecatrienoic acid (13-hydroperoxide of γ-linolenic acid) was used, the Vmax value was 18% of that of VI. 15-Hydroperoxy-(5Z,8Z,11Z,13E)-eicosatetraenoic acid (15-hydroperoxide of arachidonic acid) did not act as a substrate. Conversion of the carboxyl group to a methyl ester or alcohol function reduced reactivity. The 12-keto-13-hydroxy derivative was not a substrate. As to configuration of hydroperoxy group of the 13-hydroperoxides (VI and VII), the (S)-configuration is favored by the tea leaf enzyme (34). These data thus indicate the optimum requirements for the substrate of tea leaf hydroperoxide lyase; (1) C_{18}-straight chain fatty acid with free carboxyl group, (2) attachment of hydroperoxy group to carbon-13 with (S)-configuration and (3) (Z)-double bond at carbon-9 and (E)-double bond at carbon-11 in the chain. Introduction of a (Z)-double bond at carbon-15 and at carbon-6

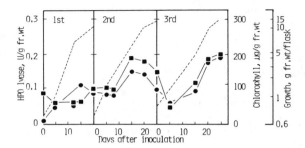

Fig. 3.   Effect of transfer from dark to light on chlorophyll
content and hydroperoxide lyase activity as tobacco
cells. The nongreen cells cultured in dark were
inoculated in the light (1st panel). The cells were
again inoculated and cultured in the light (2nd and
3rd panels). Symbols are as in Fig. 2. Reproduced
with permission from reference 26. Copyright 1984
Pergamon Press.

Table I. Substrate Specificity of HPO Lyase Solubilized from Tea Chloroplast

Compounds	Relative V_{max} (%)	K_m (mM)
13-Hydroperoxy-(9Z,11E)-octadecadienoic acid (VI)	100	1.88
13-Hydroperoxy-(9Z,11E)-octadecadienoic acid methyl ester	26	0.10
13-Hydroperoxy-(9Z,11E)-octadecadienol (XIII)	55	0.37
13-Hydroperoxy-(9Z,11E,15Z)-octadecatrienoic acid (VII)	127	2.50
13-Hydroperoxy-(6Z,9Z,11E)-octadecatrienoic acid	18	0.48
15-Hydroperoxy-(5Z,8Z,11Z,13E)-eicosatetraenoic acid	0	-
9-Hydroperoxy-(10E,12Z)-octadecadienoic acid (VIII)	0	-
9-Hydroperoxy-(10E,12Z,15Z)-octadecatrienoic acid (IX)	0	-
12-Keto-13-hydroxy-(9Z)-octadecenoic acid	0	-

increases reactivity by 27% and decreases it by 45%, respectively.
These conclusions are supported by substrate specificity studies
with soybean hydroperoxide lyase.   Soybean hydroperoxide lyase
utilized (S̲)-VI but not the (R̲)-enantiomer, (R̲,S̲)-VI and (S̲)-VIII
(23̲).

Other Properties

Optimum pH ranges from 5.5 to 8 for hydroperoxide lyase from green
and non-green tissues. K_m values for VI and VII range from 2 x
10^{-5} - 2 x 10^{-3} M (5̲,9̲,23̲,24̲,25̲,26̲,28̲).   Hydroperoxide lyase from
cucumber fruit (28̲), watermelon seedling (14̲) and tomato fruit (25̲)
was inhibited by p̲-chloromercuribenzoate, but not the enzyme from
tea leaf (28̲).   This observation indicates that chloroplastic and
non-chloroplastic hydroperoxide lyase may differ in structure.

Mechanism of Enzyme Cleavage

13-Hydroperoxide-specific hydroperoxide lyase has been shown to
cleave only the (S̲)-enantiomer of the 13-hydroperoxide into C_6-
aldehydes and C_{12}-oxo-acid in tea leaf (34̲) and in soybean seed
(23̲).   The 13-hydroperoxy-dienol (XII) was also cleaved stereo-
specifically to I and 12-oxo-(3Z)-dodecenol (XIII) by the hydroperoxide lyase (35̲).   Incorporation of the hydroperoxy oxygen of the
13-hydroperoxide (XII) into the carbonyl groups of cleavage products
was investigated in tea chloroplasts using ^{18}O-labeling.   The ^{18}O-
oxygen of the hydroperoxy group of the 13-hydroperoxide (XII) was
found in XIII, not in I (35̲), when they were reduced to the corresponding alcohols immediately after the enzyme incubation to prevent
rapid exchange of carbonyl oxygen with water (36̲,37̲).
     The origin of oxygen in cleavage products (I and XIII) of the
hydroperoxide lyase reaction presumably occurs in a manner (Fig. 4)
similar to the mechanism for acid-catalyzed rearrangement of the 13-

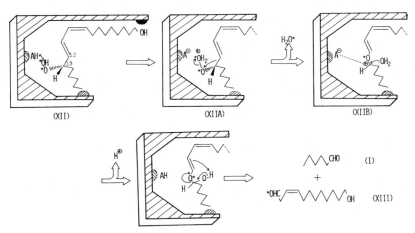

Fig. 4. Proposed mechanism for tea chloroplast hydroperoxide lyase.
*O =^{18}O.

hydroperoxide (VI) in aprotic solvent (38). In the first step, hydroperoxide lyase catalyzes cyclization of the protonated hydroperoxide (XIIA) to a 12,13-epoxide which loses a molecule of water to form an allylic ether-cation (XIIB) localized at carbon-13 via a 1,2-shift of the bond from carbon-13 to the electron-deficient oxygen. Subsequently, addition of water to the stabilized carbonium ion occurs, ultimately yielding I and XIII.

Acknowledgment

This work was supported in part by grants-in-aid from the Ministry of Education, Science and Culture of Japan (A.H., T.K.) and from Agricultural Chemical Foundation of Japan (J.S).

Literature Cited

1. Curtius, T.; Frunzen, H. Justus Liebigs Ann. Chem. 1912, 390, 89.
2. von Romburgh, P. Chem. Zentralbl. 1920I, 83.
3. Drawert, F; Heiman, W.; Emberger, R.; Tressl, R. Justus Liebigs Ann. Chem. 1966, 694, 200-8.
4. Galliard, T. In "Biochemistry of Wounded Plant Tissues"; Kahl, G., Ed.; Walterde Gruyter: Berlin, 1978; P 155.
5. Galliard, T.; Matthew, J. A. Phytochemistry 1977, 16, 339-43.
6. Sekiya, J.; Kajiwara, T.; Hatanaka, A. Plant Cell Physiol. 1984, 25, 269-80.
7. Fross, D. A.; Dunstone, E. A.; Ramshow, E. H.; Stark, W. J. J. Food Sci. 1962, 27, 90.
8. Fleming, H. P.; Cobb, W. Y.; Etchells, I. L.; Bell, T. A. J. Food Sci. 1968, 33, 572.
9. Galliard, T.; Phillips, D. R.; Reynolds, J. Biochim. Biophys. Acta 1976, 441, 181-92.
10. Hatanaka, A.; Kajiwara, T.; Harada, T. Phytochemistry 1975, 14, 2589-92.
11. Galliard, T.; Chan, H. W. -S. In "The Biochemistry of Plants, Vol. 4"; Stumpf, P. K. Ed.; Academic: New York, 1980; pp 131-61.
12. Eskin, N. A. M.; Grossman, S.; Pinsky, A. Crit. Rev. Food Sci. Nutr. 1977, 9, 1-40.
13. Zimmerman, D. C.; Coudron, C. A. Plant Physiol. 1979, 63, 536-41.
14. Vick, B. A.; Zimmerman, D. C. Plant Physiol. 1976, 57, 780-8.
15. Zimmerman, D. C. Biochem. Biophys. Res. Commun. 1966, 23, 398-402.
16. Vick, B. A.; Feng, P.; Zimmerman, D. C. Lipids 1980, 15, 468-71.
17. Sekiya, J.; Kajiwara, T.; Hatanaka, A. Agric. Biol. Chem. 1979, 43, 969-80.
18. Kazeniac, S. J.; Hall, R. M. J. Food Sci. 1970, 35, 519-30.
19. Hatanaka, A.; Kajiwara, T.; Sekiya, J.; Imoto, M.; Inouye, S. Plant Cell Physiol. 1982, 23, 91-9.
20. Kim, I. -S.; Grosch, W. J. Agric. Food Chem. 1981, 29, 1220-5.
21. Matoba, T.; Hidaka, H.; Kitamura, K.; Kaizuma, N.; Kito, M. J. Agric. Food Chem. 1985, 33, 852-5.
22. Galliard, T.; Phillips, D. R. Biochim. Biophys. Acta 1976, 431, 278-87.

23. Matoba, T.; Hidaka, H.; Kitamura, K.; Kaizuma, N.; Kito, M. J. Agric. Food Chem. 1985, 33, 856-8.
24. Phillips, D. R.; Galliard, T. Phytochemistry 1978, 17, 355-8.
25. Schreier, P.; Lorenz, G. Z. Naturforsch. 1982, 37c, 165-73.
26. Sekiya, J.; Tanigawa, S.; Kajiwara, T.; Hatanaka, A. Phytochemistry 1984, 23, 2439-43.
27. Matthew, J. A.; Galliard, T. Phytochemistry 1978, 17, 1043-4.
28. Hatanaka, A.; Kajiwara, T.; Sekiya, J.; Inouye, S. Phytochemistry 1982, 21, 13-7.
29. Sekiya, J.; Kajiwara, T.; Munechika, K.; Hatanaka, A. Phytochemistry 1983, 22, 1867-9.
30. Wardale, D. A.; Lambert, E. A.; Galliard, T. Phytochemistry 1978, 17, 205-12.
31. Wurzenbergen, M.; Grosch, W. Biochim. Biophys. Acta. 1984, 794, 25-30.
32. Galliard, T.; Matthew, J. A.; Wright, A. J.; Fishwick, M. J. J. Sci. Food. Agric. 1977, 28,863-8.
33. Sekiya, J.; Tanigawa, S.; Hatanaka, A. Proc. 1983 Ann. Mtg. Agric. Chem. Soc. Japan 1983, P 228.
34. Kajiwara,T.; Sekiya, J.; Asano, M.; Hatanaka, A. Agric. Biol. Chem. 1982, 46,3087-8.
35. Hatanaka, A; Kajiwara, T.; Sekiya, J.; Toyota, H. in preparation.
36. Wurzenberger, M.; Grosch, W. Biochim. Biophys. Acta. 1984, 794, 18-24.
37. Byrn. M.; Calvin, M. J. Am. Chem. Soc. 1966, 88, 1916-22.
38. Gardner, H. W.; Plattner, R. D. Lipids 1984, 19, 294-99.

RECEIVED February 3, 1986

14

Enzymic Formation of Volatile Compounds in Shiitake Mushroom (*Lentinus edodes* Sing.)

Chu-Chin Chen[1,2], Su-Er Liu[1], Chung-May Wu[1], and Chi-Tang Ho[2]

[1]Food Industry Research and Development Institute (FIRDI), P.O. Box 246, Hsinchu, 30099, Taiwan, Republic of China
[2]Department of Food Science, Rutgers University–The State University of New Jersey, New Brunswick, NJ 08903

Volatile compounds of Shiitake mushroom (Lentinus edodes Sing.) are composed of eight-carbon containing alcohols and sulfur compounds. 1-Octen-3-ol and 2-octen-1-ol are the major C8-compounds comprising the "mushroom" character of Shiitake mushroom. The characteristic "sulfurous" note of Shiitake mushroom is composed of cyclic S-compounds, such as lenthionine ($C_2H_4S_5$, 1,2,3,5,6-pentathiepane), 1,2,4,5-tetrathiane ($C_2H_4S_4$) and 1,2,4-trithiolane($C_2H_4S_3$). Formation of C8-compounds and S-compounds result from enzymic activities during rupture and/or drying of the tissue. C8-compounds are formed enzymically from linoleic acid. The formation of S-compounds involves two processes, enzymic reactions of lentinic acid as substrate and non-enzymic polymerization of methylene disulfide.

Shiitake (*Lentinus edodes* Sing.) is an edible mushroom highly prized in the Orient, especially Japan and China. Traditionally, Shiitake mushrooms were grown on segmented rotten wood which was placed in a cool, humid place. Several months are required before harvest. This ancient technique is still used in some countries. In Taiwan, an accelerated technique has been developed which employs treated and compressed sawdust as the growth medium. This method significantly reduces the time to harvest.

Due to the difficulties of postharvest storage, most of the mushrooms are preserved by heat drying. The drying process has to be conducted slowly in order to produce the characteristic sulfurous note of this mushroom.

Sulfurous Compounds in Shiitake Mushroom

Fresh Shiitake mushrooms exhibit only a slight odor, but upon drying and/or crushing, a characteristic sulfurous aroma gradually develops. Lenthionine (1,2,3,5,6-pentathiepane, $C_2H_4S_5$), a cyclic S-compound known to possess the characteristic aroma of Shiitake mushroom, was first identified in the dry product and subsequently synthesized by reaction of methylene chloride and sodium sulfide (1, 2, 3). Other cyclic S-compounds identified in dry mushroom and subsequently synthesized include: 1,2,4-trithiolane ($C_2 H_4 S_3$), 1,2,4,6-tetrathiepane ($C_3 H_6 S_4$) and 1,2,3,4,5,6-hexathiepane ($CH_2 S_6$) (2). The above mentioned cyclic S-compounds, with the exception of 1,2,3,4,5,6-hexathiepane, were also identified in a species of red algae (Chondria californica) (4). 1,2,4-Trithiolane was identified in the steam distillate of crushed Shiitake mushrooms as the only cyclic S-compound (5, 6). This compound has also been reported in the volatile compounds of egg (7) and as a reaction product of $H_2 S$ with D-glucose (8).

Iwami et al. (9, 10, 11) and Yasumoto et al. (12) proposed that cyclic S-compounds in Shiitake mushroom originated from a common precursor, lentinic acid, which is a derivative of γ-glutamyl cysteine sulfoxide. There are two enzymes which are responsible for the cnversion of lentinic acid into volatile S-compounds, γ-glutamyl transpeptidase and cysteine sulfoxide lyase (C-S lyase). Figure 1 shows the pathway originally proposed by Yasumoto et al. (12) for the formation of S-compounds in Shiitake mushroom.

Eight-Carbon Compounds in Shiitake Mushroom

1-Octen-3-ol occurs in many mushroom species (13, 14, 15) and contributes significantly to the "mushroom" character of species such as Agaricus campestris (16, 17) and Agaricus bisporus (18, 19). Kameoka and Higuchi (5) were the first to report the presence of C8-compounds in the steam distilled volatiles of crushed mushrooms and 1-octen-3-ol was the most abundant C8-compound identified. Other C8-compound identified were 3-octanol, 1-octanol and 2-octen-1-ol.

Enzymic Formation of Eight-Carbon Compounds

It has been established that in many edible mushrooms such as Agaricus campestris and Agaricus bisporus, the C8-compounds are formed enzymically during the oxidation of linoleic acid (16-20). In the present study, the identification of 1-octen-3-ol and 2-octen-1-ol as major C8-compounds in macerated fresh Shiitake mushrooms suggested a similar biosynthetic origin. The amount of C8-compounds (primarily 1-octen-3-ol and 2-octen-1-ol) in the blanched (97 °C, 8 min) or the hot-air dried (commercial process) Shiitake mushrooms was only 1 - 4 % of that of macerated fresh Shiitake mushrooms, confirming again the enzymic origin of these compounds.

It is well known that the amount of volatile C8-compounds produced in edible mushrooms can be greatly enhanced by adding pure linoleic acid to the enzymic reaction mixture (17, 20). Practically, the base hydrolysate of edible oils rich in linoleic acid is a less expensive substitute for pure linoleic acid.

Table I shows the volatile compounds identified in the enzymic reaction mixture containing fresh Shiitake musrooms and the base hydrolysate of sunflower oil. As compared to the control sample, a

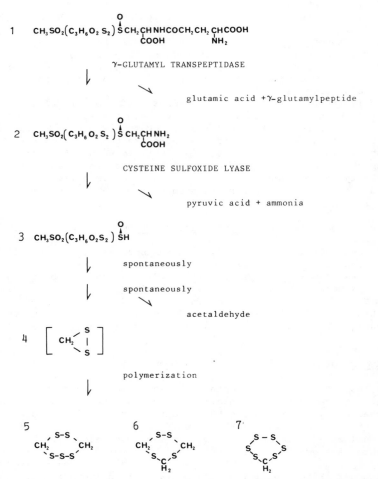

Figure 1. Proposed pathway of formation of S-compounds in Shiitake mushroom. (1) lentinic acid; (2) des-glutamyl lentinic acid; (3) a thiosulfinate, SE-3; (4) methylene disulfide; (5) lenthionine; (7) 1,2,3,4,5,6-hexathiepane.

5.5 fold increment of volatile compounds was observed in sample to which sunflower oil hydrolysate was added. This is the first report of enzymic conversion of linoleic acid into C8-compounds in Shiitake mushroom. The major volatile C8-compounds identified in this mixture were 1-octen-3-ol (79.83%) and 2-octen-1-ol (5.86%).

Table I. Composition of volatile components of Shiitake mushroom blended with base hydrolysate of sunflower oil.

No.	Compound	I^1 CW-20M	Identification	%
1.	dimethyl disulfide	1044	GC, MS	0.69
2.	hexanal	1056	GC, MS.	0.05
3.	2-alkanone	1146	MS.	trace
4.	3-octanone	1224	GC, MS.	2.70
5.	1-octen-3-one	1267	GC, MS.	0.56
6.	dimethyl trisulfide	1325	GC, MS.	3.72
7.	3-octanol	1382	GC, MS.	0.19
8.	1-octen-3-ol	1427	GC, MS.	79.83
9.	2-octenal	1448	MS.	0.09
10.	2-decanone	1470	GC, MS.	0.16
11.	1-octanol	1522	GC, MS.	1.08
12.	2-octen-1-ol	1577	GC, MS.	5.86
13.	1-methylthio-dimethyl disulfide	1587	MS.	0.22
14.	1,2,4-trithiolane	1660	GC, MS.	0.19

1.linear retention indices on fused silica capillary column (Carbowax 20M), using n-paraffins (C8 - C22) as references.

Tressl et al. (16, 17) had shown that linolenic acid (C 18:3) could also be enzymically converted to volatile C8-compounds. However, 1,5-octadien-3-ol and 2,5-octadien-1-ol would be two major C8-compounds formed. No significant amount of 1,5-octadien-3-ol and 2,5-octadien-1-ol could be detected in the volatile components of Shiitake mushrooms blended with sunflower oil hydrolysate (Table I). Instead, significant amount of 1,5-octadien-3-ol or 2,5-octadien-1-ol was observed in the sample to which soybean oil hydrolysate was added (20). This is in good agreement with the linolenic acid content in sunflower oil (trace) and soybean oil (ca. 7%).

Tressl et al. (16) proposed a enzymic pathway of C8-compounds from linoleic acid. Enzymes involved in the pathway are : lipoxygenase, hydroperoxide lyase and oxidoreductase. The 13- and 9-hydroperoxides of linoleic acid were proposed as the products of lipoxygenase action and the precursors of C8-compounds. Enzymic reduction of 1-octen-3-one to 1-octen-3-ol in Agaricus bisporus has been demonstrated (21), which is similar to the reaction of oxidoreductase mentioned by Tressl et al. (16). Wurzenberger and

Grosch (22, 23) proposed another pathway which also involved
lipoxygenase and hydroperoxide lyase. However, in their proposed
mechanism, 10-hydroperoxide of linoleic acid was suggested as the
intermediate of 1-octen-3-ol.

Enzymic Formation of Sulfurous Compounds

The enzymic activities involved in forming S-compounds can be
clearly shown by comparing the S-compounds formed at pH 9.0 (fresh
mushrooms blended in pH 9.0 buffered solution) and in the control
(fresh mushrooms blended in chloroform), as shown in Table II.
Since the enzymes were totally inactivated by chloroform, only
trace amounts of S-compounds were detected in the control sample.
Analyses of S-compounds in commercial dry Shiitake mushrooms proved
similar results (6). On the other hand, enzymic activities in
forming S-compounds were affected by the pH during blending (6).
The maximal pH during enzymic formation of S-compounds is around
9.0. The methods used in this study will be published in detail
(Chen and Ho, 1986).

There are 19 S-compounds which have been identified in
Shiitake mushrooms (Table II), and Figure 2 shows the structures of
these compounds. These S-compounds can be classified into 4 major
groups, containing 1, 2, 3 and 6 carbons. Fourteen out of the 19
S-compounds reported in this study are new to the volatiles of
Shiitake mushroom (compounds with a * in Table II).

Carbon disulfide, dimethyl trisulfide, 1,2,4-trithiolane,
1,2,4,5-tetrathiane, 1,2,3,5-tetrathiane and lenthionine were the
dominant S-compounds detected when the mushrooms were blended in pH
9.0 buffered solution. It is interesting to note that the isomers,
1,2,4,5-tetrathiane and 1,2,3,5-tetrathiane, have been tentatively
identified in red algae (Chondria californica) (4). Together with
1,2,4-trithiolane, 1,2,4,6-tetrathiepane and lenthionine, Shiitake
mushroom and red algae have in common five S-compounds. It is
quite possible that these two species share the same mechanism
which leads to the formation of above mentioned S-compounds.

With the exception of carbon disulfide (CS_2), all the S-
compound listed in Figure 2 have either the $-CH_2-S-$ or $-S-CH_2$ -S-
grouping in their structures. The $-S-CH_2-S-$ functional grouping
is similar to methylene disulfide, a proposed building block for
the polymerization of S-compounds (2).

It is worth noting that all the S-compounds identified in the
enzymic reaction mixture could also be detected in the products of
synthetic reaction of lenthionine or 1,2,4-trithiolane. Therefore,
it is reasonable to assume that chemical reactions may be the
dominant forces in the final stages of S-compounds formation. The
report of Ito et al. (24) supports the above assumption. They
found that the formation of lenthionine in dry Shiitake mushrooms
was affected by pH and temperature during rehydration. Since dry
mushrooms should be void of enzymic activities, the findings of Ito
et al. (24) might actually result from non-enzymic reaction of an
intermediate (such as methylene disulfide).

Figure 3 shows the effect of pH on the formation of 4 major
cyclic S-compounds (1,2,4,5-tetrathiane, lenthionine, 1,2,4-
trithiolane and 1,2,3,5-tetrathiane). The results show that
product formation is favored around pH 9.0. Consistent with these
findings, previous reports by Iwami et al. (9, 10, 11) showed that

Table II. Volatile sulfurous compounds identified
in Shiitake mushroom

No[1]	Compound	I^2 OV-1	M.W.	% 9.0	% contl	ID	
1.	methane thiol*	613	48	0.62	–	MS.	
2.	carbon disulfide*	625	76	4.92	+	GC,	MS.
3.	methyl hydrodisulfide*	686	80	0.17	–	MS.	
4.	dithiomethane	692	80	0.16	–	MS.	
5.	dimethyl disulfide	745	94	+	–	GC,	MS.
6.	1,3-dithietane*	786	92	+	–	MS.	
7.	dimethyl trisulfide	949	126	2.72	+	GC,	MS.
8.	methylthiomethyl- hydrodisulfide*	1011	122	1.04	+	MS.	
9.	1,2,4-trithiolane	1065	124	14.04	1.09	GC,	MS.
10.	dimethyl tetrasulfide*	1194	158	+	–	GC,	MS.
11.	1,3,5-trithiane*	1259	138	+	–	MS.	
12.	1,2,4,5-tetrathiane*	1310	156	42.34	0.93	MS.	
13.	2,3,5,6-tetrathioheptane*	1327	172	+	–	MS.	
14.	1,2,3,5-tetrathiane*	1338	156	2.18	+	MS.	
15.	1,2,4,6-tetrathiepane	1477	170	+	–	MS.	
16.	lenthionine	1590	188	39.70	0.63	GC,	MS.
17.	1,2,4,7,9,10- hexathiododecane*	1701	278	+	–	MS.	
18.	1,2,4,5,7-pentathiocane*	1749	202	+	–	MS.	
19.	1,2,3,5,6,8-hexathionane*	1901	234	+	–	MS.	

1. numbers refer to the structures in Figure 2.
2. linear retention indices on OV-1 fused silica capillary
   column, using n-paraffins (C8 - C22) as references.
* first reported in Shiitake mushroom volatiles.

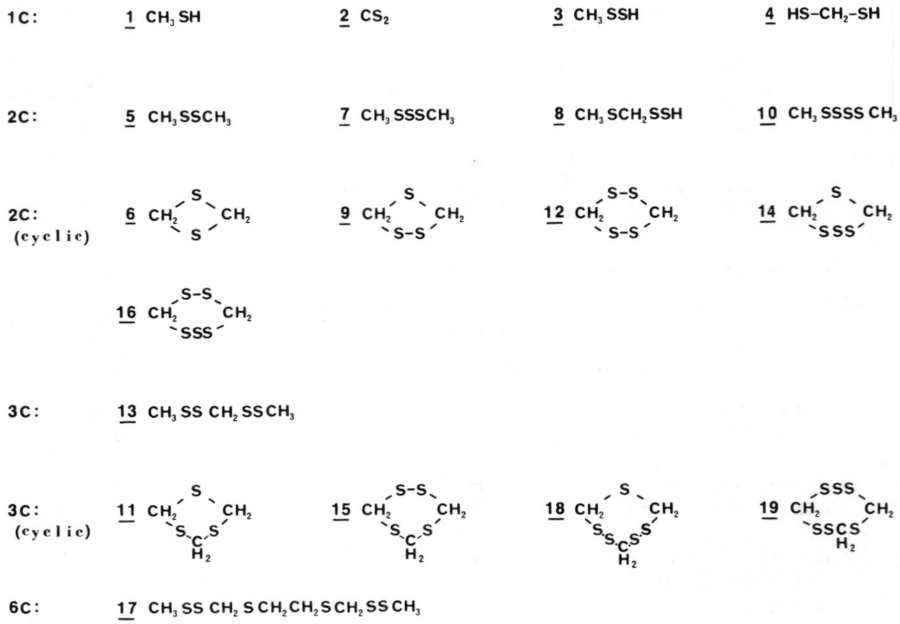

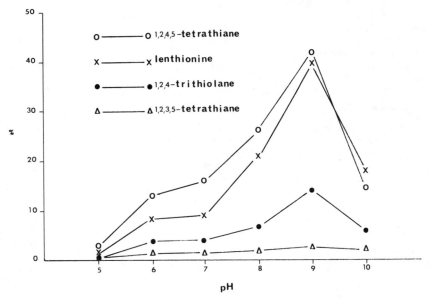

1C:
1 CH3SH
2 CS2
3 CH3SSH
4 HS-CH2-SH
2C:
5 CH3SSCH3
7 CH3SSSCH3
8 CH3SCH2SSH
10 CH3SSSSCH3
2C: (cyclic)
6
9
12
14
16
3C:
13 CH3SSCH2SSCH3
3C: (cyclic)
11
15
18
19
6C:
17 CH3SSCH2SCH2CH2SCH2SSCH3

Figure 2. Structures of S-compounds identified in the enzymic reaction mixture of Shiitake mushrooms. Numbers refer to those listed in Table II.

1,2,4,5-tetrathiane
lenthionine
1,2,4-trithiolane
1,2,3,5-tetrathiane
%
pH
0
10
20
30
40
50
5
6
7
8
9
10

Figure 3. Formation of 4 major cyclic S-compounds from pH 5.0 to 10.0 at 1.0 unit per interval.

the optimal pH for γ-glutamyl transpeptidase was around 7.6 and the optimal pH for C-S lyase was around 9.0 or higher.

Acknowledgment

Great thanks are extended to Tsen-Chien Lee for supplying the fresh Shiitake mushroom and Timothy J. Pelura for reviewing the manuscript. Chu-Chin Chen is supported by the Council of Agriculture, Republic of China. New Jersey Agricultural Experiment Station, Publication No. D-10205-1-86 supported by State Funds and Hatch Regional Fund NE-116

Literature Cited

1. Morita, K.; Kobayashi, S. Tetrahedron Lett. 1966, 6, 573.
2. Morita, K.; Kobayashi, S. Chem. Pharm. Bull. 1967, 15, 988.
3. Wada, S.; Nakatani, H.; Morita, K. J. Food Sci. 1967, 32, 559.
4. Wratten, S. J.; Faulkener, D. J. J. Org. Chem. 1976, 41, 2456.
5. Kameoka, H.; Higuchi, M. Nippon Nogei Kagaku Kaishi 1976, 50, 185.
6. Chen, C.-C.; Chen, S.-D.; Chen, J.-J.; Wu, C.-M. J. Agric. Food Chem. 1984, 32, 999.
7. Gil, V.; MacLeod, A. J. J. Agric. Food Chem. 1981, 29, 484.
8. Sakaguchi, M.; Shibamoto, T. J. Agric. Food Chem. 1978, 26, 1260.
9. Iwami, K.; Yasumoto, K.; Nakamura, K.; Mitsuda, H. Agric. Biol. Chem. 1975, 39, 1933.
10. Iwami, K.; Yasumoto, K.; Nakamura, K.; Mitsuda, H. Agric. Biol. Chem. 1975, 39, 1941.
11. Iwami, K.; Yasumoto, K.; Nakamura, K.; Mitsuda, H. Agric. Biol. Chem. 1975, 39, 1947.
12. Yasumoto, K.; Iwami, K.; Mitsuda, H. Mushroom Sci. 1976, 9, 371.
13 Picardi, S. M.; Issenberg, P. J. Agric. Food Chem. 1973, 21, 959.
14. Pyysalo, H. Acta Chem. Scand., Ser. B 1976, B30, 235.
15. Maga, J. A. J. Agric. Food Chem. 1981, 29, 1.
16. Tressl, R.; Bahri, D.; Engel, K.-H. In "Lipid Oxidation in Fruits and Vegetables"; Teranishi, R.; Barrera-Benitez, H. Eds.; ACS SYMPOSIUM SERIES No. 170, American Chemical Society: Washington, D.C., 1980; pp. 213-232.
17. Tressl, R.; Bahri, D.; Engel, K.-H. J. Agric. Food Chem. 1982, 30, 89.
18. Wurzenberger, M.; Grosch, W. Z. Lenbensm.-Unters.-Forsch. 1982, 175, 186.
19. Wurzenberger, M.; Grosch, W. Z.-Lenbensm.-Unters.-Forsch. 1983, 176, 16.
20 Chen, C.-C.; Wu, C.-M. Chinese Patent, 1983, No. 19234.
21 Chen, C.-C.; Wu, C.-M. J. Agric. Food Chem. 1984, 32, 1342.
22. Wurzenberger, M.; Grosch, W. Biochim. Biophy. Acta 1984a, 794, 18.
23. Wurzenberger, M.; Grosch, W. Biochim. Biophy. Acta 1984b, 794, 25.
24. Ito, Y.; Toyada, M.; Suzuki, H.; Iwaida, M. 1978, J. Food Sci. 43, 1287.

RECEIVED February 3, 1986

15

Biogenesis of Blackcurrant (*Ribes nigrum*) Aroma

R. J. Marriott

Barnett & Foster Ltd., Wellingborough, Northants NN8 2QJ, United Kingdom

The terpene fraction of blackcurrants is largely
responsible for its characteristic flavour and aroma.
The biogenesis of terpenes present in blackcurrant fruit
has been studied by analysis of the volatile oils during
ripening of the fruit and by feeding experiments using
$(2-^{13}C)$ mevalonate. Three blackcurrant cultivars, Ben
Lomond, Baldwin and Wellington XXX were studied, Ben
Lomond and Baldwin fruit showed very similar composition
of terpene compounds, whereas in Wellington XXX fruit,
the terpenes derived from the thujane skeleton were at
greatly reduced levels. Changes in concentration of
monoterpene hydrocarbons, alcohols and esters occurred
mainly during the period of sugar accumulation, when a
sharp decrease in the level of terpene hydrocarbons and
an increase in monoterpene alcohols and esters was
observed. Feeding experiments with $(2-^{13}C)$ mevalonate
indicated parallel incorporation into all monoterpene
hydrocarbons and monoterpene alcohols. The presence of
monoterpene and aromatic glycosides have also been
tentatively identified in blackcurrants for the first
time.

The analysis of the terpene fraction in ripening fruit was carried
out by extraction with dichloromethane followed by gas chromato-
graphy/mass spectrometry using single ion monitoring, allowing
detection of terpene compounds down to 1 µg/kg wet weight. The
aqueous residue from the dichloromethane extract was treated with
mixed glycosidases to hydrolyse non-volatile terpene glycosides.
After hydrolysis, the free terpenes were extracted with dichloromethane and identified by GC/MS/SIM. Baldwin and Wellington XXX
fruits were picked from the same bushes at two week intervals
between the end of May and the end of August, and the volatile oils
extracted immediately. Feeding experiments were carried out by
immersing freshly cut, defoliated stems in a solution containing
$(2-^{13}C)$ mevalonate and glucose, followed by a water chase and then a
nutrient solution supplemented by glucose. After feeding, fruits

0097–6156/86/0317–0184$06.00/0
© 1986 American Chemical Society

were picked at regular intervals and extracted with dichloromethane
followed by GC/MS/SIM. The labelled terpenes were detected by
monitoring m/z 138 and m/z 94, the molecular ion and M-44 fragment
respectively. Although significant work has been carried out on the
identification of individual terpenes in blackcurrant leaves, buds
and fruit (1), no studies have been carried out on their biosynthesis, although it has been assumed to be similar to that
reported in other plants. Non-volatile terpene glycosides have been
reported in a number of fruits and plants (2-4), but this is the
first time they have been tentatively identified in blackcurrants.

Experimental

Reagents. Analar dichloromethane (B.D.H. Chemicals) was redistilled
just prior to use. Pure terpene standards were obtained variously
from Aldrich Chemicals Co., Fluka A.G., S.C.M. Organics, Bush Boake
Allen Ltd and International Flavours and Fragrances. β-Terpineol
and γ-terpineol were generous gifts from Firmenich & Co. The purity
of all standards was greater than 98% and were used as supplied.
All standards were stored under nitrogen at -25°C. The glycosidic
enzyme used was Pectinol C obtained from **Rohm** and Haas. $(2-^{13}C)$ MVA
lactone (99 atom %) was obtained from MSD Isotopes. All other
reagents were at least analar purity and used as supplied.

Fruit samples. Fruits were picked from 3 year old bushes grown on a
commercial fruit farm, and were used immediately for aroma isolation.
Stems used for labelling experiments were obtained from the same
bushes.

Isolation of aroma compounds. 20g of blackcurrants were homogenised
with redistilled dichloromethane (200ml) at -10°C to minimise enzyme
reactions or terpene re-arrangements. After allowing the homogenate
to separate, the dichloromethane layer was filtered through a
previously dichloromethane washed silicone impregnated paper, and
finally dried over anhydrous sodium sulphate. The aroma isolate
was stored at -25°C under nitrogen in hypovials. In order to detect
some of the labelled compounds, it was necessary to concentrate the
dichloromethane extract by vacuum distillation.

Enzyme hydrolysis of non-volatile precursors. The aqueous residue
from the dichloromethane extraction was washed twice with 10 volumes
of redistilled dichloromethane to remove any residual terpenes. The
residue was suspended in pH 5.0 phosphate buffer and homogenised.
The homogenate was rotary film evaporated briefly to remove any
residual dichloromethane. Pectinol C (0.1%) was added in pH 5.0 buffer
and the homogenate incubated at 30°C for 24 hours. After incubation,
the terpene aglycons were extracted with 10 volumes of re-distilled
dichloromethane as previously described.

Feeding methods. Stems (ca. 150g) with 20 to 30 fruits were
defoliated, cut under sterile water and immediately immersed in a
solution of $(2-^{13}C)$ mevalonate (0.1 mmol), and glucose (0.3 mmol) in
sterile water (10 ml). Immediately before use, MVA lactone was
converted into the potassium salt of the acid by incubation with an

excess of aqueous potassium bicarbonate at 37°C for 1 hour. Water (0.5 ml) was then fed to the stems followed by feeding nutrient solution (6) supplemented with glucose (1 mg ml^{-1}) and adjusted to pH 7.8. Fruit were picked at regular intervals and extracted as described previously.

Capillary Gas Chromatography - Mass Spectrometry. Capillary GC/MS was carried out using a 50 m fused silica capillary column (0.25 mm i.d.) coated with OV-101 or Carbowax 20M in a Pye Unicam series 104 chromatograph with linear temperature program from 50°C to 200°C at 2°C/min. The column was directly coupled to the ion source of a V.G. Micromass 12B mass spectrometer set to the following conditions: accelerating voltage 5 KV; ionization voltage 70 eV; ion source temperature 200°C; resolution 1000 (10% valley). The mass spectrometer was calibrated manually using reference compounds.

Comparison of cultivars. Fully, ripe fruit was picked from 3 year old bushes of Baldwin, Ben Lomond and Wellington XXX cultivars, and the aroma isolated as previously described. The aroma isolate was analysed by GC/MS/SIM at m/z 93 and m/z 136. Identification was carried out by comparison of retention times against authentic standards using both OV-101 and Carbowax 20M columns. Quantification of identified compounds was carried out by comparison with standard solutions of known concentration. Comparison of the terpene fraction of the corresponding leaf oils was also carried out in the same way. Although it was found that the terpene fraction of the leaf oils of the three cultivars was essentially identical, the terpene fraction of the fruit was found to be similar in Baldwin and Ben Lomond cultivars, but not Wellington XXX which had greatly reduced levels of α-thujene and sabinene. Baldwin and Wellington XXX were thus chosen to study the changes in terpene composition during ripening.

Changes in terpene composition during ripening. Fruits were picked at 2 week intervals from the end of May to the end of August from the same bushes grown at the same site. The aroma isolation was carried out within 6 hours of picking, and the aroma extracts analysed as previously described. Feeding experiments with $(2-^{13}C)$ mevalonate were carried out at three stages of ripening using stems from the same bushes and using the method previously described.

Non-volatile precursors. It was observed during routine processing of blackcurrants that the level of monoterpene alcohols increased after the addition of pectin hydrolysing enzymes. This suggested the presence of terpene glycosides, and the level of these compounds in all samples of fruit picked during ripening was determined in the aqueous residue after dichloromethane extraction of the volatile aroma compounds using the method previously described.

Results and Discussion

Comparison of cultivars. The analysis of the monoterpene fraction of the aroma isolated from the three cultivars is given in table I. The analysis was carried out using both OV-101 and Carbowax 20M

Table I.   Monoterpenes found in 3 blackcurrant cultivars (levels in
μg/Kg)

Monoterpene	Ben Lomond	Baldwin	Wellington XXX
α-thujene	49	37	5
α-pinene	36	42	37
camphene	2	2	10
sabinene	<u>1450</u>	<u>1867</u>	28
β-pinene	46	59	21
myrcene	48	72	50
α-phellandrene	29	28	53
Δ-³-carene	649	666	<u>1454</u>
α-terpinene	2	<1	38
d-(+)-limonene	174	205	188
cis-β-ocimene	97	141	79
trans-β-ocimene	58	65	52
γ-terpinene	39	38	13
terpinolene	<1	<1	2
allo ocimene	<1	51	77
linalol	44	59	16
cis-β-terpineol	11	18	22
trans-β-terpineol	11	14	16
terpinene-4-ol	40	46	18
α-terpineol	211	237	213
γ-terpineol	<1	<1	2
citronellol	139	185	138
nerol	<1	6	8
geraniol	92	96	96

phases and using m/z 93 and m/z 136. Good correlation was obtained by using any combination of column or single ion and wherever possible, complete spectra were obtained for confirmation of structure. The results clearly show that both Baldwin and Ben Lomond cultivars have very similar monoterpene profiles, but Wellington XXX has greatly reduced levels of monoterpenes of the thujane group. Baldwin and Ben Lomond aroma volatiles contain sabinene as the major monoterpene hydrocarbon, but this is almost totally absent in Wellington XXX, as is the related monoterpene α-thujene. There is, however, a marked increase in the level of Δ³-Carene in Wellington XXX volatiles. Examination of the leaves of these cultivars, showed that this difference is not reflected in the leaf oils, in fact all three leaf oils are almost identical. The same observation has also been made by Latrasse (7) regarding the bud oils. Since all the cultivars examined were grown at the same location, the effects of physio-logical and environmental factors can be ignored, and therefore it is concluded that the monoterpene profile reflects the genetic control of the monoterpene synthesising enzyme systems, and that these systems differ in the leaf, bud and fruit. One can also conclude that in Wellington XXX, the enzyme system involved in the synthesis of monoterpenes of the thujane type is either largely absent or inactive.

Changes in terpene composition during ripening. In order to examine the biosynthetic pathway of terpenes in blackcurrant fruit, regular analysis of the fruit was carried out during development and ripening which is from May to August. Changes in the concentration of total monoterpene hydrocarbons, alcohols and esters during ripening are shown in figure 1. The result of this work was rather disappointing since no clear change in the ratio of any terpene hydrocarbons or alcohols could be defined, in fact the ratios remained almost constant during the entire period of development and maturation. Changes in concentration of monoterpene hydrocarbons, alcohols and esters occurred mainly during the period of sugar accumulation when it is observed that the level of monoterpene hydrocarbons drops sharply and the level of alcohols and esters slowly increase, but not proportionally to the hydrocarbon decrease. A similar rapid reduction of sesquiterpene hydrocarbons is also observed. Our experiments with $(2-^{13}C)$ mevalonate have confirmed many of these observations. The incorporation of the labelled mevalonate was extremely poor (1% maximum) and this presented the following difficulties with the analysis. In order to detect the labelled compounds, it was necessary to concentrate the dichloromethane extract by vacuum distillation, which we had previously avoided in order to minimise any re-arrangements which might occur, and further-more, the most abundant ion (m/z 94) also occurs in unlabelled terpenes from naturally occurring ^{13}C. However, the results obtained from these studies showed the incorporation of $(2-^{13}C)$ mevalonate into the same monoterpene hydrocarbons irrespective of the stage of ripening, and that the percentage rate of incorporation was approx-imately the same for each group of monoterpenes. During these studies, it was found that the label took at least 24 hours to appear in any monoterpene and that the portion of the monoterpene derived from isopentenyl pyrophosphate (IPP) was labelled prior to that

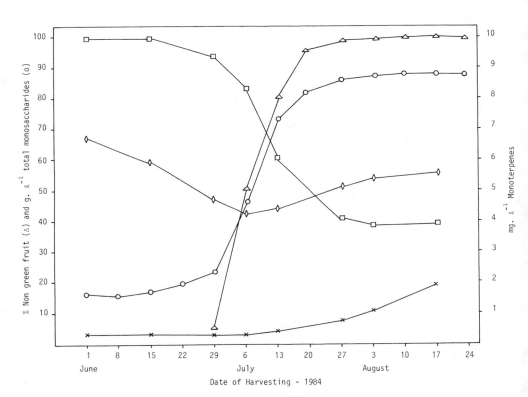

Figure 1.   Changes in concentration of total monoterpene hydrocarbons, (-□-) alcohols (-◊-) and esters (-✗-) during ripening of Wellington XXX blackcurrants in 1984.

portion derived from dimethylallyl pyrophosphate (DMAPP) (figures 2
and 3).   Monoterpene alcohols were found to increase in concentration
parallel to the hydrocarbons.   In none of our studies using labelled
mevalonate was any label found in any monoterpene esters or glycosides, but this may have been because they were at a concentration
lower than our detection limit, rather than being truly absent.

Non-volatile precursors.   During the processing of blackcurrants to
juice, the pulp is treated with a pectin hydrolysing enzyme, prior
to pressing, clarification and concentration, and we have observed
that the level of p-menthane and acyclic monoterpene alcohols
increased by up to 100%, after this treatment.   Since terpene
glycosides have been widely reported in fruit, we investigated the
products of glycosidic hydrolysis in blackcurrants after the
volatile terpenes had been extracted with dichloromethane.   Enzymic
hydrolysis at pH 5.0 with a mixed glycosidase yielded monoterpene
alcohols and their corresponding olefins in a ratio of 5:1.   The
predominant alcohols were terpinen-4-ol and α-terpineol.   The level
of glycosides during ripening did not increase until sugar
accumulation was almost complete (figure 4), but then continued to
rise steadily during over-ripening.   Thus, it would appear that an
excess of sugars is required before glycoside synthesis commences.

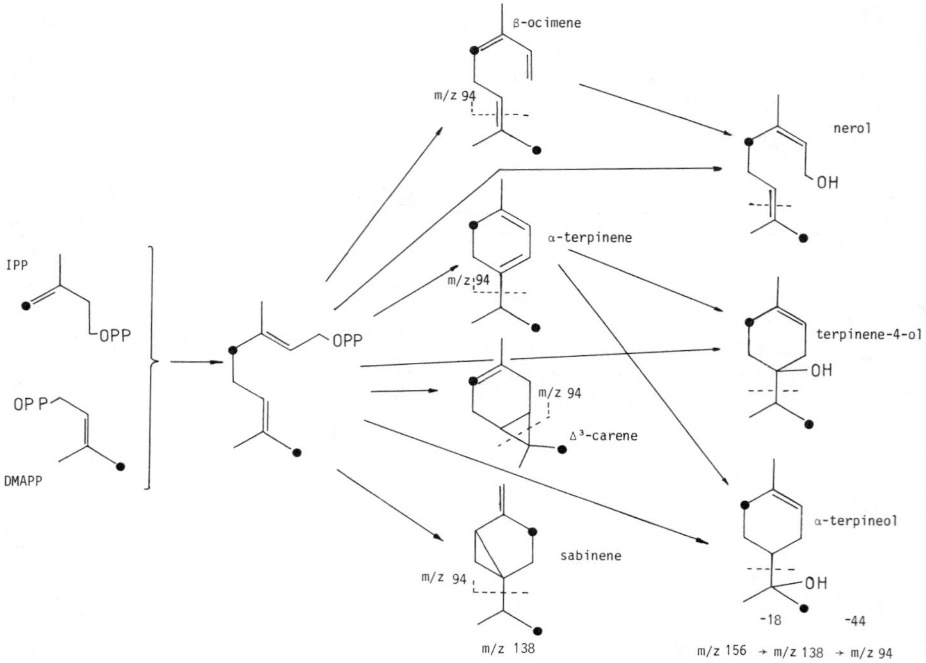

Figure 2.   Fragmentation patterns of ^{13}C labelled monoterpenes
derived from $(2-^{13}C)$ mevalonate.

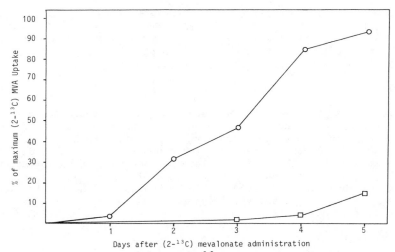

Figure 3.   Incorporation of $(2-{}^{13}C)$ mevalonate into labelled monoterpene hydrocarbons monitored at m/z 94 (-o-) and m/z 138 (-□-).

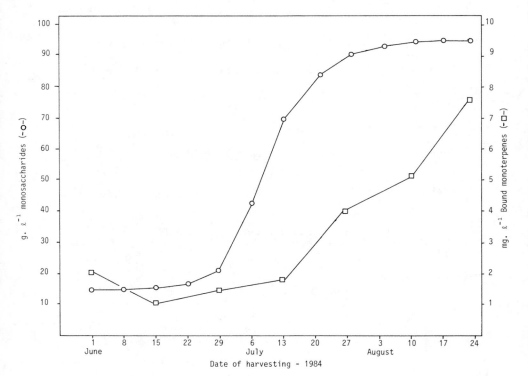

Figure 4.   Changes in concentration of bound monoterpenes during ripening of Wellington XXX blackcurrants in 1984.

The structure of the sugar moieties in the terpenol glycosides have not yet been investigated, but further work in this area is in progress. In a separate study, we have also found evidence of aromatic glycosides in blackcurrants, in particular, glycosides of phenylethanol, benzyl alcohol and p-cymene-8-ol.

Conclusions

The results of this work demonstrate the genetic control exerted on this group of compounds even between cultivars of the same species. The changes in composition during ripening, and labelling studies have suggested a common precursor for monoterpene olefins and alcohols, and the relatively low level of secondary transformations in this group of compounds. The presence of terpene and aromatic glycosides has been indicated in blackcurrants for the first time and may contribute to the variation in aroma observed during processing of blackcurrants.

Literature Cited

1) "Volatile Compounds in Food"; van Straten, S.; Maase, H. Ed.; TNO : Zeist, 1983; 7.1-7.3

2) Wilson, B.; Strauss, C.R.; Williams, P.J. J. Agric. Food Chem. 1984, 32, 919-924

3) Engel, K.H.; Tressl, R. J. Agric. Food Chem. 1983, 31, 998-1002

4) Kodama, H.; Kato, K. Phytochemistry, 1984, 23, 690-692

5) Konoshima, T.; Sawada, T. Chem. Pharm. Bull. 1984, 32, 2617-2621

6) Banthorpe, D.V.; Wirz-Justice, A. J. Chem. Soc. (C). 1969, 541

7) Latrasse, A.; Lantrin, B. Annales de Technol, Agric. 1974, 23, 65-74

8) Ruzicka, L.; Eschenmoser, A.; Heusser, H. Experimentia, 1953, 9, 357

9) Croteau, R. in "Biosynthesis of Isoprenoid Compounds"; Vol. I; Porter, J.W.; Spurgeon, S.L. Ed.; Wiley : New York, 1981; 227-270

RECEIVED March 25, 1986

Volatile Compounds from Wheat Plants: Isolation, Identification, and Origin

Thomas R. Hamilton-Kemp[1] and R. A. Andersen[2]

[1]Department of Horticulture, University of Kentucky, Lexington, KY 40546
[2]Agricultural Research Service, U.S. Department of Agriculture, and Department of Agronomy, University of Kentucky, Lexington, KY 40546

Volatile compounds have been reported to stimulate germination, in vitro and in vivo, of spores of the fungal pathogens which cause wheat rust diseases. Since growth and development of the pathogens disrupt plant tissue with resultant formation of volatiles, we have isolated and identified volatile compounds from disrupted wheat leaves and stems. Volatiles were isolated by steam distillation-extraction at reduced pressure and thirty-five compounds were identified by GC-MS. Among these volatiles are compounds reported to stimulate rust spore germination. Comparisons have been made among volatiles from leaves and culms, immature and mature plants and among different stages of tissue disruption. The possible origin of compounds isolated and the effects of tissue source are discussed.

One of the areas in nature where volatile compounds from plants may play a role which has not been extensively investigated is the chemical interaction between plants and microorganisms, especially fungi. Many species of fungi are known which cause serious diseases in economically important plants. These diseases frequently result in the death of a plant or the destruction of a crop if conditions for their development are favorable. Early examples of the effects of volatile compounds on the development of fungi were reported by Soulanti (1) and Fries (2) who showed that the growth of wood-decomposing fungi was stimulated by volatiles. We found that volatiles from cut or crushed leaves of cucurbitaeous plants such as melon and squash increased the degree and severity of infection of stem blight disease (fungal origin) in the cucurbits (3).

The most thorough studies of promotion of fungal development by volatiles was done by French and co-workers (4-6) who showed that volatile compounds in solution or their vapors were effective promotors of rust spore germination. Rusts are diseases found on many species including crop plants such as wheat, corn, beans and others. They are caused by fungi whose life cycle includes multiple development steps with the formation of rust colored uredospores in

0097-6156/86/0317-0193$06.00/0

pustules beneath the epidermis of a host plant. Many synthetic
volatiles have been tested and several, including nonanal and nonanol,
were effective at sub-parts per million levels in stimulating germination of wheat rust uredospores (4,5).

A factor to be considered in studying the volatiles from a host
plant is that the development of the fungus on its host causes damage
or disruption of the tissue. That is, in rust diseases the invasive
vegetative growth of the fungal hyphae and the formation of spore
pustules on leaves and stems of an infected plant disrupts tissue and
cells and results in the biogenesis of volatile compounds. The study
of injured or disrupted tissue seems contrary to the objectives of
most physiological studies in which the "normal" physiological
functions of the plants take place. However, under the circumstances
described for the interactions of a plant and an invasive fungus it
seems pertinent to determine the effects of injury or disruption.
Under these conditions the compounds formed can be expected to arise
from enzymatic oxidations and other degradative reactions.

Experimental

Winter wheat plants cultivar 'Arthur 71' were harvested from field
plots at various stages of maturity (Figure 1) as described below
and refrigerated at 5°C or stored frozen at -20°C.

Isolation and Separation. Volatiles were isolated from 1 kg of plant
tissue by steam distillation-extraction in a water recycling still
(7) containing 2.8L of water and operated at reduced pressure.
Hexane (4ml) was placed on the water in the side arm and the still
was operated at 65°C and approximately 200mm of Hg for 3 hrs. The
hexane was removed after a run and the combined hexane layers from
eight distillations were dried over Na_2SO_4 and concentrated to
approximately 100 µl under a stream of nitrogen. The concentrate was
separated initially on a 1.8m x 4mm column packed with 20% SP 2100
(nonpolar phase) on Supelcoport. The column was temperature
programmed from 100°C to 180°C at a rate of ∿1°C/min. Fractions
corresponding to chromatographic peaks were collected in U-shaped
tubes cooled in a Dry-Ice-acetone bath. Fractions were further
separated on either a 30m x 0.25mm Carbowax 20M or a 30m x 0.32mm
Supelcowax 10 fused silica capillary column.

Mass spectral studies. Mass spectral analyses were performed using
a Finnegan 330-6100 GC-MS instrument equipped with the same fused
silica capillary columns described above. Electron impact studies
were carried out at 70 eV and chemical ionization studies were
performed using methane.

Quantitative analysis. For all quantitative comparisons two analyses
were performed by dividing a distillate into two equal portions prior
to concentration and GC analysis. The peak areas obtained from
capillary column chromatograms were normalized to an internal
standard, pentadecane, added to each fraction. The means of two
analyses are presented and the values obtained can be interpreted as
approximately equivalent to parts per billion of volatile compound
per tissue wet weight.

Results and Discussion

Identification of compounds and origin. The volatile compounds isolated by steam distillation-extraction of a wheat sample are listed
in Table I.

Table I.   Volatile Compounds Isolated from Vegetative Wheat Tissue
           by Reduced Pressure Steam Distillation Extraction

Compound	Compound
Hexanal	1-Hexanol
t-2-Hexenal	t-2-Hexen-1-ol
Heptanal	c-3-Hexen-1-ol
t-2-Heptenal	1-Heptanol
t-2-Octenal	1-Octanol
Nonanal	1-Nonanol
t-2-Nonenal	t-2-Nonen-1-ol
t,c-2,6-Nonadienal	c-3-Nonen-1-ol
Decanal	t,c-2,6-Nonadien-1-ol
t-2-Decenal	c,c-3,6-Nonadien-1-ol
Undecanal	1-Decanol
t-2-Undecenal	
Tridecanal	c-3-Hexenyl Acetate
Tetradecanal	Eugenol
Pentadecanal	B-Ionone
c,c-8,11-Heptadecadienal	Pentadecane
c,c,c-8,11,14-Heptadecatrienal	Hexadecane
Benzaldehyde	2-Octadecanone

Compound identifications were based on comparison of mass spectral
data and co-chromatography of plant components with standards as
reported (8,9). A sample of 3,6-nonadien-1-ol was isolated from
melon (10) whereas 8,11-heptadecadienal and 8,11,14-heptadecatrienal
were obtained from cucumber fruit (11). Of primary interest in the
present study is the identification of C_9 aldehydes and alcohols
including nonanal and nonanol which are effective promotors of wheat
rust spore germination. The unsaturated C_9 compounds have not been
evaluated for activity but their close structural relationships to
nonanal and nonanol make them candidates for germination promotors.
     The unsaturated C_9 aldehydes and alcohols probably arise in
wheat from cleavage of the 9-10 double bond of unsaturated C_{18} fatty
acids, linoleic and linolenic acids (Figure 2). Galliard and co-
workers (12,13) have partially purified enzyme systems capable of
catalyzing these transformations. Lipoxygenase initially converts
linoleic and linolenic acids to 9-hydroperoxides which are subsequently cleaved by hydroperoxide lyase to volatile C_9 unsaturated
aldehydes and 9-oxo-nonanoic acid. The 3-enals are the primary
volatile cleavage products from the fatty acids and these are
transformed by an isomerase to the more stable 2-enals (14). The
3-enals are rather unstable but Hatanaka et al. (15) have confirmed
their presence in plant tissue with authentic samples. The C_9

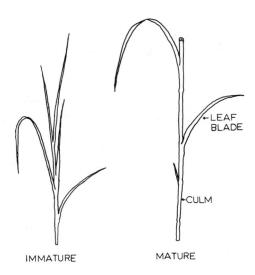

Figure 1.   Development stages of winter wheat.

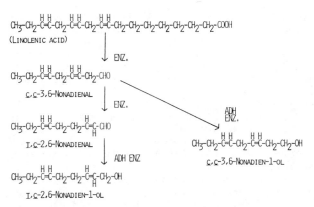

Figure 2.   Pathways for biosynthesis of C_9 aldehydes and alcohols from linolenic acid.

alcohols probably arise from reduction of the aldehydes by alcohol dehydrogenase enzymes.

Another major group of compounds found in wheat are the C_6 aldehydes and alcohols. These compounds are formed from the C_{18} fatty acids by pathways similar to those for the C_9 compounds except that the 12-13 double bond of the fatty acids is cleaved (13,14,16).

Comparison of fresh and frozen immature plants. To obtain an adequate amount of field grown wheat for initial identification of compounds, plants were harvested and stored frozen prior to distillation-extraction. After identifications were confirmed a comparison was made between the volatiles from fresh and frozen tissue to determine if the compounds isolated from frozen samples were also present in fresh wheat tissue. For these comparisons immature plants (Figure 1) 20-30cm tall were harvested and macerated in a Waring blender prior to distillation-extraction. From Table II it can be seen that the

Table II. Comparison of Relative Quantities of Volatiles from Fresh Versus Frozen Vegatative Tissue of Immature Wheat Plants

	Wheat Tissue	
Compound	Fresh	Frozen
Hexanal	3.2	0.4
1-Hexanol	290	1.2
2-Hexenal	2090	29.0
2-Hexen-1-ol	2050	3.2
3-Hexen-1-ol	1820	0.4
Nonanal	4.4	10.6
1-Nonanol	1.0	2.5
2-Nonenal	0.4	1.4
2-Nonen-1-ol	0.2	2.0
3-Nonen-1-ol	0.6	0.9
2,6-Nonadienal	0.9	4.5
2,6-Nonadien-1-ol	0.4	5.7
3,6-Nonadien-1-ol	1.2	1.4
Tetradecanal	0.1	0.2
Pentadecanal	3.0	4.0
8,11-Heptadecadienal	0.04	0.2
8,11,14-Heptadecatrienal	0.1	1.0
B-Ionone	32.9	11.5
Eugenol	0.2	0.2

same compounds including the C_9 compounds were present in the fresh and frozen wheat. By far the greatest quantitative differences were found for the C_6 compounds which were isolated in amounts up to several thousand fold greater from fresh plants than from frozen plants of the same age. Thus freezing causes a marked loss of C_6 compounds. It appears that one or more enzyme systems responsible for the formation of these compounds is inactivated by freezing.

As part of a study of the compounds that attract insects to intact wheat plants, Buttery, et al. (17) recently identified volatiles obtained from undisrupted immature plants by headspace

analysis (purge and trap technique). 3-Hexen-1-ol and 3-hexenyl
acetate were found to be major components of the young whole plants
(15-20 cm tall). 3-Hexen-1-ol and other C_6 compounds have "green"
grass-like odors and appear to be qualitatively and quantitatively
important components which contribute to the characteristic odor of
the immature plants studied in the present work also.

Analysis of the C_9 compounds indicated that freezing caused an
increase of 2-3 fold of most of these volatiles with the exception
of the 2-enols which increased more than 10-fold. Since the 2-enols
were similarly affected there may be an enzyme responsible for the
reduction of the 2-enals to alcohols which is distinct from the
system responsible for reduction of 3-enals.

Comparison of leaves and culms of mature plants. Comparisons were
made between volatiles steam distilled from leaves and culms (hollow
stems) of mature plants which were 40-50 cm tall (Figure 1). Rusts
manifest different affinities for these plant parts. Plants were
harvested, stored refrigerated overnight and leaf blades were cut
from the culms and both parts were cut into 3-6 cm segments. The
results from these comparisons are presented in Table III. With

Table III. Comparison of Relative Quantities of Volatiles from
Leaves and Culms of Mature Wheat Plants

Compound	Leaves	Culms
Hexanal	0.5	0.3
1-Hexanol	0.9	0.5
2-Hexenal	1.1	0.5
2-Hexen-1-ol	0.6	0.1
3-Hexen-1-ol	3.3	1.0
Nonanal	39.2	7.9
1-Nonanol	17.7	9.8
2-Nonenal	0.2	0.8
3-Nonen-1-ol	0.5	11.3
2,6-Nonadienal	0.4	1.8
2,6-Nonadien-1-ol	0.2	1.1
3,6-Nonadien-1-ol	1.0	19.7
Tetradecanal	0.2	0.5
Pentadecanal	2.1	6.6
8,11-Heptadecadienal	0.1	0.2
B-Ionone	4.4	1.4
3-Hexenyl Acetate	1.3	0.3
Hexadecane	0.5	0.3

regard to the C_9 compounds, there was a decrease in the quantity of
the major components, nonanal and nonanol, in the culms. There was
a marked increase of 10-20 fold in the 3-enols in the culms. This
suggests that the 3-enols are biosynthetically related since they
both increased whereas 2,6-nonadien-1-ol did not increase as markedly. (The 2-nonen-1-ol was obscured by an adjacent peak.)

The amounts of C_6 compounds were greatly reduced in mature plants when compared to the young plants used in the previous comparisons. The precipitous drop in these compounds may be due to a decrease in fatty acid substates, enzymic activity or both in mature plants. There was also a trend toward reduction in amounts of C_6 compounds in culms compared to leaves.

Comparison of whole and cut plants. An experiment was also done to compare volatiles from cut plants with those from whole or intact plants. Immature plants 30-50 cm tall were harvested and refrigerated overnight and then cut into 3-6 cm segments. From Table IV it

Table IV. Comparison of Relative Quantities of Volatiles from Whole and Cut Segments of Immature Wheat Plants

Compound	Whole	Cut
Nonanal	2.1	5.0
1-Nonanol	2.2	6.0
2-Nonenal	0.1	0.2
3-Nonen-1-ol	0.1	1.4
2,6-Nonadienal	0.1	0.2
2,6-Nonadien-1-ol	----	0.1
3,6-Nonadien-1-ol	0.1	2.6
B-Ionone	0.7	2.2
Pentadecanal	1.6	3.0

can be seen that the yields of most of the C_9 compounds increased 2-3 fold. In contrast, 3-nonen-1-ol and 3,6-nonadien-1-ol increased approximately 15-25 fold on cutting indicating that they are biosynthetically related. The 2-enols were isolated in such small amounts that changes could not be quantitated. It should be noted that the procedures necessary for harvesting and distillation of whole plants may have resulted in bruising which contributed to the volatiles isolated. However, there were clear differences in yield of 3-enols between cut and whole plants.

Conclusions

The results obtained provide information on the types of compounds which are obtained from wheat plants by a single isolation method, steam distillation-extraction, and how the relative amounts of these compounds vary with certain treatments. Further studies are needed to expand knowledge of the types and amounts of compounds present and their relevance to the interaction of a fungal pathogen with a host plant during disease development. French and Gallimore (5) have suggested that certain volatiles such an nonanol, might be applied to plants to cause premature germination of spores and reduce the spread of disease. Similarly, it may be possible to obtain high levels of certain endogenous volatile promotors of spore germination, in the future, through genetic manipulation after the volatiles present in host plants and their metabolic origins have been thouroghly investigated.

Acknowledgments

We are grateful to Lois Kemp, Pierce Fleming, and John Loughrin for
assistance and to George Lovelace for mass spectral analysis. We
thank Pam Wingate for typing the manuscript.

Literature Cited

1. Soulanti, O. Publ. Tech. Forschungsanst. Finland. 1951, 21,
    1-95.
2. Fries, N. Sven. Bot. Tidskr. 1961, 55, 1-16.
3. Pharis, V. L.; Kemp, T. R.; Knavel, D. E. Scientia Hortic. 1982,
    17, 311-317.
4. French, R. C.; Gallimore, M. D. J. Agric. Food Chem. 1971, 19,
    912-915.
5. French, R. C.; Gallimore, M. D. J. Agric. Food Chem. 1972, 20,
    421-423.
6. French, R. C.; Gale, A. W.; Graham, C. L.; Rines, H. W. J.
    Agric. Food Chem. 1975, 23, 4-8.
7. Kemp, T. R.; Stoltz, L P.; Smith, W. T.; Chaplin, C. E., Proc.
    Amer. Soc. Hort. Sci. 1968, 93, 334-339.
8. Hamilton-Kemp, T. R.; Andersen, R. A. Phytochemistry 1984, 23,
    1176-1177.
9. Hamilton-Kemp, T. R.; Andersen, R. A. Phytochemistry, in press.
10. Kemp, T. R.; Knavel, D. E.; Stoltz, L. P.; Lundin, R. E.
    Phytochemistry 1974, 13, 1167-1170.
11. Kemp, T. R. J. Amer. Oil Chem. Soc. 1975, 52, 300-302.
12. Galliard, T.; Phillips, D. R. Biochem. Biophys. Acta. 1976, 431,
    278-287.
13. Galliard, T.; Phillips, D. R.; Reynolds, J. Biochem. Biophys.
    Acta. 1976, 441, 181-192.
14. Phillips, D. R.; Mathew, J. A.; Reynolds, J.; Fenwick, G. R.
    Phytochemistry. 1979, 18, 401-404.
15. Hatanaka, A.; Kajiwara, T.; Harada, T. Phytochemistry. 1975,
    14, 2589-2592.
16. Hatanaka, A.; Kajiwara, T.; Sekiya, J. Phytochemistry 1976, 16,
    1125-1126.
17. Buttery, R. G.; Xu, C-j; Ling, L. C. J. Agric. Food Chem. 1985,
    33, 115-117.

RECEIVED January 3, 1986

Enzymic Generation of Volatile Aroma Compounds from Fresh Fish

David B. Josephson and R. C. Lindsay

Department of Food Science, University of Wisconsin–Madison, Madison, WI 53706

Characterizing aromas for freshly-harvested fish are
derived from polyunsaturated fatty acids through
lipoxygenase-mediated reactions, and include both
volatile alcohols and carbonyls. Short-chain
alcohols and aldehydes were shown to suppress the
activity of lipoxygenase which was derived from
trout gill tissue, and may serve as a means to
regulate the formation of fresh fish aroma
compounds. Exploratory experiments revealed that
plant-derived lipoxygenases have potential for the
biogenesis of fresh fish flavors.

Flavors and aromas commonly associated with seafoods have been
intensively investigated in the past forty years (1-7), but the
chemical basis of these flavors has proven elusive and difficult
to establish. Oxidized fish oils can be described as painty,
rancid or cod-liver-oil like (8), and certain volatile carbonyls
arising from the autoxidation of polyunsaturated fatty acids have
emerged as the principal contributors to this type of fish-like
aroma (3, 5, 9-10). Since oxidized butterfat (9, 11-12) and
oxidized soybean and linseed oils (13) also can develop similar
painty, fish-like aromas, confusion has arisen over the compounds
and processes that lead to fish-like aromas. Some have believed
that the aromas of fish simply result from the random
autoxidation of the polyunsaturated fatty acids of fish lipids
(14-17). This view has often been retained because no single
compound appears to exhibit an unmistakable fish aroma. Still,
evidence has been developed which indicates that a relatively
complex mixture of autoxidatively-derived volatiles, including
the 2,4-heptadienals, the 2,4-decadienals, and the
2,4,7-decatrienals together elicit unmistakable, oxidized
fish-oil aromas (3, 9, 18). Additionally, reports also suggest
that contributions from (Z)-4-heptenal may add characteristic
notes to the cold-store flavor of certain fish, especially cod
(4-5).

0097–6156/86/0317–0201$06.00/0
© 1986 American Chemical Society

The definition of oxidized fish oil-like aromas still leave
fresh fish aromas undefined. Various freshly harvested fish have
distinguishing aromas, but they also are characterized by a
common plant-like, seaweed-like aroma. Thus, compounds and
reaction pathways different from random autoxidation appear
likely and reasonable. Even conflicting descriptions of fishy
odors, i.e., including roles for volatile amines (2, 19) and
sulfur compounds (20-22), can be accommodated by the hypothesis
that previously unrecognized biochemical reactions yield
characterizing fresh fish aromas. These premises led to
investigations (23-26) which have resulted in the identification
of a group of enzymically-derived volatile aroma compounds that
contribute fresh, plant-like aromas to freshly harvested fish
(Table I).

Eight-carbon alcohols and ketones have been previously
identified in mushrooms (27-28), and occur in all species of fish
surveyed to this point (23-25, 29), in crustaceans (30-31), and
in shellfish (26). Although the eight-carbon volatile compounds
individually possess mushroom, geranium-like aroma notes, they
contribute distinct heavy, plant-like aromas to freshly harvested
fish. The eight-carbon volatile alcohols appear to occur in
greater abundance than the corresponding ketones, and this is
consistent with a similar origin for these volatiles in mushrooms
(27).

Table I. Enzymically derived carbonyls and alcohols associated
with freshly harvested fish.[1]

ALCOHOLS	Conc. Range (ppb)[2]	CARBONYLS	Conc. Range (ppb)[2]
1-Penten-3-ol	3-30	Hexanal	10-100
(Z)-3-Hexen-1-ol	1-10	(E)-2-Hexenal	1-10
1-Octen-3-ol	10-100	1-Octen-3-one	0.1-10
1,5-Octadien-3-ol	10-100	1,5-Octadien-3-one	0.1-5
2-Octen-1-ol	1-20	2-Octenal	0.1-5
2,5-Octadien-1-ol	1-20	(E)-2-Nonenal	0-25
6-Nonen-1-ol	0-15	(E,Z)-2,6-Nonadienal	0-35
3,6-Nonadien-1-ol	0-15	6-Nonenal[3]	trace
		3,6-Nonadienal[3]	trace

[1] Josephson et al. (23-24).
[2] Present in surface slime/water extracts.
[3] Reported for the first time.

Nine-carbon volatile alcohols and aldehydes, responsible for
much of the characterizing aromas of cucumber and melon fruits
(32-35), occur only in some species of fish, such as whitefish
(Coregonus clupeaformis), ciscoe (Coregonus artedi), smelt
(Osmerus mordax) (23-24, 36), and spawning Pacific salmon
(Oncorhynchus sp.) (unpublished data). In these species the
nine-carbon compounds provide characterizing cucumber-,
melon-like aromas. Pacific oysters also produce both the eight-
and nine-carbon alcohols and carbonyls identified in fresh fish,
but Atlantic oysters biosynthesize only the eight-carbon alcohols
and ketones (26). In seafoods the nine-carbon volatile aldehydes

have been found to occur in greater abundance than the
nine-carbon alcohols, and this is consistent with their enzymic
formation in cucumbers (37).

Six-carbon volatile alcohols and aldehydes have been found in
all freshwater fish surveyed (23-24). However, these compounds
have not been found in either salmon residing in saltwater
(unpublished data) or in oysters (26). Hexanal has been found in
modestly fresh (5-6 days old) saltwater fish (24), but its
formation may be the result of autoxidation rather than via
enzyme-mediated reactions. Thus, data for the occurrence of
hexanal in freshly harvested saltwater fish remains to be
developed. Hexanal and (E)-2-hexenal contribute coarse,
green-plant-like, aldehydic aroma notes to freshly harvested
finfish where their aroma dominates the overall odors within
seconds after the death of the fish. (Z)-3-Hexen-1-ol
contributes a clean, green-grass-like aroma note. Hexanal always
occurs in substantially greater abundance than 1-hexanol in fish.

In addition to the six-carbon volatile compounds,
1-penten-3-ol is also found in all freshwater fish. However,
concentrations of 1-penten-3-ol in fish remain below its
recognition threshold (400 ppb; 38), and therefore it is unlikely
that this volatile contributes strongly to the characteristic
aroma of freshly harvested fish.

Generally, the volatile carbonyls found in fish exhibit
coarse, heavy aromas whereas the volatile alcohols contribute
smoother qualities. Lower threshold values for the volatile
carbonyls, especially 1-octen-3-one (0.005 ppb; 39);
1,5-octadien-3-one (0.001 ppb; 12); (E)-2-nonenal (0.08ppb; 39),
and (E,Z)-2,6-nonadienal (0.01ppb; 39), result in greater
contributions to overall fresh fish-like odors than do the
corresponding alcohols (1-octen-3-ol, 1,5-octadien-3-ol, and
3,6-nonadien-1-ol) whose threshold values are each 10 ppb (28,
31, 39).

In addition to the enzymically-derived volatile aroma
compounds (Table I), low levels of autoxidatively-derived
carbonyls can also be detected in harvested fish held a day on
ice, and these volatiles are listed in Table II. The
oxidatively-derived carbonyls modify the fresh plant-like aromas
of fresh fish by providing oxidized-oil-like, staling fish-type
odor notes (3-5, 9). The formation of hexanal in freshly
harvested fish appears to be enzymic because the concentration of
this compound can be diminished by lipoxygenase inhibitors (25).
However, when fresh fish are stored on ice or are held under
frozen storage, hexanal concentrations also increase because of
autoxidative processes (40).

Trimethylamine and other amines have often been variously
associated with the aromas of fish (2, 19, 41). Much of the
trimethylamine found in fresh fish arises from the microbial
reduction of trimethylamine oxide (42-44) which is found
abundantly in only marine fish (45-47). On the other hand,
dimethylamine is an abundant product of an endogenous enzymic
action on trimethylamine oxide in marine fish muscle, and it is
readily produced even under high-sub-freezing conditions in
marine fish (48). Both trimethylamine and dimethylamine

contribute significently to the stale, fishy aromas of aging
fresh and frozen marine fish as well as to overall boiling-crab,
fish house-type aromas. However, freshly harvested fish contain
little of this compound.

Table II.   Oxidatively-derived carbonyls identified in fresh
                fish stored on ice.

CARBONYLS	Conc. Range (ppb)[1]	CARBONYLS	Conc. Range (ppb)[1]
Hexanal	***[2]	(E,E)-3,5-Octadiene-2-one	0.1-2
(E,Z)-2,4-Heptadienal	1-10	(E,Z)-2,4-Decadienal	1-5
(E,E)-2,4-Heptadienal	1-10	(E,E)-2,4-Decadienal	1-5
(E,Z)-3,5-Octadien-2-one	0.1-2		

[1]Present in surface slime/water extracts.
[2]Hexanal is formed both enzymically and autoxidatively, which
 makes contributions difficult to delineate.

     Sulfur-containing compounds have also been shown to
contribute to deteriorative aromas associated with seafood
spoilage, and other instances of off-flavor occurrences.
Dimethyl sulfide has been identified as the causative agent in
one case of non-spoilage off-flavor in seafoods where it was
formed either from the enzymic or thermal degradation of
dimethyl-ß-propiothetin (49-50). Still, dimethyl sulfide
provides a characterizing top-note aroma to cooking or stewing
clams and oysters where its flavor contribution is expected and
desirable (51-53). In another instance, Shiomi et al. (22) have
shown that methyl mercaptan and dimethyl disulfide are
responsible for the offensive odor of freshly-captured flat-head
(Calliurichthys doryssus), and in this case these two volatiles
appear to be formed by endogenous enzymes. The apparent enzymic
formation of methyl mercaptan and dimethyl disulfide in the
flat-head differs substantially from the usual microbial origin
of these compounds in spoiling fresh fish (54-55). A garlic-like
off-flavor caused by bis-(methylthio)-methane has been reported
for several species of prawns and the sand lobster (Ibacus
peronii), and evidence also supports endogenous biosynthesis in
these cases (21).

Biogenesis of Fresh Fish Carbonyls and Alcohols

Although for many years lipoxygenase activity in fish was
discounted (14-17), plant lipoxygenases were easily demonstrated,
and became well-accepted. Research into the biosynthesis of the
six-, eight-, and nine-carbon volatile aroma compounds of
mushrooms (27, 56) and cucumber/melon-fruits (37, 57-58) showed
the concerted activities of both site-specific lipoxygenases and
hydroperoxide lyases. Because animal lipoxygenases exhibit
self-inactivation properties (59-60), early experimentation
failed to detect an active lipoxygenase from fish, and this led
to conclusions that autoxidation was responsible for the

production of all volatiles derived from lipids that were found in fish tissue.

Tsukuda and Amano (61-62) and Tsukuda (63-64) believed they had observed lipoxygenase activity in fish by showing that fading or color deterioration associated with cartenoid pigment destruction in the skin of red fish could be delayed by introducing a variety of non-specific enzyme inhibitors. Yet, previous skepticisms continued to prevail (17). Difficulties in clearly demonstrating enzymic oxidative involvement in the biogenesis of fresh fish volatile aroma compounds were also encountered in our work with freshly-harvested fish held on ice. To circumvent obscure initial enzyme activity as well as subsequent autoxidation in even very fresh fish, techniques were developed which employ sacrificing live fish at the initiation of experiments which are designed to demonstrate lipoxygenase activity. Thus, by inhibiting activity at the time of death through the use of lipoxygenase-specific inhibitors, such as esculetin (65) and Sn(II)Cl (66), it was possible to more clearly separate enzymic and autoxidative lipid degradations. The integration of this approach with highly-sensitive Tenax GC headspace analysis (67) has provided a powerful biochemical probe technique for investigating the biogenesis of fresh fish flavors.

Recently, German and Kinsella (68-69) have demonstrated activity of a 12-lipoxygenase in the gill and skin tissues of trout (Salmo sp.), and these findings identify an appropriate precursor to some of the fresh fish aroma volatiles. This allows association of an appropriate precursor with the occurrence of certain volatiles, and provides very strong support for the view that the volatile aroma compounds characterizing the fresh, plant-like aromas of freshly harvested fish result from lipoxygenase-mediated bioconversions of polyunsaturated fatty acids.

A summary of the current hypothesis for the formation of volatile aroma compounds in freshly harvested fish is shown in Figure 1. These proposed pathways differ in part from those previously proposed (25), but they are consistent with all data. Evidence in support of the mechanism shown in Figure 1 can be derived from parallel lipoxygenase-mediated reaction pathways which have been proposed for the formation of the eight-carbon volatiles in mushrooms (71) and the nine-carbon volatiles in cucumber fruits (37). Even more direct evidence, comes from the recent identification of a 12-lipoxygenase in the skin and gill tissue of trout by German and Kinsella (68-69). However, an unknown remains at this time regarding the existence of the 15-lipoxygenase. If the carbon number and functional group of fresh fish volatile aroma compounds are viewed as indicative of a specific, required hydroperoxy-fatty acid precursor, the biogenesis of the five- and six-carbon volatile aroma compounds would depend on the existence of 15-lipoxygenase. In such a case, the reaction pathways for the formation of the five- and six-carbon volatile compounds would parallel those for the eight- and nine-carbon volatiles once the site-specific hydroperoxidation has occurred.

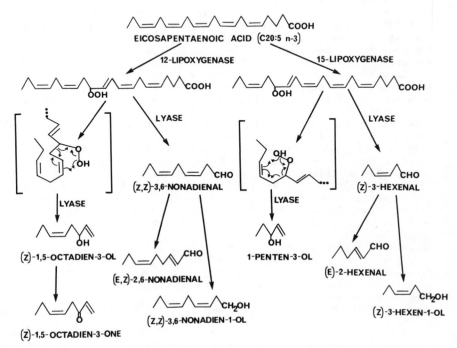

Figure 1.   Proposed mechanism for the biogeneration of some
                 fresh seafood aroma compounds.

An important difference exists in terms of fatty acid
precursors that are available as substrates for plant and animal
lipoxygenases. Plant lipoxygenases primarily have available as
substrates only linoleic (C18:2, n-6) and linolenic (C18:3, n-3)
acids (72), whereas fish lipoxygenases have access to the more
unsaturated fatty acids, including arachidonic (C20:4, n-6),
eicosapentaenoic (C20:5, n-3), and docosahexaenoic (C22:6, n-3)
acids, which are quite abundant in most fish (73-74). Thus, in
the proposed scheme eicosapentaenoic acid would be available for
enzymic conversion to monounsaturated five-, and six-carbon, and
diunsaturated eight-, and nine-carbon volatile aroma compounds
(Figure 1, Table I). Docosahexaenoic acid could also serve as a
precursor for these volatiles. Alternatively, arachidonic acid
(an n-6 fatty acid) could serve as a precursor for the saturated
six, and monounsaturated eight-, and nine-carbon volatiles
(Table I). Biogenesis of the five- and eight-carbon volatiles
appears to involve lipoxygenase-mediated hydroperoxidations which
are likely followed by a reductive rearrangement and cleavage to
result in the formation of the secondary alcohols. Enzymic
oxidation of the secondary alcohols would result in the formation
of the five- and eight-carbon ketones. The six- and nine-carbon
volatile compounds would be formed via the concerted activities
of lipoxygenase and a hydroperoxide lyase to form
(Z,Z)-3,6-nonadienal [or (Z)-3-hexenal], followed by either
enzymic isomerization to form (E,Z)-2,6-nonadienal [or
(E)-2-hexenal] or dehydrogenase activity to form
(Z,Z)-3,6-nonadien-1-ol [or (Z)-3-hexen-1-ol].

Proposed Physiological Role for the Enzymic Formation of Fresh Fish Volatile Carbonyls and Alcohols

The biogeneration of volatile aroma compounds in fish can be
hypothetically related to mechanisms for the regulation of
recently-recognized physiologically-active compounds, especially
leukotrienes and hydroxy-fatty acids (75-79). This view provides
the basis for the diagram shown in Figure 2 where the skin-water
interface has been chosen as an example of a site for this type
of enzymic activity. Stress-related lipid bioconversions (80)
would be expected to occur at this interface to maintain
physiological conditions in the fish, and lipoxygenase activity
has been detected recently in skin and gill tissues of trout
(69). Further, enzymically-derived volatile aroma compounds
appear to occur in higher concentrations in the outer skin and
mucus layer of fish when compared to muscle tissue per se
(23-24). If the biochemical processes are viewed as initial
enzymic formations of physiologically-active or regulatory
compounds (i.e., leukotrienes) from polyunsaturated fatty acids
that are followed by their inactivation through indicated enzymic
mechanisms (i.e., lyases), the result is a release of the
volatile aroma compounds that are associated with the aromas of
freshly-harvested fish. Differing physiological demands of
individual genus/species of finfish would account for at last
some of the varying concentrations and occurrences of the fresh
fish volatiles that have been observed (23-26, 36).

Hydroperoxides formed by animal lipoxygenases serve as
intermediates of physiologically-active compounds in fish (79),
but these hydroperoxides also appear to catalyze the initiation
of free radical autoxidation. Such activity has been recently
proposed by German and Kinsella (69) who isolated hydroxy-fatty
acids formed by reduction of hydroperoxides generated via
12-lipoxygenase in the skin of trout. Further evidence for the
hydroperoxide-induced formation of certain volatiles in fresh
fish is provided in Table III where occurrences of the products
of both enzymic and non-enzymic oxidation are presented. Data
were obtained for juvenile hybrid muskellunge (4-5 gm each; Esox
masquinongy) that were sacrificed by homogenization either in
distilled water (control-blend) or in the presence of esculetin
(65). Also included with these data are results obtained when
surface extracts (control-mucus) were taken from recently
sacrificed (suffocated) fish. Enzymically-derived volatiles were
produced along with low levels of autoxidatively-derived
carbonyls in distilled water homogenates. In the presence of the
lipoxygenase inhibitor, esculetin, neither fresh fish volatiles
nor the autoxidatively-derived volatiles were present.
Therefore, generation of autoxidatively-derived volatiles appears
to occur at least initially from the reactions of susceptible
lipid fractions with enzymically-derived hydroperoxides.

Table III. Effect of a lipoxygenase-specific inhibitor on the
            occurrence of enzymically- and oxidatively-derived
            volatiles in fish.

	Treatment		
Volatile	Control blend	Esculetin (10 mM)	Control mucus
ENZYMICALLY-DERIVED			
1-Octen-3-one	+[1]	−[1]	+
1,5-Octadien-3-one	+	−	+
1-Octen-3-ol	+	−	+
1,5-Octadien-3-ol	+	−	+
2-Octen-1-ol	+	−	+
2,5-Octadien-1-ol	+	−	+
(E)-2-Nonenal	+	−	+
(E,Z)-2,6-Nonadienal	+	−	+
6-Nonen-1-ol	+	−	+
3,6-Nonadien-1-ol	+	−	+
AUTOXIDATIVELY-DERIVED			
(E,Z)-2,4-Heptadienal	+	−	−/+[2]
(E,E)-2,4-Heptadienal	+	−	−/+
(E,Z)-2,4-Decadienal	+	−	−/+

[1] + = Positively Identified; − = Not Identified.
[2] Not found initially; later found after 2-days on ice.

Surface extracts (mucus) of the recently sacrificed fish
contained only enzymically-derived volatiles. However,

autoxidatively-derived carbonyls along with enzymically-derived
carbonyls and alcohols were subsequently found in surface (mucus)
extracts from the sacrificed fish after storage on ice for two
days. This indicates slower reactivity via autoxidation compared
to enzymically-induced oxidations. These results are consistent
with the observations of Yamaguchi et al., (81-82) and Yamaguchi
and Toyomizu (83) who suggested that accelerated rates of lipid
oxidation in fish skin extracts were caused by lipoxygenase
activity.

Current Investigations on the Contributions of Various Oxidative Enzymes to the Generation of Fresh Fish Volatile Carbonyls and Alcohols

While lipoxygenases (EC# 1.13.11.12) catalyze the site-specific
insertion of molecular oxygen into polyunsaturated fatty acids,
cytochrome P-450 monooxygenases (EC# 1.14.14.1) and other
oxidative enzymes share a common trait of catalyzing non-specific
oxidation of organic compounds, including polyunsaturated fatty
acids. The later oxygenases result in the formation of variously
positioned epoxy-, and hydroxy-fatty acids (84). Thus, enzymes
catalyzing site-specific hydroperoxidations of polyunsaturated
fatty acids should be regarded as lipoxygenases, and measurements
of volatile aroma compounds which result from the cleavage of
predictable hydroperoxy-fatty acids appear appropriate as probes
into this type of biochemical reaction in fish. The high
selectivity of fused silica capillary gas chromatography allows
use of this information to differentiate between volatiles
generated via lipoxygenase-mediation from volatiles generated
through random autoxidation.

Earlier we reported that the biogenesis of fresh fish
volatile aroma compounds were suppressed by the addition of
asprin (25), a potent inhibitor of cyclooxygenase (EC# 1.14.99.1;
85). Cyclooxygenase converts polyunsaturated fatty acids into
physiologically-active prostaglandins. However, recent
experiments have demonstrated that lower pH effects from acetyl
salicylic acid rather than specific cyclooxygenase inhibition
were primarily responsible for the suppression observed (70).
Thus, at this time unambiguous data for the involvement of
prostaglandins in the biogenesis of fresh fish volatile aroma
compounds either through enzymic regulation mechanisms or
directly from prostaglandins themselves as precursors to some
fresh fish volatile compounds are not available. However,
hydroperoxy-fatty acids are now implicated in the production of
fresh fish aroma volatiles, and these compounds are also
converted to leukotrienes or other physiologically-active
hydroxy-fatty acid derivatives (75-79).

Potential Technical Difficulties in Applying Lipoxygenases for the Controlled Biogeneration of Fresh Fish Aromas

Although little is known about many of the important enzymic
parameters of the flavor generating systems that are endogenous
in fish, the self-inactivation or suicidal nature of animal

lipoxygenases (59-60) is also characteristic of the lipoxygenases
from fish (68-69). Lipid hydroperoxides formed by animal
lipoxygenase are required to catalytically activate the enzyme,
but the hydroperoxides are also responsible for causing the
self-inactivation of the enzyme (60).

The very rapid rate of self-inactivation of animal
lipoxygenases (60, 68), which delayed recognition of these
enzymes in fish (14-16), can be suppressed by addition of reduced
glutathione (1 mM; 68; Figure 3). In these systems, glutathione
functions by chemically reducing hydroperoxy-fatty acid products
(animal-lipoxygenase inhibitors) to hydroxy-fatty acids. Without
such additions, it has been difficult to maintain activity in
order to demonstrate effects of various treatments on actual
enzyme activity. When crude trout gill lipoxygenase preparations
were stored at 0°C, activities could be sustained over a six to
eight hour period if glutathione was present, but this appeared
to vary with different preparations. However, gradual losses of
enzyme activity were still noted over this time span. Since all
experiments on gill lipoxygenase to date have employed crude
supernatant preparations (centrifuged at 15,000 x g), it remains
to be determined whether further purification will enhance or
diminish the overall activity of the enzyme.

The involvement of lipoxygenases in the overall
flavor-generating scheme for fish encompasses only the
hydroperoxidation of free fatty acids, and little is known about
the additional enzymes that are likely involved. At a minimum,
these probably include lyases, isomerases, and dehydrogenases.

Another major consideration regarding the
lipoxygenase-mediated biogeneration of volatile aroma compounds
in fish is the hydroperoxidative nature of the enzyme products
themselves. Since any lipid hydroperoxide is capable of
participating in subsequent random autoxidative deterioration of
polyunsaturated fatty acids following dismutation to alkoxy
(RO·) and hydroxy (·OH) radicals, the uncontrolled generation
of lipid hydroperoxides by lipoxygenases will lead to the
production of undesirable amounts of classical oxidized fish-like
aroma compounds (Table III; 3-5, 9, 11-12). However,
microencapsulation (86-87) of enzyme systems which separates the
flavor-generating reactions from the bulk of the food and allows
passage of the volatile compounds that are generated into the
food may find application for controlled seafood-flavor
generations (88). In foods where uncontrolled
hydroperoxide-initiated autoxidation might not limit the quality
of the overall flavor of the product, then perhaps direct
incorporation of a lipoxygenase flavor-generating system might be
exploited.

The relative ratios of alcohols and carbonyls for the six-,
eight- and nine-carbon volatiles in fish (23-24) and oysters (26)
parallel those encountered in cucumber fruits (37) and mushrooms
(27, 56) if the two systems are combined. Therefore, the use of
plant-based enzyme systems for the controlled generation of fresh
seafood flavors and aromas has been under consideration in our
laboratory as a means to overcome some of the self-inactivating
problems associated with fish lipoxygenases.

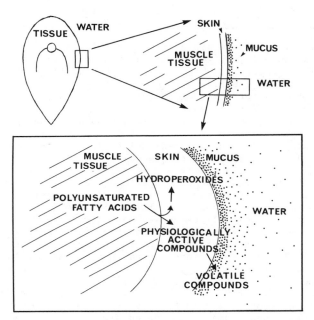

Figure 2.   Proposed physiological role for the enzymic formation of volatile aroma compounds in fresh seafoods.

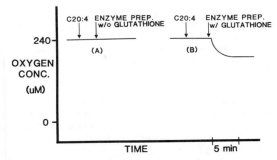

Figure 3.   The effect of reduced glutathione on fish lipoxygenase activity as measured by oxygen consumption employing a Johnson-type oxygen electrode.

Exploratory Experiments on the Means to Assist in the Regulation
of Lipoxygenases in Tissue Extracts

Mitsuda et al. (89) have demonstrated inhibition of soybean
lipoxygenase activity by the addition of monohydric alcohols
(n-butanol, n-pentanol, n-hexanol, and n-heptanol), and noted a
marked increase in inhibition with increasing chain-length.
Evidence obtained for other alcohols, including straight-chain,
secondary and tertiary alcohols, strongly suggested the
inhibition was caused by non-specific hydrophobic bonding between
the alcohol and enzyme.
        These observations stimulated exploration of the possibility
that naturally occurring fresh fish volatiles might be involved
in regulating enzymic activity. Further, it was of interest to
determine whether the presence of certain fresh fish volatiles
might exert regulating action on lipoxygenase in flavor
generating systems. To test the effect of fresh fish volatile
compounds on trout gill lipoxygenase activity, a crude
preparation of the enzyme was obtained from rainbow (Salmo
gairdneri; 400 gm each) or brown (Salmo trutta; 300 gm each)
trout (68). These preparations proved to be sensitive to classic
lipoxygenase inhibitors, i.e., esculetin and tin (II) chloride,
as reported by German and Kinsella (69).
        For each enzyme preparation, the gills were removed from five
live trout and homogenized in 50 ml of tris buffer (0.05M; pH
7.8; 0°C) which contained 1 mM reduced glutathione. Homogenates
were subsequently centrifuged at 15,000 x g for 10 min., and
filtered through cheesecloth to remove free-floating lipids (69),
and then used directly in enzyme assays. For assays of trout
gill lipoxygenase, 1.5 ml of the preparation was equilibrated (5
min) with 50 ul of either a solution of ethyl acetate containing
an inhibitor or neat ethyl acetate (control). One ml of the
equilibrated extract was then added to a vial containing 2 ml of
Tris buffer (0.05M; pH 7.8), 100 nmoles of sodium arachidonate
(in ethanol), and a Johnson-type lead-silver electrode (90-91)
was used for the measurement of oxygen consumption. All
extractions and measurements were performed in a walk-in cooler
at 4°C.
        When the n-alkanols, n-butanol, n-hexanol, and n-heptanol
(10-100 mM), were preincubated with trout gill lipoxygenase
preparations, some degee of inhibition of oxygen consumption was
observed as compared to control determinations. These data
suggested that the alcohols inactivated fish lipoxygenase in a
manner similar to that observed for soybean lipoxygenase (89).
Additionally, a number of enzymically-derived volatiles
identified in fresh fish were tested for inhibitory activity on
trout gill lipoxygenase, including 1-penten-3-ol, 1-octen-3-ol,
6-nonen-1-ol, hexanal, and 1-octen-3-one. In all cases the fresh
fish volatile compounds also exhibited some inhibitory influence
on trout gill lipoxygenase preparations. 1-Octen-3-ol and
1-octen-3-one at a concentration of 10 mM each provided about 50%
inhibition of trout gill lipoxygenase activity. The similar
inhibitory properties of 1-octen-3-ol and 1-octen-3-one suggest a
non-specific binding for these volatiles to lipoxygenase.

Hexanal (to 150 mM) and 6-nonen-1-ol (to 100 mM) were less
effective in inhibiting gill lipoxygenase, while 1-penten-3-ol
(to 300 mM) proved to be the least effective.

The range of concentrations of volatiles which caused some
degree of inhibition of trout gill lipoxygenase were in the mM
range (i.e., 1000-3000 ppm). Therefore, relatively high
concentrations of these compounds are necessary to substantially
reduce the activity of trout gill lipoxygenase in crude enzyme
preparations. However, solubility factors may influence the
apparent effectiveness of the longer-chain compounds.
Experiments also have shown that ethanol and ethyl acetate do not
influence the activity of trout gill lipoxygenase up to about
1.2M (6.6%) and 0.75M (6.6%), respectively. Thus, ethanol and
ethyl acetate provided attractive solvents for diluting both
substrates and inhibitory compounds in lipoxygenase inhibition
studies. Although selected volatiles might be exploited for
providing contributions of beneficial flavor effects as well as
lipoxygenase-control, the high concentrations required for
lipoxygenase control will require special manipulation in
microencapsulated systems to allow practical applications.

Exploratory Model Employing Surimi to Evaluate Plant-Based Seafood Flavor Generation

Since plant lipoxygenases lack the self-inactivation properties
of animal lipoxygenases, plant-based enzyme flavor generating
systems have a greater potential for early application in the
biogeneration of fresh seafood aromas and flavors than those of
fish. At this time some of the most desirable, characterizing
fresh seafood-like aroma/flavor compounds appear to be provided
by combinations of the eight-carbon volatiles, and these are also
produced in mushrooms and geranium leaves.

Thus, the models for initial experiments included crude
enzyme preparations from plants that were incorporated into
pollack surimi which is a washed, minced fish flesh product
widely used as a base for seafood analogs such as imitation
crab. It exhibits relatively mild fish-like aromas, and its
aroma seems to respond well to added plant-based enzyme
preparations.

Flavor-generating systems have included those from mushrooms
(mascreated; Agaricus bisporus) and cucumber fruits (mascreated;
Cucumis sativus) which were each added to surimi at 1%.
Additionally, single geranium leaves (either crushed or
uncrushed; Pelargonium sp.) were placed into 100 gm of surimi.
The descriptions of aromas produced in these systems are shown in
Table IV. The principal compounds which contribute to the
initial fish-like aromas of surimi appear to be the
enzymically-derived eight-carbon carbonyls and alcohols in
combination with some oxidized fishy aroma undertones that are
caused by very low levels of autoxidatively-derived carbonyls,
including the 2,4-heptadienals and the 2,4-decadienals
(unpublished data). When geranium leaves were macerated before
addition to surimi, the six-carbon volatile compounds dominated
the overall aroma, and the desirable contributions associated

Table IV. Exploratory applications of plant-derived lipoxygenases
           for flavor modifications of pollack surimi.

| Surimi Sample Treatment | Aroma Quality | |
	After Incubation (12 h at 4°C)	Probable Dominant Contributing Compounds
No treatment	Mild, clean fishy	C_7-, and C_{10}-dienals, C_8 Alcohol/ketones
Uncrushed geranium leaves	Initially "marine green"; Then potent fresh trout	1,5-Octadien-3-one, Possibly unknowns
Mascerated mushroom	Masked fishiness; Oyster-like	C_8 Alcohols/ketones
Mascerated cucumber	Lacks masking effect; Green-vine-like	C_9 Aldehydes

with the presence of 1,5-octadien-3-one were overwhelmed.
However, when uncrushed geranium leaves provided
1,5-octadine-3-one to the aroma of fresh surimi, a marine-green
aroma often found in fresh lake or ocean breezes was observed
upon opening the container. As the sample warmed from 4°C, the
marine-green-like aroma was replaced by an aroma that was
characteristic of fresh trout, and this persisted for several
hours.
    When mushroom homogenates were incubated with surimi,
enhanced plant-like aromas somewhat reminiscent of oysters were
produced, and this treatment also resulted in the masking of some
of the fish-like aromas of the surimi. Cucumber homogenates
developed strong cucumber, cardboard-like aromas which appear to
be contributed principally by 2-nonenal and 2,6-nonadienal. As a
result, the cucumber homogenates caused undesirable and
unbalanced aromas that did not suppress unpleasant fishiness.
Watermelon fruit extracts behaved similarly, and also provided
unbalanced sweet aromas to surimi. Tests to date have been
limited to short-term incubations of crude enzyme preparations
with surimi. Further exploration of more purified and controlled
plant-based flavor-generating enzyme systems for the production
of fresh seafood-like aromas, and especially those for the
eight-carbon volatile aroma compounds, appear warrented.

Other Future Directions In Seafood Aroma Research

Although polyunsaturated fatty acids have historically received
the most attention as precursors for the formation of volatiles
associated with the aromas of seafoods, another source of
characterizing flavors for certain seafoods appears to exist in
the naturally-occurring carotenoid pigments found in some fish
tissue. Lipoxygenase can co-oxidize carotenoid pigments while
catalyzing the hydroperoxidation of fatty acids (92-94), and this
reaction has been routinely exploited in bleaching wheat flour

for breadmaking (95). In this respect, some of the unique flavor of cooked salmon now appears to result from selected oxidative reactions involving carotenoid pigments and polyunsaturated fatty acids in the flesh of these fish.

Overall, the development of microencapsulation techniques which allow biogenerations of fresh seafood aroma compounds while separating the potentially damaging lipoxygenase-derived hydroperoxides from the bulk food consituents appears to be a promising area of future research. Combinations of such techniques with controllable plant-based enzyme systems seem within technological reach at this time. If fish lipoxygenases can be stabilized against self-inactivation, these enzymes potentially could be exploited also. However, regulating the rate of selective production of important volatile aroma compounds will remain one of the biggest hurdles in developing commercially feasible processes, and every means of controlling the formation of enzymic products should be carefully considered in the next stages of research and development.

Acknowledgments

This research was supported by the College of Agricultural and Life Sciences and the Sea Grant College Program, Univeristy of Wisconsin-Madison.

Literature Cited

1. Obata, Y., Yamanishi, T. and Ishida, M. Bull. Jap. Soc. Sci. Fish. 1950. 15, 551-553.
2. Jones, N.R. In "Symposium on foods: The Chemistry and Physiology of Flavors"; Schultz, H.W., Day, E.A. and Libbey, L.M. Eds.; AVI; Westport, Conn., 1967; pp. 267-295.
3. Meijboom, P.W. and Stroink, T.B.A. J. Am. Oil Chem. Soc. 1972, 49, 555-558.
4. McGill, A.S., Hardy, R., Burt, J.R., and Gunstone, F.D. J. Sci. Food Agric. 1974, 25, 1477-1489.
5. McGill, A.S., Hardy, R., and Gunstone, F.D. J. Sci. Food Agric. 1977, 28, 200-215.
6. Aitken, A. and Connell J.J. In "Effects of Heating on Foodstuffs"; Priestly, J.R. Ed.; Applied Sciences: London, 1979; pp. 219-254.
7. Ikeda, S. In "Advances in Fish Science Technology"; Connell, J.J., Ed.; Fishing News Book; Farnham, Eng. 1980; pp. 111-123.
8. Yu, T.C., Day, E.A., and Sinnhuber, R.O. J. Food Sci. 1961, 26, 192-197.
9. Badings, H.T. J. Am. Oil Chem. Soc. 1973, 50, 334.
10. Josephson, D.B., Lindsay, R.C., and Stuiber, D.A. J. Food Sci. 1983, 48, 1064-1067.
11. Badings, H.T. Neth. Milk Dairy J. 1970, 24, 147-256.
12. Swoboda, P.A.T. and Peers, K.E. J. Sci. Food Agric. 1977, 28, 1010-1018.

216

BIOGENERATION OF AROMAS

13. Seals, R.G. and Hammond, E.G. J. Am. Oil Chem. Soc. 1970, 47, 278-280.
14. Tappel, A.L. Food Res. 1952, 17, 550-559.
15. Tappel, A.L. Food Res. 1953, 18, 104-108.
16. Tappel, A.L. In "Autoxidation and Antioxidants, Vol. 1"; Lundberg, W.O. Ed.; Wiley: New York, 1961, p. 325.
17. Gardner, H.W. J. Agric. Food Chem. 1975, 23, 129-136.
18. Ke, P.J., Ackman, R.G. and Linke, B.A. J. Am. Oil Chem. Soc. 1975, 52, 349-353.
19. Moncrieff, R.W. In "The Chemical Senses"; John Wiley and Sons: New York, 1944; 424p.
20. Ackman, R.G., Hingley, J. and MacKay, K.T. J. Fish Res. Board Can. 1972, 25, 267-284.
21. Whitfield, F.B., Freeman, D.J., Last, J.H., and Bannister, P.A. Chem. Ind. (London) 1981, 5, 158-159.
22. Shiomi, K., Noguchi, A., Yamanaka, H., Kikuchi, T. and Iida, H. Comp. Biochem. Physiol. 1982, 71B, 29-31.
23. Josephson, D.B., Lindsay, R.C., and Stuiber, D.A. J. Agric. Food Chem. 1983, 31, 326-330.
24. Josephson, D.B., Lindsay, R.C., and Stuiber, D.A. J. Agric. Food Chem. 1984, 32, 1344-1347.
25. Josephson, D.B., Lindsay, R.C., and Stuiber, D.A. J. Agric. Food Chem. 1984, 32, 1347-1352.
26. Josephson, D.B., Lindsay, R.C., and Stuiber, D.A. J. Food Sci. 1985, 50, 5-9.
27. Tressl, R., Bahri, D. and Engel, K. -H. J. Agric. Food Chem. 1982, 30, 89-93.
28. Pyysalo, H. and Suihko, M. Leb. Wiss. Tech. 1976, 9, 371-373.
29. Josephson, D.B., Lindsay, R.C., and Stuiber, D.A. "Abstracts of Papers"; The 185th National Meeting of the American Chemical Society, Seattle, WA 1983; American Chemical Society: Washington, DC. AGFD # 28.
30. Whitfield, F.B. and Freeman, D.J. Wat. Sci. Tech. 1983, 15, 85-95.
31. Whitfield, F.B., Freeman, D.J., Last, J.H., Bannister, P.A. and Kennett, B.H. Aust. J. Chem. 1982, 35, 373-383.
32. Kemp, T.R., Knavel, D.E. and Stoltz, L.P. Phytochem. 1971, 10, 1925-1928.
33. Kemp, T.R., Knavel, D.E. and Stoltz, L.P. J. Agric. Food Chem. 1974, 22, 717-718.
34. Kemp, T.R., Knavel, D.E., Stoltz, L.P. and Lundin, R.E. Phytochem. 1974, 3, 1167-1170.
35. Kemp, T.R. Phytochem. 1975, 14, 2637-2638.
36. Suyama, M., Hirano, T. and Yamazaki, S. Bull. Jap. Soc. Sci. Fish. 1985, 51, 287-294.
37. Hatanaka, A., Kajiwara, T. and Harada, T. Phytochem. 1975, 14, 2589-2592.
38. Frazzalari, F.A. (Ed.) In "Compilations of Odor and Taste Threshold Value Data"; American Society for Testing and Materials: Philadelphia, PA, 1978; p. 130.
39. Buttery, R.G. In "Flavor Research-Recent Advances"; Teranishi, R., Flath, R.A.; Sugisawa, H. Ed.; Marcel Dekker: New York, 1981; pp. 193-210.

40. Josephson, D.B., Lindsay, R.C., and Stuiber, D.A. Can. Inst. Food Sci. Technol. J. 1984, 17, 178-182.
41. Stansby, M.E. Food Technol. 1962, 16, 28-32.
42. Beatty, S.A. J. Fish Res. Board Can. 1938, 4, 63-68.
43. Beatty, S.A. J. Fish Res. Board Can. 1939, 4, 229-232.
44. Watson, D.W. J. Agric. Food Chem. 1939, 25, 124-127.
45. Jacquot, R. In "Fish as Food: Vol. 1. Production, Biochemistry, and Microbiology" Borgstrom G. Ed.; Academic Press: New York, 1961; pp. 145-209.
46. Simidu, W. In "Fish as Food. Vol. 1. Production, Biochemistry, and Microbiology." Borgstrom, G. Ed.; Academic Press: New York, 1961; pp. 353-384.
47. Reay, G.A. and Shewan, J.M. Advances Food Res. 1949, 2, 343-398.
48. Hebard, C.E., Flick, G.J. and Martin, R.E. In "Chemistry and Biochemistry of Marine Food Products." Martin, R.E., Flick, G.J., Hebard, C.E. and Ward, D.R. Ed.; AVI: Westport, Conn., 1982, pp. 155-175.
49. Motohiro, T. Memoirs of the Faculty of Fisheries, Hokkaido University. 1962, 10, 1-65.
50. Granroth, B. and Hattula, T. Finn. Chem. Lett. 1976., 148-150.
51. Yueh, M.H. Dissert Abs. 1961, 22, 730.
52. Ronald, A.P. and Thomson, W.A.B. J. Fish Res. Board Can. 1964, 22, 1481-1487.
53. Mendelsohn, J.M. and Brooke, R.O. Food Technol. 1968, 22, 1162-1166.
54. Kadota, H. and Ishida, Y. Ann. Review Micro. 1972, 26, 127-138.
55. Miller, A., Scanlan, R.A., Lee, J.S. Libbey, L.M. and Morgan, M.E. Appl. Microbiol. 1973, 26, 18-21.
56. Tressl, R., Bahri, D. and Engel, K. -H. In "Quality of Selected Fruits and Vegetables of North America", Teranishi, R. and Barrera-Benitez, H. Eds.; American Chemical Society: Wash. D.C. 1981. pp. 213-232.
57. Vick, B.A. and Zimmerman, D.C. Plant Physiol. 1976, 57, 780-788.
58. Galliard, T. and Phillips, D.R. Biochim. Biophys. Acta. 1976, 431, 278-287.
59. Hartel, B., Ludwig, P., Schewe, T. and Rapoport, S.M. Eur. J. Biochem. 1982, 126, 353-357.
60. Lands, W. Prostaglandins Leukotrienes and Medicine 1984, 13, 35-46
61. Tsukuda, N. and Amano, K. Bull. Japan. Soc. Sci. Fish. 1967, 33, 962-969.
62. Tsukuda, N. and Amano, K. Bull. Japan. Soc. Sci. Fish. 1968, 34, 633-639.
63. Tsukuda, N. Bull. Japan. Soc. Sci. Fish. 1970, 36, 725-733.
64. Tsukuda, N. Bull. Japan. Soc. Sci. Fish. 1970, 36, 806-811.
65. Sekiya, K., Okuda, H. and Arichi, S. Biochem. Biophys. Acta. 1982, 713, 68-72.
66. Wallach, D.P. and Brown, V.R. Biochem. Biophys. Acta. 1981, 663, 361-372.

67. Olafsdottir, G., Steinke, J.A., and Lindsay, R.C. J. Food
    Sci. 1985, 50, 1431-1437.
68. German, J.B. and Kinsella, J.E. Biochem. Biophys. Acta.
    1986, 875, 12-20.
69. German, J.B. and Kinsella, J.E. J. Agric. Food Chem. 1985,
    33, 680-683.
70. Josephson, D.B., Lindsay, R.C., and Stuiber, D.A. J. Food
    Sci. 1986, 51:in press.
71. Wurzenberger, M. and Grosch, W. Z. Lebensm. Unters. Forsch.
    1982, 175, 186-190.
72. Hitchock, C. and Nichols, B.W. In "Plant Lipid Biochemistry"
    Sutcliffe, J.F. and Mahlberg, P.; Ed.; Academic Press: New
    York, 1971, pp. 1-42.
73. Kinsella, J.E., Shimp, J.L., Mai, J. And Weihrauch, J. J.
    Am. Oil Chem. Soc. 1977, 54, 424-429.
74. Viswanathan Nair, P.G. and Gopakumar, K. J. Food Sci. 1978,
    43, 1162-1164.
75. Brain, S.D., Camp, R.D.R., Dowd, P.M., Black, A., Woollard,
    P.M., Mallet, A.I. and Greaves, M.W. In "Leukotrienes and
    Other Lipoxygenase Products"; Piper, P.J. Ed.; Research
    Studies Press: New York, 1982; pp. 248-254.
76. Pace-Asciak, C.R. and Smith, W.L. In "The Enzymes. Lipid
    Enzymology Vol 16."; Boyer, P.D. Ed.; Academic Press: New
    York, 1983; pp. 543-603.
77. Parker, C.W. In "The Leukotrienes Chemistry and Biology.";
    Chakrin, L.W. and Bailey, D.M. Eds.; Academic Press: New
    York, 1984, pp. 125-137.
78. Viliegenthart, J.F.G., Veldink, G.A.; Verhagan. J. and
    Slappendel, S. In "Biological Oxidations, Colloquium der
    Gesellschaft fur Biologische Chemie" Sund. H. and Ullrich,
    V. Eds.; Springer-Verlag: Berlin, 1983, pp. 203-223.
79. Piomelli, D. Naturwissenschaften 1985, 72, 276-277.
80. Fletcher, T.C. In "Stress and Fish"' Pickering, A.D. Ed.;
    Academic Press: New York, 1981. pp. 171-183.
81. Yamaguchi, K., Toyomizu, M., and Nakamura, T. Bull. Japan.
    Soc. Sci. Fish. 1984, 50, 1245-1249.
82. Yamaguchi, K., Nakamura, T., and Toyomizu, M. Bull. Japan.
    Soc. Sci. Fish. 1984, 50, 869-874.
83. Yamaguchi, K. and Toyomizu, M. Bull. Japan. Soc. Sci. Fish.
    1984, 50, 2049-2054.
84. Capdevila, J., Saeki, Y. and Falck, J.R. Xenobiotica 1984,
    14, 105-118.
85. Roth, G.J., Machuga, E.T. and Ozols, J. J. Biochem. 1983, 22,
    4672.
86. Bakan, J.A. Food Technol. 1973, 27(11), 34-44.
87. Balassa, L.L. and Fanger, G.O. CRC Critical Reviews in Food
    Technol. 1971, 2(2), 245-265.
88. Magee, E.L., Olson, N.F. and Lindsay, R.C. J. Dairy Sci.
    1982, 64, 616-621.
89. Mitsuda, H., Yasumoto, K. and Yamamoto, A. Archiv. Biochem.
    Biophys. 1967, 118, 664-669.
90. Johnson, M.J., Borkowski, J. and Engblom, C. Biotechnol.
    Bioeng. 1964, 6, 457-468.

91. Borkowski, J.D. and Johnson, M.J. Biotechnol. Bioeng. 1967, 9, 635-639.
92. Grosch, W., Weber, F. and Fischer, K.H. Ann. Technol. Agric. 1977, 26, 133-137.
93. Galliard, T., and Chan, H. W.-S. In "The Biochemistry of Plants Vol. 4, Lipids: Structure and Function"; Stumpf, P.K. Ed.; Academic Press: New York, 1980, pp. 149-150.
94. Ikediobi, C.O. and Snyder, H.E. J. Agric. Food Chem. 1977, 25, 124-127.
95. Schwimmer, S. In "Source Book of Food Enzymology" AVI: Westport, Conn, 1981, p. 425.

RECEIVED May 2, 1986

METABOLISM OF SPECIFIC COMPOUNDS

18

Role of Monoterpenes in Grape and Wine Flavor

Christopher R. Strauss, Bevan Wilson, Paul R. Gooley, and Patrick J. Williams

The Australian Wine Research Institute, Private Mail Bag, Glen Osmond, SA, 5064, Australia

This review brings up to date developments in the field of grape and wine monoterpenes with emphasis on formation of volatile flavorants from flavorless precursors. Data are presented substantiating the importance of monoterpenes to grape and wine flavor and the extent to which these compounds contribute to varietal character. The various forms of monoterpenes present in grapes and the significance of the different forms to winemaking are examined. Sections on pre- and postharvest influences as well as analytical techniques are included along with suggestions for future research in the field.

Monoterpenes were first recognized in muscat grapes almost thirty years ago when linalool was characterized by thin layer chromatography (1). The number of compounds identified has now been greatly expanded and the list includes alcohols, ethers, aldehydes and hydrocarbons as well as polyfunctional derivatives (Figures 1 and 2). The number of grape varieties in which monoterpenes have been detected has also increased, although the findings have been largely confined to the species Vitis vinifera and its hybrids.

Earlier reviews on grape and wine flavor have covered the occurrence and significance of monoterpenes in varying degrees of detail. Terrier and Boidron (2) comprehensively reviewed the literature up to 1972 and discussed experimental techniques used for the analysis of the then recognized grape and wine monoterpenes. In later reviews on wine flavor, Schreier (3) and Williams (4) each included a section on terpene compounds. The former work drew attention to the importance of monoterpene alcohols and oxides for the differentiation of grape and wine varieties. Two recent reviews (5,6) have been devoted exclusively to the subject of terpenes in grapes and wines. Both of these works include referenced tabulations of the incidences of occurrence of individual

0097–6156/86/0317–0222$06.25/0
© 1986 American Chemical Society

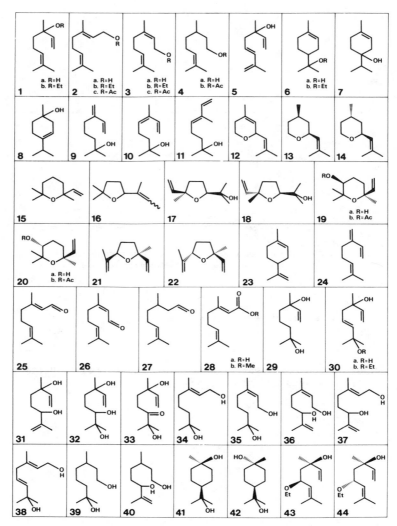

Figure 1. Monoterpenes identified in Vitis vinifera grapes and wines. For references to identification of compounds **1a, 2a, 2c, 3a, 3c, 4a, 4b, 5, 6a, 7, 9–18, 19a, 20a, 21–27, 28a, 28b, 29, 30a, 31, 32, 36–42** see reviews (5) and (6); **1b**, publication (95); **2b, 3b, 6b** (53); **33** (96); **19b** and **20b,** (97). By an oversight **19b** and **20b** were erroneously named as isomers of 2,6,6-trimethyl-2-vinyl-4-acetoxytetrahydropyran instead of 2,6,6-trimethyl-2-vinyl-5-acetoxytetrahydropyran (97). The error has been reproduced in reviews (3,4,6). Diols **34** and **35** have not been previously reported in grapes and wines but have been observed by the authors as components of Muscat of Alexandria juice.

No absolute stereochemistry is implied for the monoterpenes shown.

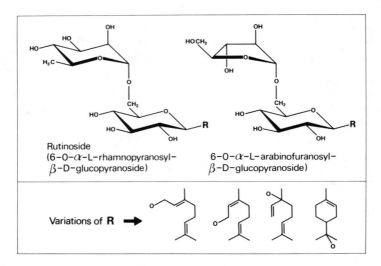

Figure 2. Monoterpene disaccharides identified in <u>Vitis</u>
<u>vinifera</u> grapes and wines (<u>24</u>).

monoterpenes in grapes and wines. The survey of Marais (5) collects published information on the aroma threshold values of the major alcohols and oxides and gives the concentrations of these compounds in different grape varieties. The work of Rapp et al. (6) was presented at a symposium along with two other papers (7,8) which also contained sections on the role of terpene constituents in the formation of wine aroma.

The present review reports on developments in the field of grape and wine monoterpenes. Where appropriate, early work is clarified by interpretation of data in terms of the most recent results. New developments have led to an understanding of the mechanisms of formation of monoterpenes from flavorless precursors in grapes and these concepts are emphasized in this work.

The Importance of Monoterpenes to the Flavor of *Vitis vinifera* Grapes and Wines

It has been a long standing aim of enological research to distinguish analytically between grape varieties, or between wines of similar styles, on the basis of compositional parameters. Such a differentiation is essential to an understanding of the factors responsible for varietal flavors of wines. It has now been established that analysis of fruit-derived monoterpenes will allow such an aim to be realized. This group of compounds, alone among the many and varied constituents of grapes and wines, shows a relationship with varietal flavor characteristics. Wagner et al. (9) demonstrated a correlation between the content of linalool **1a** and geraniol **2a** (see Figure 1) in grapes and the flavor intensity of the fruit. This applied to muscat varieties and to experimental cultivars bred by crossing muscat and non-muscat varieties.

Ribereau-Gayon et al. (10) determined the aroma threshold values of eight monoterpenes of muscat grapes (i.e. **1a,2a,3a,6a, 17,18,19a** and **20a**) and found that the major monoterpenes of the fruit, linalool and geraniol, were present in concentrations higher than their aroma thresholds. Furthermore, it was observed that none of the individual compounds studied had sensory properties identical with muscat character but a combination of the eight volatiles was essential to muscat grape aroma. Other aromatic but non-muscat grape varieties, such as those of the Alsace region (i.e. Riesling, Sylvaner, Müller-Thurgau and Gewürztraminer) also contained moderate to high levels of monoterpenes. Terrier et al. (11) concluded that the flavor of these varieties appeared to be controlled by monoterpenes. Non-aromatic and bland varieties gave juices in which terpenoids could either not be detected or were present in trace amounts (11).

Schreier et al. (12,13) used differences between the concentrations of **1a,2a,3a,5,6a,17,18,19a** and **20a** to distinguish six wine grape cultivars (i.e. Gewürztraminer, Scheurebe, Ruländer, Riesling, Morio-Muskat and Müller-Thurgau). Stepwise discriminant analysis was then applied to quantitative data for the same monoterpenes in wines prepared from the six varieties (14,15). All except for the wines prepared from the closely related cultivars Riesling and Müller-Thurgau could be distinguished.

These last works were highly significant since, in spite of the abundance of volatiles formed by yeast during fermentation, these non-terpene compounds did not vary systematically among the varieties. Furthermore, the yeast metabolites did not mask the distinguishing aroma properties of the grape monoterpenes.

Statistical correlations of grape and wine volatiles, particularly monoterpenes, have also been used by Rapp et al. (16,17). However, these workers have noted the ability of monoterpene diols and triols, which are also present in musts and wines, to act as precursors of volatiles (6). Accordingly, variations in monoterpene abundance between samples could be reflective of the method used in preparing the sample or of the analysis itself. A similar questioning of the emphasis given to the analytical characterization of grapes and wines based on quantification of free monoterpenes had been expressed previously when the lability of various precursor forms of these compounds was first recognized (18,19). Nevertheless these caveats questioned only the detailed interpretation of the statistical data and the need for appropriate techniques to avoid artefact formation. The fundamental soundness of the correlations remains intact.

Non-Volatile or Bound Forms

Cordonnier and Bayonove (20) first suggested the possible existence of bound, acid labile, non-volatile terpenoids in muscat grapes. A crude enzyme preparation from muscat grapes, when applied to muscat pulp and juices, produced up to six-fold increases in concentrations of geraniol and nerol 3a. It was tentatively concluded that the so-called "bound" monoterpenes may have been β-glucosides, which for unknown reasons, were stable towards attack by exogenous β-glucosidases such as almond emulsin (20). Di Stefano (21) later observed in White Muscat from Piemonte, (Moscato bianco del Piemonte) a terpene precursor which produced, by acid hydrolysis, linalool, α-terpineol 6a and geraniol, along with a pentose sugar, arabinose. Ultimately it was shown that several different categories of monoterpenes existed in muscat grapes (19); namely free volatile monoterpenes, free polyhydroxylated monoterpenes (polyols) and glycosidic derivatives of the two former monoterpenoid types.

The free volatiles consist mainly of linalool, geraniol, nerol, furan and pyran forms of the linalool oxides (**17, 18, 19a, 20a** respectively), α-terpineol, hotrienol 5 and citronellol 4a, which are also the major terpene aroma compounds of the fruit.

The odorless polyols appear to be derived from four "parent" monoterpenes - linalool, citronellol, nerol and geraniol by hydration and oxidation reactions (6,22). Figure 3 lists the polyols identified in grapes and also shows structures of components not yet found but likely to be observed should polyol formation from each parent molecule follow an analogous route.

Techniques were developed for the isolation of monoterpene glycosides (23) and structural studies on those with aglycons at the linalool oxidation state were carried out (24). These compounds

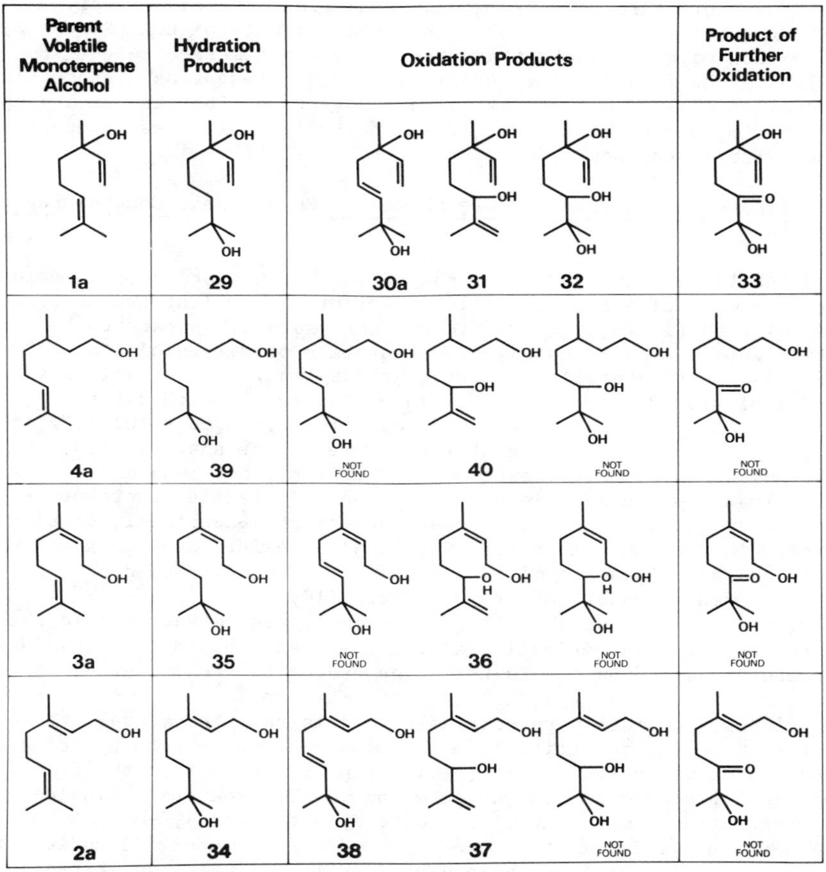

Figure 3. Polyols related to parent monoterpene alcohols. Compounds not yet identified in Vitis sp. products are so designated. For references to all other products see Figure 1.

consisted of a glycosidic mixture of β-rutinosides (i.e.
6-O- α-L-rhamnopyranosyl-β-D-glucopyranosides) and
6-O- α-L-arabinofuranosyl-β-D-glucopyranosides of predominantly
geraniol, nerol, and linalool with traces of α-terpineol (Figure 2).
Subsequent work has shown that other monoterpenes free in the fruit,
including the four linalool oxides, are also present as disaccharide
glycosides (25). Some of the polyols have been liberated from grape
isolates by action of glycosidases, indicating that these compounds
also exist in the bound form in the fruit (26,27).

Significance of Polyols and Glycosides to Monoterpene Composition of the Fruit

Acid lability. Several of the polyols although odorless, are acid
labile and readily form volatile flavorants at ambient temperature
and juice pH (18,28,29). Hotrienol, for example, appears to be
formed wholly by acid catalyzed dehydration of dienediol **30a**
(26,30). Four naturally occurring hydroxylinalool derivatives i.e.
29,30a,31 and **32** were heated for 15 min at 70°C and pH 3.2 and
thirteen volatile monoterpene products (i.e. **1a,5,6a,9,10,11,12,15,
16,17,18,21** and **22**) were identified by headspace analysis (18).
When muscat juices were heated under similar conditions a
significant enhancement in concentration of volatile monoterpenes
was observed. With the notable exceptions of α-terpineol, linalool,
nerol, geraniol and the pyran ring linalool oxides most of the
heat-induced terpenoids of the juices could be attributed to
rearrangement products of grape polyols (18).
    The complexity of the mixture of monoterpene glycosides in the
grape prevented investigation of the chemical behavior of individual
precursor constituents. Instead, the hydrolytic reactions of
synthetic geranyl, neryl, linalyl and α-terpinyl
β-D-glucopyranosides were studied as models for the natural product
mixture in order to rationalize the observed monoterpene hydrolysis
products of wines and juices (31). At pH 3.2 the first three
mentioned glucosides each gave the same major products - linalool
and α-terpineol, with nerol, 3,7-dimethyloct-1-ene-3,7-diol **29**,
2,6,6-trimethyl-2-vinyltetrahydropyran **15** and limonene **23** as lesser
components. Geraniol was also a product from linalyl and geranyl
glucosides. The α-terpinyl glucoside gave no acyclic monoterpenes
and α-terpineol was the predominant product.
    Arising from this study (31) of the glycosides were a number of
points of importance regarding the free monoterpene composition of
the fruit. It was clear that hydrolysis of glycosides was a major
pathway to the free compounds and the previously unexplained
enrichment of linalool, geraniol, nerol and α-terpineol in heated
juice could now be accounted for by this mechanism.
    A detailed investigation into the chemical mechanisms of
monoterpene glycoside hydrolysis in acidic medium has not been so
far reported, but it seems that the water solubility of the
compounds, as well as the presence of an allylic glycosidic linkage
and/or an O-glycosidic bond to a tertiary carbon centre may
determine the reactivity of these substrates.

Enzymatic hydrolysis. Valuable information on the composition of the naturally occurring glycosides has been obtained by use of enzymatic hydrolyses. Figure 4 shows a portion of a gas chromatogram featuring aglycons liberated enzymatically from an isolate of muscat juice (25). Included among the products were monoterpene alcohols with primary, secondary and tertiary hydroxyl functions, as well as benzyl alcohol and 2-phenylethanol. These last two compounds, which exist as disaccharide glycosides in the fruit (32), are not readily liberated by mild acid hydrolysis.

A perspective of the monoterpene composition of grapes. The previous studies provide an overview of the interrelationships between the various monoterpene forms of the grape (see Figure 5). Glycosidic derivatives and free polyols are a reserve of odorless precursors of fruit flavor but only the free aroma compounds make any direct contribution to fruit character. However, flavor is generated by hydrolysis of both the glycosidic compounds and the polyols.

Analytical Methods for Monoterpenes in Grapes and Wine

Preparation of the fruit. Since the early research of Webb et al. (33) several workers have recognized the problem of artefact formation during work-up of grapes. Accordingly, these investigators have employed additives such as NaF and ascorbic acid or SO_2 and sorbic acid to minimize microbial and oxidative degradation of the juice (2,10,34).

The findings of Cordonnier and Bayonove (20) alerted workers to the presence in grapes of enzymes potentially capable of liberating some volatile monoterpenes during work-up. Either methanol addition or rapid pasteurization have been used to inhibit such enzymes (35). Until recently, however, the combined effects of acid and heat during sample preparation had been largely unrecognized. The presence of hotrienol and nerol oxide **12** in extracts can be attributed to decomposition of dienediol **30a** during analysis (26). α-Terpineol, barely present naturally in the fruit (26,36), is readily formed from monoterpene glycosides under acid conditions (21,31). Accordingly Wilson et al. (26) employed salt-saturation to denature endogenous enzymes and also neutralized the juice to prevent hydrolysis of precursors.

Isolation of free monoterpenes from juices and wines. The sensitivity of monoterpenes to heat and acid dictates that the extraction of these compounds should be carried out under conditions which are as mild as possible. Early isolation procedures involving steam distillation, or distillation under reduced pressure as a preconcentration step (33,35,37,38) have now largely been abandoned in favor of more gentle methods.

Liquid extraction techniques using a variety of solvents have been reviewed (2,39) and the criteria for an ideal partition system for alcoholic beverages discussed by Clutton and Evans (40). These workers examined nine solvent systems for extraction of monoterpenes

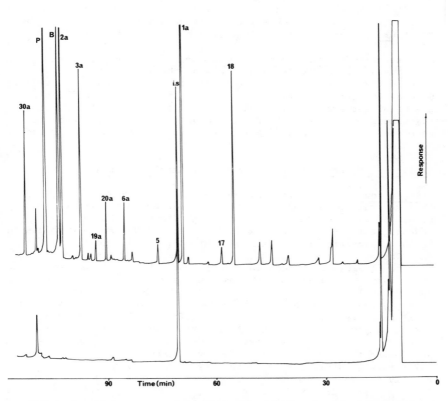

Figure 4. Top trace: Gas chromatogram of aglycons generated by enzyme hydrolysis of C_{18} retained monoterpene glycosides. Lower trace: The same glycosidic material treated with denatured enzyme. For peak assignments refer to Figure 1. B=benzyl alcohol; P=2-phenylethanol and i.s.=internal standard (octan-1-ol).

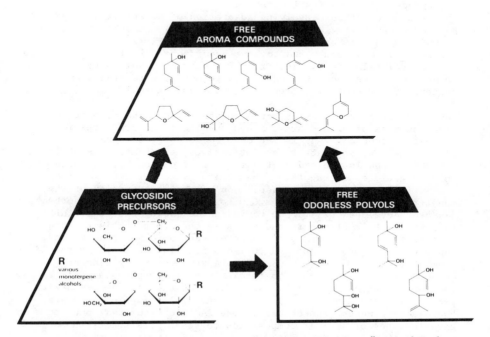

Figure 5. Categories of monoterpenes in grapes. "Reproduced with permission from Ref. (98). Copyright 1985, Practical Winery".

from a model alcoholic solution. The suitability of trichlorofluoromethane (Freon F11) for isolation of wine volatiles has been
investigated (39) and this solvent has been used for exhaustive
extraction of polyhydroxylated linalool derivatives from grape juice
(19). Addition of pyridine prior to concentration of Freon F11
extracts containing dienediol **30a**, was important to the attainment
of reproducible quantitative results (26).

Headspace methods involving adsorption of volatiles onto porous
polymer traps have also been developed (41-45). Another technique,
in which volatiles are entrained into a Freon F11/water partition
system (46), has been found to have good reproducibility for a range
of wine volatiles (47). The Freon F11 extract obtained by this
technique can be concentrated at low temperature to give aroma
concentrates essentially free from water and ethanol.

Isolation of glycosidically bound monoterpenes. The monoterpene
disaccharides of grapes are too hydrophilic for direct organic
solvent extraction. Any isolation procedure for the grape
glycosides must separate these derivatives, which are present in
concentrations in the order of only mg/l, from the bulk of other
polar organic constituents of the fruit.

An isolation procedure based upon selective retention of the
monoterpene disaccharides on a C_{18} bonded reversed-phase adsorbent
has been developed (23). The bonded phase showed little affinity
towards the bulk of sugars and acids in the fruit. This technique,
which gave a 20,000 fold enrichment of the compounds from grapes,
has been subsequently applied to isolation of monoterpene glycosides
from other plant materials (30,48). A styrene-divinylbenzene
co-polymer XAD-2, was used to retain free monoterpenes as well as
monoterpene glycosides (34). Successive elution of the resin with
pentane and ethyl acetate gave fractions containing the free and
bound monoterpenes respectively. The XAD technique was found to
give isolates of similar composition to those obtained by the C_{18}
reversed-phase procedure (34).

Another technique for the purification of monoterpene glycosides
utilizing gel permeation and hydrophobic-interaction chromatography
on polyacrylamide resin has recently been published (49). This
gentle yet highly selective procedure for purification was applied
on a micro-scale to model monosaccharides of eight monoterpene
aglycons.

Analysis of free monoterpenes. The review of Terrier and Boidron
(2) summarized a range of analytical techniques including gas and
silica gel chromatography, for determination of free monoterpenes.
Schreier et al. (35) employed an initial fractionation of extracts
on silica gel, followed by gas chromatographic analysis.

Retention indices for several monoterpenes relevant to grapes
and wine have been published (31,50-52), as have mass spectral data
(22,24,28,29,52-58). Thomas et al. (59) have discussed some causes
of anomalous mass spectra of terpenes.

Analysis of monoterpene glycosides. Owing to high molecular weight,
lack of suitable chromophore, water solubility and nonvolatility,
the monoterpene disaccharides of grapes have proven difficult to

analyze directly. Monoterpene glycoside concentrates have been silylated and then analyzed by gas chromatography on SE30 liquid phase (24, 34). This procedure, however, presents difficulties in identification of individual components in the mixture. Many of the essential reference monoterpene disaccharides have not yet been prepared or isolated in high purity and there are similarities in the mass spectra of several of the silylated components. Williams et al. (24) acetylated the glycoside mixture and using silica gel chromatography, separated arabinoglucosides from rutinosides. Mass spectra of the acetylated fractions gave more structural information than did spectra of the silyl derivatives but gas chromatography was difficult owing to the limited volatility of the acetates.

Another procedure involves enzymatic hydrolysis of monoterpene glycoside mixtures to liberate monoterpenes which can then be analyzed by gas chromatography (26,34). Such methods provide identifications and give good quantitative data for the important aglycon, but not the sugar moieties. When polyols are released by the enzyme, no evidence can be obtained about the site of attachment of the carbohydrates to the terpenols.

The above methods are slow, highly labor intensive and cannot be used to identify and quantify individual monoterpene glycosides of grapes. More rapid techniques have been developed but these also are non-specific. Di Stefano (60) used acid hydrolysis of the glycosides to yield an array of volatile monoterpenes which could then be analyzed by gas chromatography. By this method, the volatiles formed are not diagnostic of the monoterpene aglycon composition of the glycosides (31). Dimitriadis and Williams (61) developed a technique for rapid assay of free and potentially volatile monoterpene flavorants of grapes. This procedure is based on a colorimetric assay and treats the compounds as 'linalool equivalents'. Steam distillation of juice at neutrality yields free volatile aroma components, while at low pH, monoterpenes derived from the polyols and glycosides are collected. The assay has value in that it is simple to establish, rapid (62) and can readily be applied to a range of viticultural treatments such as irrigation (63).

Preharvest Influences on Grape Monoterpenes

Monoterpene levels in different grape cultivars. A number of surveys have been made of monoterpene concentrations in different grape varieties (11,12,16,34,60,61). Some of the works report on free monoterpenes only while others give concentrations of both free and glycosidically bound forms. As the quantitative data were obtained by differing techniques and from samples of fruit from diverse areas, direct comparison is not feasible. Nevertheless, a general classification of those varieties which have been screened is possible (see Table I) allowing division into (a) intensely flavored muscats where total monoterpene concentrations can be as high as 6 mg/l, (b) aromatic but non-muscat varieties with total monoterpene concentrations of 1-4 mg/l, and (c) more neutral varieties not dependent upon monoterpenes for their flavor.

TABLE I.

Classification of Some Vitis vinifera Varieties Based on Monoterpene
Content.

Muscat varieties	Non-muscat aromatic varieties	Varieties independent of monoterpenes for flavor
Canada Muscat*	Gewürztraminer	Bacchus
Muscat of Alexandria	Huxel	Cabernet-Sauvignon
(Syn. Muscat Gordo	Kerner	Carignan
Blanco)	Morio-Muskat	Chardonnay
Muscat a petits grains	Müller-Thurgau	Chasselas
blanc (Syn. White	Riesling	Cinsault
Frontignac, Muscat	Scheurebe	Clairette
Canelli, Moscato bianco)	Schonburger	Dattier de Beyrouth
Moscato bianco del	Siegerebe	Doradillo
Piemonte	Sylvaner	Forta
Muscat Hambourg	Wurzer	Grenache
Muscat Ottonel		Merlot
Italia		Nobling
		Ruländer
		Shiraz
		(Syn. Syrah)
		Terret
		Thompson Seedless
		Ugni Blanc

*Canada Muscat is a hybrid of Vitis vinifera and Vitis labrusca

Changes in monoterpene composition with grape development. Several
studies have been conducted in which changes were recorded in the
concentration of free monoterpenes in ripening grapes (11,38,64-66).
However, most of these works were carried out before the presence of
glycosidically bound or polyhydroxylated forms of monoterpenes in
fruit were recognized.

In research on Muscat of Alexandria (26,34) as well as Moscato
bianco (67,68) the concentrations of free linalool were observed to
increase to a maximum and then to decrease in the overripe fruit.
In one of these studies (26) five phases of monoterpene development
were distinguished during berry ontogeny. At berry set high
concentrations of free and bound geraniol were observed; approaching
veraison, levels of all terpenes decreased; during sugar
accumulation terpene concentrations increased and several
monoterpenes reached peak levels in the overripe fruit. After
veraison linalool, geraniol, nerol, α-terpineol and the furan
linalool oxides were present mainly as glycosides. While the total
concentration of glycosides increased with ripening, only free
linalool showed development paralleling that of its glycoside. At
grape maturity the level of dienediol **30a** was greater than the total
concentration of all other monoterpenes. The concentrations of
glycosides and odorless polyols exceeded those of the free flavor

compounds and maximal terpene levels were found in the fruit after
it had attained commercial maturity.

**The distribution of monoterpenes among separate fractions of the
grape.** As a prerequisite to any biosynthetic studies on terpenes in
grapes the sites of production and storage within the berry must be
known. Such knowledge may have considerable practical consequences
by providing valuable guidance in the application of skin contact
and press conditions to optimize flavorants in juice.

Bayonove et al. (69) observed a highly uneven distribution of
some free monoterpenes in juice, skins and pulp of muscat grapes. A
similar non-uniform distribution was found in Riesling (Riesling
Renano) (66) and in Moscato bianco del Piemonte (60). A more
detailed study has been made by Gunata (34) who quantified free and
glycosidically bound linalool, α-terpineol, citronellol, nerol and
geraniol in juice, skins and pulp of Muscat of Alexandria and Muscat
a petits grains. Wilson et al. (27) examined the location of free
and glycosidically bound forms of monoterpenes at the oxidation
state of geraniol as well as those at the linalool oxide oxidation
state in Muscat of Alexandria, Muscat a petits grains blanc and
Gewürztraminer.

A feature of all of these surveys (27,34,60,66,69) was the
observation that free geraniol and nerol were concentrated in the
skins of each grape variety studied. It was found that free
linalool was more uniformly distributed than geraniol and nerol in
the two muscats and glycosylation effectively redistributed all
monoterpenes throughout the berry fractions (27,34). Wilson et al.
(27) suggested that the hypodermal cells of the berry may be sites
of biosynthesis as well as storage of geraniol and that this
compound could play a fundamental role in monoterpene biosynthesis
in the grape. In addition, the co-occurrence of linalool with
dienediol **30a** implies that linalool is a substrate for the oxidation
to the polyol **30a**. The water solubility of the glycosidic forms
could account for their wide distribution throughout the grape. The
experimental data demonstrate that the long held view that aroma
substances of the grape are concentrated largely in the skins (70)
must be qualified. High concentrations of free linalool in muscat
juices as well as the presence of potential flavorants in parts
other than the epidermis of the grape, emphasize the importance of
processes alternative to skin contact for aroma enhancement.

Effect of Botrytis cinerea. The fungus Botrytis cinerea, which is
endemic in some European vineyards, profoundly influences the flavor
of wines made from the grapes it infects.

Boidron (71) showed that there was a decrease in the
concentration of acyclic monoterpene alcohols and oxides in grape
juice which was fortified with these compounds and then inoculated
with B. cinerea. After 25 days of fungal growth none of the added
monoterpenes could be detected in the juice. Two proposals were put
forward to account for these observations – (a) B. cinerea
metabolized acyclic monoterpenes to oxides and α-terpineol, then
degraded these postulated metabolites or (b) a chemical oxidation
paralleled the attack by the fungus (71).

Shimizu et al. (72) established that B. cinerea did not produce
any monoterpenes in must under culture conditions but led to an
oxidation of linalool. Although the authors identified the four
linalool oxides among the products, they did not identify the major
metabolite, which from their data, was probably dienediol 31. Thus
a parasitic organism commonly found on grapes can metabolize one of
the major monoterpene flavorants of the fruit to a flavorless
polyol. Of note in this regard were the reported losses of both
aroma and volatile terpenes in botrytised grapes (73).

Viticultural effects. The rapid assay technique of Dimitriadis and
Williams (61) has been used to study the influences of some major
viticultural variables (e.g. crop load and irrigation) on grape
monoterpene concentrations. McCarthy and Nicholas (74) reported
lower concentrations of total monoterpenes, particularly potential
volatile terpenes, in both Riesling and Muscat of Alexandria grown
on high yielding rootstocks relative to fruit grown on low yielding
roots. Irrigation did not affect development of free volatile
terpenes in ripening fruit but resulted in a significantly lower
rate of increase in potential volatile terpenes (63).

Postharvest Effects

Glycosidases in winemaking. Enzyme treatment of juices or wines to
increase the concentrations of volatile monoterpene flavorants
could, if successful, provide considerable benefit to winemaking.
In muscat grapes an endogenous glycosidase activity was demonstrated
when increased levels of free geraniol and nerol, but not of
linalool, were observed in juice held at pH 5 and 30°C (75).
However under conditions of temperature and pH appropriate to
winemaking, no significant increases in free terpenes were obtained
(75).

Aryan et al. (76) have also isolated and partially purified an
endogenous grape glucosidase and have confirmed the above findings
(75). However, the partially purified grape enzyme was inhibited
by both glucose and ethanol and was also found to be inactive
towards glycosides of tertiary alcohols such as linalool, and the
furan ring linalool oxides. Glycosidases were also isolated from
yeasts and found to be similarly inhibited by glucose (but not
ethanol) and only partially active towards glycosides of tertiary
alcohols (76).

The findings thus far indicate that there are several
constraints to be overcome before commercial use of enzymes for
volatile monoterpene enhancement in winemaking can be developed.

Effect of pressings on monoterpene composition of musts and wines.
Kinzer and Schreier (77) found that, compared with free run juice,
pressings contained higher concentrations of monoterpene alcohols.
When a pressing aid was used, wines made from the juice contained
34-38% less monoterpenes than wines produced from the same system
without pressing aid. Bayonove et al. (78,79) also studied effects
of pressing and of skin contact in muscat musts.

Formation and metabolism of monoterpenes by yeasts. Although certain species of yeasts are capable of producing monoterpenes (80,81), the wine yeast species Saccharomyces cerevisiae appears not to share this capacity (82). Accordingly, it has been concluded that terpene composition in various grape cultivars and varietal wines is not influenced by fermentation (6,82). Although apparently unable to biosynthesize monoterpenes, wine yeasts may carry out certain transformations of these compounds (e.g. double bond reduction) (83).

Effect of wine ageing on monoterpene composition. Several studies have been made in which alterations in monoterpene distribution in wines over time or with increases in temperature were recorded (84-89). The effect on wine flavor of prolonged ageing or exposure to elevated temperatures is a loss of fruit bouquet. By contrast, brief heating of juices of aromatic grapes can enhance the fruit flavor by increasing the concentration of free volatile monoterpenes through hydrolysis of precursors.
In Riesling wines held at 50°C for varying periods to mimic the ageing process, linalool concentration increased initially and then decreased (84). α-Terpineol was also found to increase in some of the samples (84). In another study of Riesling wines of different ages the only monoterpenes observed were trans-furan linalool oxide and nerol oxide and these two appeared to increase with wine ageing (85). Recognition of the roles of monoterpene polyols and glycosides as precursors of volatiles in juices facilitated interpretation of data from wine ageing experiments. Simpson and Miller (86) detected 17 monoterpenes in aged Riesling wine and of these, 13 were oxides. The data indicated a significant overall increase in the concentrations of volatile monoterpenes in the aged wines and these increases were accounted for in terms of the above degradations. The fruit flavor loss of the aged wines, which occurred in spite of overall increases in levels of volatile monoterpenes, was attributed to preferential formation of compounds with high flavor thresholds, i.e. 6a,12,17,18,41.
Damage to the flavor of "Asti Spumante" wines stored at 20°C when compared with samples held at 10°C was related to losses of linalool and glycosidic precursors of linalool, geraniol and nerol (87). Also the concentrations of α-terpineol, hotrienol, nerol oxide and the furan linalool oxides increased in wines held at ambient temperature. Di Stefano and Castino (87) concluded that low temperature storage retarded these reactions and wines so held retained the characteristic muscat flavor for years.
Rapp et al. (88) and Di Stefano (89) have summarized changes observed in the terpene compositions of aged Riesling wines. These changes were similar to those reported in the studies discussed above.
The differing flavor effects of hydrolytic reactions in aged wines and heated juices can be understood in terms of the conditions under which different volatile monoterpenes are formed. In aged wines, slow transformations of both free monoterpenes and glycosides give the more thermodynamically stable products which are mainly cyclic compounds of high flavor threshold. Additionally, monoterpene ether formation would be expected (53,54) and this too

would represent a flavor loss. In juices, brief exposure to high
temperature effects a rapid hydrolysis of precursors to give a
kinetically controlled product distribution of acyclic and cyclic
volatiles (18,31). It has been observed (31) that prolonged heating
of juice at pH 3 ultimately impaired the sensory character by
imparting a eucalyptus-like aroma which is attributed to the
presence of excessive quantities of 1,8-cineole in the headspace
composition of the juice.

Future Developments in the Field

Enhancement of volatile monoterpene composition. In the light of
results in hand, the genetic manipulation of wine yeasts to produce
non-selective glycosidases which would be active during fermentation
represents a biotechnological challenge. Alternatively, flavor
enhancement might be obtained through use of immobilized enzymes
from sources such as fungi. Nevertheless, limitations to the
ultimate extent of flavor induction possible in grape juices through
the action of glycosidases must always be considered. The abundance
of polyols, especially dienediol **30a** (26,27), in some grapes
represents a rich potential source of volatile monoterpenes in the
processed juice. However, only acid hydrolysis is applicable to the
release of flavor compounds from the polyols and it is therefore of
little consequence whether these compounds are present free or as
glycosides.

Induced muscat aroma (IMA). When the heating process is applied to
muscat grape juices an intense, luscious, raisin-like aroma is
induced. The source of this aroma can not be attributed to release
of known monoterpenes from precursors so far elucidated. This
separate enchanced flavor, which was recognized some years ago (90)
is referred to here as IMA.
    Production of IMA in juices of mature muscat grapes is
facilitated by the presence in the juice of a low concentration of
free SO_2. However, IMA develops in muscat grape juices which have
been cold stored for several months and is clearly apparent in
liqueur muscat wines. IMA may not be widely recognized as the
outcome of a processing step and may be considered by many as the
natural, specific aroma of muscat grapes.
    It has been understood for some time that none of the known free
volatile monoterpenes of the grape is responsible for the specific
flavor of muscats (18,19). Etievant et al. (91) recently drew
attention to this point, but were unsuccessful in identifying
compounds which could account for all the nuances of muscat flavor.
    Research on muscat flavor should focus on IMA and the compounds
responsible for the phenomenon identified.

Viticultural aspects. Viticultural research would benefit from
detailed investigations concerning influences of major preharvest
variables on monoterpene composition of fruit. Different pruning
techniques, trellis designs, soil characteristics and climatic
variables have yet to be assessed for their impact on grape
flavorants. Research has been carried out on some of these factors
in relation to wine quality (92,93) and now their effect on fruit

flavor composition can be investigated in those cultivars which are dependent on monoterpenes for varietal character. Monoterpene analyses should also be useful in future grape breeding programs by offering an objective measure of flavor quality.

Biosynthetic and bioregulatory studies. In recent years substantial advances have been made in understanding the biosynthesis and catabolism of monoterpenes in plants (94). In the so called 'essential oil' plants the monoterpenes are predominantly carbocyclic and hence monoterpene cyclases have been a primary focus of the research (94).

By contrast, biosynthetic studies on whole fruits have been limited and, probably because of the low levels of monoterpenes in grapes, the pathways in Vitis species remain unexplored. Since monoterpenes important to the flavor of grapes are mainly acyclic, interconversions of neryl, geranyl and linalyl derivatives may be of special significance. The oxidative steps leading to polyol production represent a shunt of flavor-active compounds to flavorless forms and require investigation.

An important aspect in understanding the metabolic processes is a knowledge of the control mechanisms involved. When the regulatory mechanisms inducing fruit to produce monoterpene flavorants become known it may be possible to alter varietal character of grapes by application of appropriate regulatory substances. It would then follow that vines could be adjusted to produce grapes tailor-made to the winemakers' flavor specifications.

LITERATURE CITED

1. Cordonnier, R. Ann. Technol. Agric. 1956, 5, 75.
2. Terrier, A.; Boidron, J.N. Conn. Vigne Vin 1972, 1, 69.
3. Schreier, P. CRC Crit. Rev. Food Sci. Nutr. 1979, 12, 59.
4. Williams, A.A. J. Inst. Brew. 1982, 88, 43.
5. Marais, J. S. Afr. J. Enol. Vitic. 1983, 4, 49.
6. Rapp, A.; Mandery, H.; Güntert, M. In "Flavour Research of Alcoholic Beverages"; Nykänen, L.; Lehtonen, P., Eds.; Foundation for Biotechnical and Industrial Fermentation Research: Helsinki, 1984; pp. 255-274.
7. Schreier, P. In "Flavour Research of Alcoholic Beverages"; Nykänen, L.; Lehtonen, P., Eds.; Foundation for Biotechnical and Industrial Fermentation Research: Helsinki, 1984; pp. 9-37.
8. Strauss, C.R.; Williams, P.J.; Wilson, B.; Dimitriadis, E. In "Flavour Research of Alcoholic Beverages"; Nykänen, L.; Lehtonen, P., Eds.; Foundation for Biotechnical and Industrial Fermentation Research: Helsinki, 1984; pp. 51-60.
9. Wagner, R.; Dirninger, N.; Fuchs, V.; Bronner, A. In "International Symposium on the Quality of the Vintage"; l'Office International de la Vigne et du Vin: Cape Town, 1977; pp 137-142.
10. Ribereau-Gayon, P.; Boidron, J.N.; Terrier, A. J. Agric. Food Chem. 1975, 23, 1042.
11. Terrier, A.; Boidron, J.N.; Ribereau-Gayon, P. C.R. Hebd. Seances Acad. Sci., Ser. D 1972, 275, 941.

12. Schreier, P.; Drawert, F.; Junker, A. Chem. Mikrobiol.
    Technol. Lebensm. 1976, 4, 154.
13. Schreier, P. In "University of California, Davis Grape and
    Wine Centennial Symposium Proceedings"; Webb, A.D., Ed.;
    University of California: Davis, 1982; pp. 317-321.
14. Schreier, P.; Drawert, F.; Junker, A.; Reiner, L. Mitt.
    Klosterneuburg 1976, 26, 225.
15. Schreier, P.; Drawert, F.; Junker, A. Chem. Mikrobiol.
    Technol. Lebensm. 1977, 5, 45.
16. Rapp, A.; Knipser, W.; Hastrich, H.; Engel, L. In "University
    of California, Davis Grape and Wine Centennial Symposium
    Proceedings"; Webb, A.D., Ed.; University of California: Davis,
    1982; pp. 304-316.
17. Rapp, A.; Güntert, M. Vitis 1985, 24, 139.
18. Williams, P.J.; Strauss, C.R.; Wilson, B. J. Agric. Food Chem.
    1980, 28, 766.
19. Williams, P.J.; Strauss, C.R.; Wilson, B. Am. J. Enol. Vitic.
    1981, 32, 230.
20. Cordonnier, R.; Bayonove, C. C.R. Hebd. Seances Acad. Sci.,
    Ser. D 1974, 278, 3387.
21. Di Stefano, R. Vignevini 1982, 9, 45.
22. Williams, P.J.; Strauss, C.R.; Wilson, B. Phytochemistry 1980,
    19, 1137.
23. Williams, P.J.; Strauss, C.R.; Wilson, B.; Massy-Westropp, R.A.
    J. Chromatogr. 1982, 235, 471.
24. Williams, P.J.; Strauss, C.R.; Wilson, B.; Massy-Westropp, R.A.
    Phytochemistry 1982, 21, 2013.
25. Strauss, C.R. PhD Thesis, University of Adelaide, South
    Australia, 1983.
26. Wilson, B.; Strauss, C.R.; Williams, P.J. J. Agric. Food Chem.
    1984, 32, 919.
27. Wilson, B.; Strauss, C.R.; Williams, P.J. Am. J. Enol. Vitic.
    1986, in press.
28. Rapp, A.; Knipser, W. Vitis 1979, 18, 229.
29. Rapp, A.; Knipser, W.; Engel, L. Vitis 1980, 19, 226.
30. Engel, K.H.; Tressl, R. J. Agric. Food Chem. 1983, 31, 998.
31. Williams, P.J.; Strauss, C.R.; Wilson, B.; Massy-Westropp, R.A.
    J. Agric. Food Chem. 1982, 30, 1219.
32. Williams, P.J.; Strauss, C.R.; Wilson, B.; Massy-Westropp, R.A.
    Phytochemistry 1983, 22, 2039.
33. Webb, A.D.; Kepner, R.E.; Maggiora, L. Am. J. Enol. Vitic.
    1966, 17, 247.
34. Gunata, Y.Z. Docteur-Ingenieur These, Universite des Sciences
    et Techniques du Languedoc, Montpellier, 1984.
35. Schreier, P.; Drawert, F.; Junker, A. J. Agric. Food Chem.
    1976, 24, 331.
36. Usseglio-Tomasset, L. 6th International Oenological Symposium,
    Mainz/Germany, 1981, p. 353.
37. Stevens, K.L.; Bomben, J.; Lee, A.; McFadden, W.H. J. Agric.
    Food Chem. 1966, 14, 249.
38. Hardy, P.J. Phytochemistry 1970, 9, 709.
39. Rapp, A.; Hastrich, H.; Engel, L.; Knipser, W. In "Flavor of
    Foods and Beverages"; Charalambous, G.; Inglett, G.E., Eds.;
    Academic Press: New York, 1978; pp. 391-417.

40. Clutton, D.W.; Evans, M.B. J. Chromatogr. 1978, 167, 409.
41. Jennings, W.G.; Wohleb, R.; Lewis, M.J. J. Food Sci. 1972, 37, 69.
42. Bertuccioli, M.; Montedoro, G. J. Sci. Food Agric. 1974, 25, 675.
43. Williams, P.J.; Strauss, C.R. J. Inst. Brew. 1977, 83, 213.
44. Williams, A.A.; May, H.V.; Tucknott, O.G. J. Sci. Fd. Agric. 1978, 29, 1041.
45. Noble, A.C.; Murakami, A.A.; Coope, G.F. J. Agric. Food Chem. 1979, 27, 450.
46. Rapp, A.; Knipser, W. Chromatographia 1980, 13, 698.
47. Guichard, E.A.; Ducruet, V.J. J. Agric. Food Chem. 1984, 32, 838.
48. Nitz, S.; Fischer, N.; Drawert, F. Chem. Mikrobiol. Technol. Lebensm. 1985, 9, 87.
49. Croteau, R.; El-Hindawi, S.; El-Bialy, H. Anal. Biochem. 1984, 137, 389.
50. Klouwen, M.H.; Ter Heide, R. J. Chromatogr. 1962, 7, 297.
51. Ter Heide, R. J. Chromatogr. 1976, 129, 143.
52. Jennings, W.; Shibamoto, T. "Qualitative Analysis of Flavor and Fragrance Volatiles by Glass Capillary Gas Chromatography"; Academic press: New York, 1980.
53. Strauss, C.R.; Williams, P.J. In "Flavour of Distilled Beverages: Origin and Development"; Piggott, J.R., Ed.; Ellis Horwood Ltd.; Chichester, 1983; pp. 120-133.
54. Strauss, C.R.; Wilson, B.; Rapp, A.; Guentert, M.; Williams, P.J. J. Agric. Food Chem. 1985, 33, 706.
55. Thomas, A.F.; Willhalm, B. Helv. Chim. Acta. 1964, 47, 475.
56. Heller, S.R.; Milne, G.W.A. "EPA/NIH Mass Spectral Data Base", Vols. 1-4; U.S. Department of Commerce and The National Bureau of Standards: Washington DC, 1978.
57. Rapp, A.; Mandery, H.; Ullemeyer, H. Vitis 1983, 22, 225.
58. Rapp, A.; Mandery, H.; Ullemeyer, H. Vitis 1984, 23, 84.
59. Thomas, A.F.; Willhalm, B.; Flament, I. In "Chromatography and Mass Spectrometry in Nutrition Science and Food Safety"; Frigerio, A.; Milon, H., Eds.; Elsevier: Amsterdam, 1984; pp. 47-65.
60. Di Stefano, R. Vini d'Italia 1981, 23, 29.
61. Dimitriadis, E.; Williams, P.J. Am. J. Enol. Vitic. 1984, 35, 66.
62. Dimitriadis, E.; Bruer, D.R.G. Aust. Grapegrower Winemaker 1984, 244, 61.
63. McCarthy, M.G.; Coombe, B.G. Acta Horticulturae 1985, in press.
64. Bayonove, C.; Cordonnier, R. Ann. Technol. Agric. 1970, 19, 79.
65. Bayonove, C.; Cordonnier, R. Ann. Technol. Agric. 1971, 20, 347.
66. Versini, G.; Inama, S.; Sartori, G. Vini d'Italia 1981, 23, 189.
67. Di Stefano, R.; Corino, L.; Bosia, P.D. Riv. Vitic. Enol. 1983, 36, 263.
68. Di Stefano, R.; Corino, L. Riv. Vitic. Enol. 1984, 37, 657.

69. Bayonove, C.; Cordonnier, R.; Ratier, R. C.R. Acad. Agric. Fr.
    1974, 60, 1321.
70. Winkler, A.J.; Cook, J.A.; Kliewer, W.M.; Lider, L.A. "General
    Viticulture"; University of California Press: Berkeley, 1974;
    p. 145.
71. Boidron, J.N. Ann. Technol. Agric. 1978, 27, 141.
72. Shimizu, J.; Vehara, M.; Watanabe, M. Agric. Biol. Chem. 1982,
    46, 1339.
73. Guerzoni, M.E. Enotecnico 1984, 20, 469.
74. McCarthy, M.G.; Nicholas, P.R. Aust. Grapegrower Winemaker
    1984, 249, 10.
75. Bayonove, C.; Gunata, Z.; Cordonnier, R. Bull. O.I.V. 1984,
    643-644; 741.
76. Aryan, A.P.; Strauss, C.R.; Wilson, B.; Williams, P.J.
    Unpublished.
77. Kinzer, G.; Schreier, P. Am. J. Enol. Vitic. 1980, 31, 7.
78. Bayonove, C.; Cordonnier, R.; Benard, P.; Ratier, R. C.R.
    Acad. Agric. Fr. 1976, 62, 734.
79. Cordonnier, R.; Bayonove, C. Rev. Franc. Oenol. (Paris) 1979,
    16, 79.
80. Drawert, F.; Barton, H. J. Agric. Food Chem. 1978, 26, 765.
81. Fagan, G.L.; Kepner, R.E.; Webb, A.D. Vitis 1981, 20, 36.
82. Hock, R.; Benda, I.; Schreier, P. Z. Lebensm. u Forsch. 1984,
    179, 450.
83. Gramatica, P.; Manitto, P.; Ranzi, M.; Delbianco, A.;
    Francavilla, M. Experientia 1982, 38, 775.
84. Simpson, R.F. Vitis 1978, 17, 274.
85. Simpson, R.F. Vitis 1979, 18, 148.
86. Simpson, R.F.; Miller, G.C. Vitis 1983, 22, 51.
87. Di Stefano, R.; Castino, M. Riv. Vitic. Enol. 1983, 36, 245.
88. Rapp, A.; Güntert, M.; Ullemeyer, H. Z. Lebensm. u Forsch.
    1985, 180, 109.
89. Di Stefano, R. Riv. Vitic. Enol. 1985, 38, 228.
90. Williams, P.J.; Strauss, C.R.; Wilson, B. Aust. Grapegrower
    Winemaker 1980, 202, 10.
91. Etievant, P.X.; Issanchou, S.N.; Bayonove, C.L. J. Sci. Food
    Agric. 1983, 34, 497.
92. Smart, R.E. In "University of California, Davis Grape and Wine
    Centennial Symposium Proceedings"; Webb, A.D., Ed.; University
    of California: Davis, 1982: pp. 362-375.
93. Carbonneau, A.; Huglin, P. In "University of California, Davis
    Grape and Wine Centennial Symposium Proceedings"; Webb, A.D.,
    Ed.; University of California: Davis, 1982: pp. 376-385.
94. Croteau, R. In "Isopentenoids in Plants, Biochemistry and
    Function"; Nes, W.D.; Fuller, G.; Tsai, L.S., Eds.; Marcel
    Dekker: New York, 1984: pp 31-64.
95. De Smedt, P.; Liddle, P.A.P. Ann. Fals. Exp. Chim. 1976, 69,
    865.
96. Dimitriadis, E.; Williams, P.J. Chem. Ind. (London) 1984. 108.
97. Schreier, P.; Drawert, F. Chem. Mikrobiol. Technol. Lebensm.
    1974, 3, 154.
98. Strauss, C.R.; Wilson, B; Williams, P.J. Practical Winery 1985,
    6, (3), 27.

RECEIVED February 3, 1986

Metabolism of Linalool by *Botrytis cinerea*

G. Bock[1], I. Benda[2], and P. Schreier[1]

[1]Lehrstuhl für Lebensmittelchemie, Universität Würzburg, Am Hubland,
 D–8700 Würzburg, West Germany
[2]Bayerische Landesanstalt für Weinbau und Gartenbau, Residenzplatz 3,
 D–8700 Würzburg, West Germany

Biotransformation of linalool was studied in grape must
using three strains of Botrytis cinerea (5901/2; 5901/1;
5899/4). Capillary gas chromatography (HRGC) and coupled capillary gas chromatography-mass spectrometry
(HRGC–MS) revealed predominant conversion (> 90 %) of
linalool to (E)-2,6-dimethyl-2,7-octadiene-1,6-diol.
In minor amounts (< 10 %) the corresponding (Z)-iso-
mer, 2-vinyl-2-methyl-tetrahydrofuran-5-one, the four
isomeric linalool oxides in their furanoid and pyranoid forms, the isomeric acetates of pyranoid linalool
oxides as well as 3,9-epoxy-p-menth-1-ene were identified as linalool metabolization products. Quantitative
variations depending on the B. cinerea strain used were observed.

Alcoholic beverages such as wine and beer are classical examples of
biologically produced foods. In these systems, the various enzymes
of yeasts catalyze the biotransformation of plant constituents. Ultimately, this leads to a complex mixture of flavor compounds, which
in a characteristic quantitative distribution determine the quality
of the final product (1). In winemaking, additionally, there are
cases where the influence of a fungus has to be considered, i.e.
Botrytis cinerea (2). The infection of unripe grapes with this
fungus is very feared since the grapes become moldy ("grey rot").
With fully ripe grapes, however, the growth of B. cinerea is desirable; at this point, B. cinerea is called "noble rot" as grapes infected in this stage deliver the famous sweet wines, such as
Sauternes of France, Tokay Aszu of Hungary, or Trockenbeerenauslese
wines of Germany.
    The high metabolic activity of B. cinerea, in particular reac-
tions involving oxidations, is well-known. Thus, the formation of
glycerol and gluconic acid as well as citric acid in grape musts
infected by B. cinerea has been observed (2-4). As to the volatiles of grapes and wines, several key components were identified by
our group (1) consisting of terpenes and derived terpenoids. Since
these components have been found to be effected by B. cinerea
(5-7), we studied the metabolization of terpene alcohols by this

0097–6156/86/0317–0243$06.00/0

fungus. In this paper, the results of biotransformation of linalool by B. cinerea are presented.

Experimental

In Figure 1, a scheme of the different steps of our investigations is outlined. Four separate experiments were conducted on grape musts containing 8.5 g/l acid (pH 3.5) and 200 g/l sugar.

Untreated control. (Experiment 1). We started the studies with the analysis of volatiles from grape must in order to perform a control assay (Figure 1-1). This analysis was carried out by capillary gas chromatography (HRGC) and combined capillary gas chromatography-mass spectrometry (HRGC-MS). Samples were prepared by extractive separation and enrichment of the volatiles. These samples were further separated into three fractions using liquid chromatography on silica gel (8). Analysis of acids was performed by bicarbonate extraction followed by diazotization.

B. cinerea metabolic studies. (Experiment 2). The volatile compounds formed by three different strains of B. cinerea (5901/2; 5909/1; 5899/4) (cf. Figure 1-2) were studied (strains obtained from Bayerische Landesanstalt für Weinbau und Gartenbau, Würzburg, culture collection).

B. cinerea plus linalool studies. (Experiment 3). In a third series of experiments linalool (50 mg/l) was added to the botrytized musts. The volatiles formed from linalool after incubation for two weeks (25°C) with the three strains of B. cinerea were analyzed by the above-mentioned techniques (Figure 1-3).

Linalool control experiment. (Experiment 4). All grape musts were adjusted to pH 3.5 and at this low pH the degradation of linalool must be considered (9). Thus a control experiment was performed by adding 50 mg/l linalool to the must (Figure 1-4).

Results and Discussion

In experiment 4, linalool (1) underwent a variety of well-known chemical reactions (hydrolysis, deprotonation, hydration, cyclization) leading to a series of hydrocarbons (2)-(14) as well as α-terpineol (15), 3,7-dimethyl-1-octene-3,7-diol (16), 1,8-cineole (17) and 2,6,6-trimethyl-2-vinyl-tetrahydropyran (18) shown in Figure 2.

In experiment 2, higher alcohols originating from amino acid metabolism, such as 2-methyl-1-propanol, 3-methyl-1-butanol and 2-phenylethanol were found as metabolization products of B. cinerea. Figure 3 shows distinct quantitative differences depending on the strain used. Control experiments demonstrated that these alcohols were exclusively formed by B. cinerea and not by contamination by yeasts. This fact should be stressed since contradictory results have been published (2,3).

In experiment 3 (cf. Figure 1-3), i.e. after addition of linalool to the botrytized must, a series of transformation products were identified by HRGC and HRGC-MS, which are outlined in Figure 4.

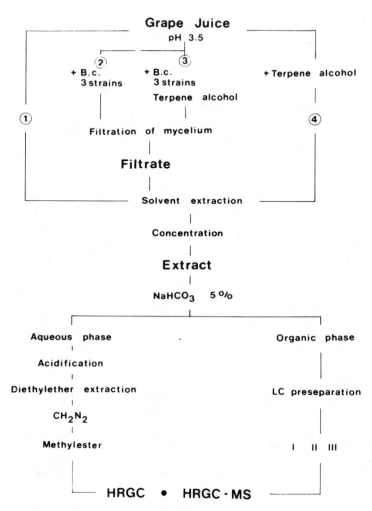

Figure 1. Scheme of sample preparation steps. 1) grape juice volatiles; 2) botrytized (3 strains, 5901/2; 5909/1; 5899/4); 3) addition of linalool to the botrytized musts; 4) addition of linalool to the must.

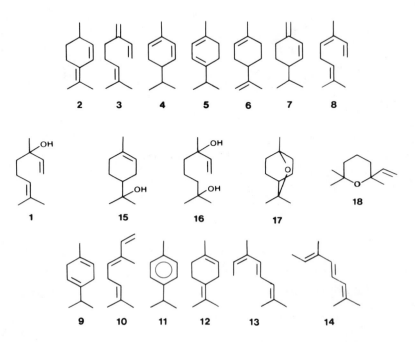

Figure 2. Structures of terpenoids chemically formed from linalool (1) at pH 3.5 (cf. Fig.1-4). (2) 2,4(8)-p-menthadiene; (3) ß-myrcene; (4) α-phellandrene; (5) α-terpinene; (6) limonene; (7) ß-phellandrene; (8) (Z)-ocimene; (9) γ-terpinene; (10) (E)-ocimene; (11) p-cymene; (12) terpinolene; (13) (E,Z)-alloocimene; (14) (E,E)-alloocimene; (15) α-terpineol; (16) 3,7-dimethyl-1-oct-ene-3,7-diol; (17) 1,8-cineole; (18) 2,2,6-trimethyl-2-vinyl-te-trahydropyran.

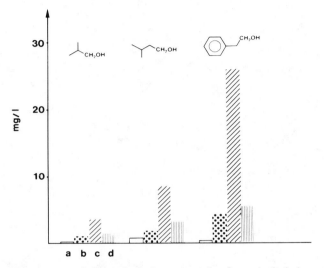

Figure 3. Formation of 2-methyl-1-propanol, 3-methyl-1-butanol and 2-phenylethanol by three strains of <u>Botrytis</u> <u>cinerea</u> (a = grape must; b = 5901/2; c = 5909/1; d = 5899/4).

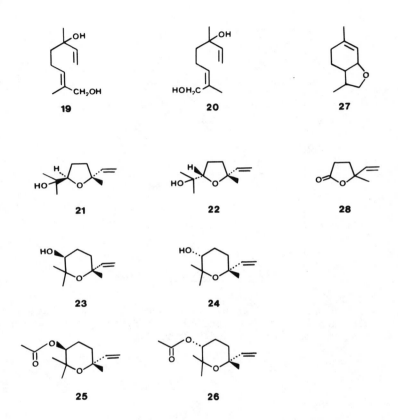

19   **20**   **27**

21   **22**   **28**

23   **24**

25   **26**

Figure 4. Structures of terpenoids formed from linalool (1) by
Botrytis cinerea. (19),(20) (E)- and (Z)-2,6-dimethyl-2,7-octadie-
ne-1,6-diol; (21),(22) (Z)- and (E)-linalool oxides, furanoid;
(23),(24) (Z)- and (E)-linalool oxides, pyranoid; (25),(26) (Z)-
and (E)-linalool oxide acetates, pyranoid; (27) 3,9-epoxy-p-menth-
1-ene; (28) 2-vinyl-2-methyl-tetrahydrofuran-5-one.

These fungal conversion products comprised (E)- (19) and (Z)-2,6-di-
methyl-2,7-octadiene-1,6-diol (20), the furanoid (Z)- (21) and (E)-
linalool oxides (22), the pyranoid (Z)- (23) and (E)-linalool oxides
(24), their isomeric acetates (25) and (26), 3,9-epoxy-p-menth-1-ene
(27) and 2-vinyl-2-methyl-tetrahydrofuran-5-one (28).

    Quantitative HRGC showed that linalool was predominantly (> 90
%) metabolized to (E)-2,6-dimethyl-2,7-octadiene-1,6-diol (19); the
compounds (20)-(28) were only found in minor amounts (< 10 %). In
these experiments, quantitative variations depending on the strain
of B. cinerea used were also observed as shown in Figure 5.

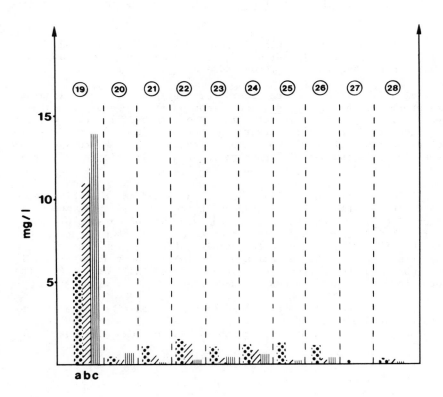

Figure 5. Quantitative distribution of terpenoids (19)-(28) (cf.
Fig.4) formed from linalool (1) by three strains of Botrytis cinerea (a = 5901/2; b = 5909/1; c = 5899/4).

Let us next consider the possible biogenetic pathways of conversion products formed from linalool by B. cinerea. One could assume that 2,6-dimethyl-3,7-octadiene-2,6-diol (29) and 2,6-dimethyl-1,7-octadiene-3,6-diol (30) (Figure 6), both detected among the natural grape must constituents (cf. Figure 1-1), function as intermediates in the formation of 2,6-dimethyl-2,7-octadiene-1,6-diols (19) and (20). However, in control experiments, in which the diols (29) and (30) were added to the botrytized must instead of linalool, no formation of (19) or (20) could be observed. Consequently, for the production of (E)-2,6-dimethyl-2,7-octadiene-1,6-diol (19) a direct enzymic ω-hydroxylation of linalool is likely, as already proposed for analogous reactions of bacterial metabolization of linalool (10-12). This scheme is shown in Figure 6.

While the above-mentioned diols (29) and (30) are odorless, 2,6-dimethyl-2,7-octadiene-1,6-diol (19) is an odoriferous compound, useful in the perfume and flavor industry (13).

Finally, let us consider the by-products of linalool transformation by B. cinerea. The formation of 3,9-epoxy-p-menth-1-ene (27), the character impact compound of fresh dill herb (14), can be understood by allylic rearrangement of (E)-2,6-dimethyl-2,7-octadiene-1,6-diol (19) to the corresponding hydroxygeraniol (-nerol). This compound is known to undergo cyclization to the epoxy derivative (27) under acidic conditions (15).

6,7-Epoxy-linalool (31) has been proposed as the biogenetic precursor of the isomeric hydroxy ethers (21)-(24) as shown in Figure 7 (16). Recently, in our studies of the precursors of papaya fruit volatiles, compound (31) was detected as a natural constituent of this fruit (17). Due to the acidity of the medium (pH 3.5), in the present study, the detection of the labile epoxy derivative (31) was not expected. From the hydroxy ethers (23) and (24), the formation of acetates is easy to understand.

Finally, the lactone (28) is generally described as a formal oxidation product of furanoid linalool oxides (21) and (22), but the biogenetic formation of this compound has not been elucidated yet.

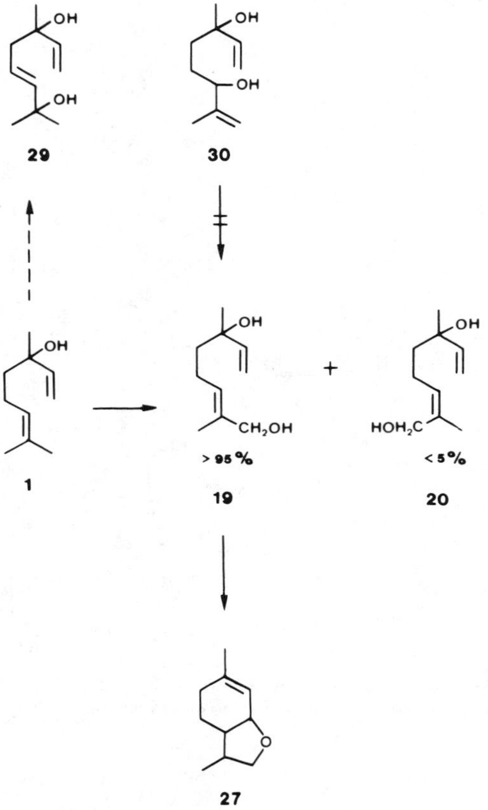

Figure 6. Scheme for biogenetic formation of (19) and (20) by direct ω-hydroxylation of linalool (1) and exclusion of diols (29) and (30) as precursors of (19) and (20). Formation of (27) by allylic rearrangement of (19) and cyclization (15).

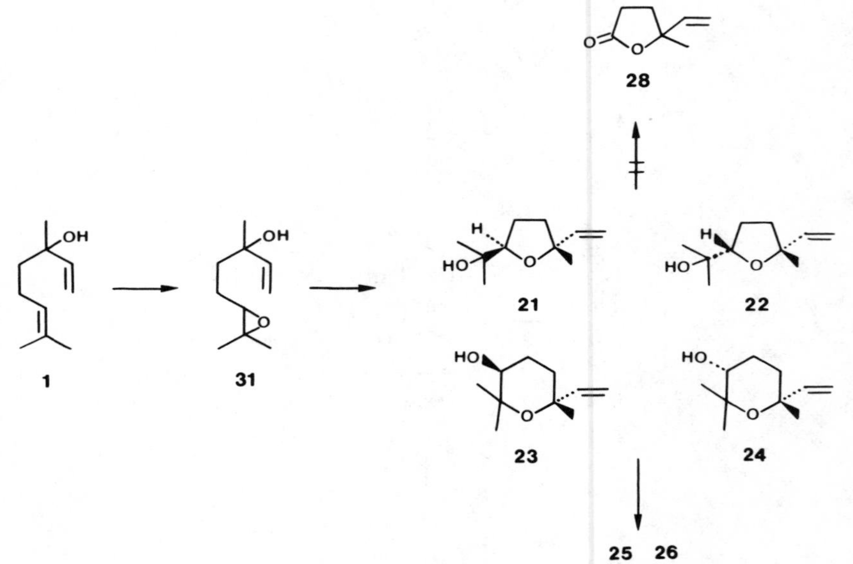

Figure 7. Biogenetic pathways for the formation of isomer linalool oxides (21)-(24) with 6,7-epoxy-linalool (31) as intermediate.

Literature cited

1. Schreier, P. In "Flavour Research of Alcoholic Beverages"; Nykänen, L.; Lehtonen, P., Eds.; Foundation of Biotechnical and Industrial Fermentation Research: Helsinki, 1984, pp. 9–37.
2. Dittrich, H.H. "Mikrobiologie des Weines"; Ulmer: Stuttgart, 1977.
3. Bertrand, A.; Pissard, R.; Sarre, C.; Sapris, J.C. Conn. Vigne Vin 1976, 10, 427–446.
4. Schreier, P.; Drawert, F.; Kerenyi, Z.; Junker, A. Z. Lebensm. Unters. Forsch. 1976, 161, 249–258.
5. Boidron, J.N. Ann. Technol. Agric. 1978, 27, 141–145.
6. Boidron, J.N.; Torres, P. Progres Agric. Vitic. 1978, 95, 612–618.
7. Shimizu, J.; Uehara, M.; Watanabe, M. Agr. Biol. Chem. 1982, 46, 1339–1344.
8. Idstein, H.; Schreier, P. J. Agric. Food Chem. 1985, 33, 138–143.
9. Morin, P.; Richard, H. In "Progress in Flavour Research 1984"; Adda, J., Ed.; Elsevier: Amsterdam, 1985, pp. 563–576.
10. Devi, J.R.; Bhat, S.G.; Bhattacharyya, P.K. Indian J. Biochem. Biophys. 1977, 14, 359–363.
11. Devi, J.R.; Bhat, S.G.; Bhattacharyya, P.K. Indian J. Biochem. Biophys. 1978, 15, 323–327.
12. Madyastha, K.M. Proc. Indian Acad. Sci. 1984, 93, 677–686.
13. Hasegawa, T. Japanese Patent 58,140,032, 1982.
14. Schreier, P.; Drawert, F.; Heindze, I. Lebensm. Wiss. u. Technol. 1981, 14, 150–152.
15. Kitagawa, I.; Tsujii, S.; Nishikawa, F.; Shibuya, H. Chem. Pharm. Bull. 1983, 31, 2639–2651.
16. Ohloff, G.; Flament, I.; Pickenhagen, W. Food Reviews Int. 1985, 1, 99–148.
17. Winterhalter, P.; Katzenberger, D.; Schreier, P. Phytochemistry 1986, in press.

RECEIVED January 3, 1986

20

Selective Production of Ethyl Acetate by *Candida utilis*

David W. Armstrong

Division of Biological Sciences, National Research Council of Canada, Ottawa, Canada, K1A 0R6

Ethyl acetate produced via bioconversion processing
has potential use as a 'natural' flavor and fragrance
compound; for this reason the physiological manipulation of the yeast C. utilis to produce significant
yields of this ester from glucose or ethanol has been
studied. Production of the ester was dependent on the
stage of growth. By use of iron-limited conditions,
under adequate aeration, ethyl acetate accumulated
from ethanol. Studies using specific metabolic
inhibitors implicated acetyl-CoA as a key intermediate
in formation of the ester. Chelators could be used to
increase yield and specific production of ethyl
acetate from glucose and ethanol. EDTA appeared to
function at the level of cell permeability whereas
EGTA or NTA may have encouraged larger acetyl-CoA
pools to accumulate.

Currently the bulk of flavor and fragrance compounds is provided
through traditional methods which include chemical synthesis or
extraction of desired components from natural sources such as
plants. Accordingly, with the great current interest in "natural"
products more pressure has been placed on expensive and labor
intensive extraction processes since the FDA specifies that only
products derived from living sources can be termed "natural".
Bowing to consumer demand for these products, the flavor industry is
beginning to enlist the help of biotechnology to produce natural
flavor and aroma compounds via fermentative routes. At this time,
the production of these fermentation compounds is a largely untapped
area of bioconversion research in which bioesterification has great
potential as esters play a key role in flavors.
     Simple organic acid esters, compounds of an alcohol with a
monobasic acid, are produced in small amounts by some microorganisms
as by-products in their utilization of organic compounds. However,
comparatively few microorganisms are able to form significant
amounts of esters (1). Various yeasts of the genera Saccharomyces
(2) and Hansenula (3) produce ethyl acetate from glucose and/or

ethanol. Saccharomyces sp. produce very low concentrations of ethyl acetate (i.e. levels found in alcoholic beverages) whereas Hansenula anomala produces significant amounts of the ester (4). Hansenula anomala (4) produced maximum amounts of the ester after about 8 d in a nutritionally complex medium while it took up to 50 d to achieve the same levels (2.6 g/L ethyl acetate from 20 g/L glucose) in a simple salts medium. More recently Thomas and Dawson (5), while studying the effect of iron limitation on energy metabolism in the yeast Candida utilis, found significant rates of ethyl acetate formation from glucose in cultures synchronized by repeated dilution with iron-limited minimal medium.

Physiological studies on the formation of esters such as ethyl acetate have been targeted towards the alcoholic beverage industry (2,6) and production of single cell protein (SCP) (7). The studies have looked at regulation of the levels of these esters by environmental factors in order to prevent off-flavors due to their organoleptic properties in beverages or to maximize SCP yields from ethanol. More recently, interest has grown in the biological production of certain esters specifically for use in flavors and fragrances. Esterification of racemic alcohols with fatty acids and other organic acids by Candida cylindracea lipase (8) and in work by Paterson and Bell using Rhizopus arrhizus esterase (9) have indicated potential for microbiological flavor and fragrance processing. The present work examines the physiological control of ethyl acetate production by C. utilis.

Materials and Methods

Culture and Medium Formulation. Candida utilis NRC 2721 (NRRL Y-900; ATCC 9950) was grown at 28°C in the minimal-salts medium of Thomas and Dawson (5). Glucose, as indicated, was added before adjusting the final volume, while 95% ethanol was added post-autoclaving where indicated. The medium was adjusted to pH 5.8 and sterilized at 121°C for 15 minutes. All chemicals were of analytical grade.

Analytical. Viable cell counts: Samples, suitably diluted, were spread on the surface of agar medium (minimal salts medium described above containing 20 g/L glucose and agar at 18 g/L). Colonies were counted after incubation at 28°C for 48 h. Cell mass density (A_{620}): Cell mass density was determined by absorbance at 620 nm in cylindrical cuvettes having a light path of 1 cm. Dry cell weight: DCW was determined gravimetrically. Analysis of fermentation products: Products were identified by gas chromatography/mass spectrometry with routine analysis by gas chromatography. Glucose analysis: The dinitrosalicylate method of Miller (10) was used to measure glucose content of the medium. Determination of iron concentration of medium: The α,α-dipyridyl method of Herbert et al. (11) was used.

Ethanol Uptake Studies. Uptake of ethanol was studied using [1-^{14}C] ethanol suitably diluted in deionized distilled water to make a stock solution of 100 µCi/mL (= stock solution A). Where indicated 50 µL of stock solution A and 450 mL of 10^{-3} M non-radioactive ethanol were mixed (= stock solution B). Cells isolated at

different points in the growth curve (A_{620} readings of 0.35, 0.61
and 1.05, Figure 3), were collected on membrane filters (0.45 μ) and
washed three times with carbon-free medium (10 mL volumes) with
final cell suspension (A_{620} = 0.4) in medium to which glucose was or
was not added (10 g/L final concentration). Aliquots (450 μL) of
the latter suspensions were placed in 100 mL Pyrex screw-cap tubes
to allow for adequate aeration. Stock ethanol solution B (50 μL)
was then added (1 μCi/mL). Samples (100 μL) were spotted on Whatman
3 mm discs at 1.5 h and subsequently rinsed with three portions of
150 mL of ice-cold 5% trichloroacetic acid (TCA) and then with three
portions of 100 mL ether/ethanol (1:1) with suction on a large
filter paper supported on a Büchner funnel. Filter discs were air
dried (on aluminum foil) then oven dried at 110°C. Radioactivity
was counted in 0.4% Omnifluor in toluene (5 mL) in a Beckman LS 7000
scintillation counter. Results are averages of duplicate samples.

Fermentation Studies. Butyl rubber-stoppered vials (Wheaton 160 mL)
were used. The ratio of headspace to culture volume (H/C) indicated
the degree of aeration; an H/C of ca. 4 represented a 'high' level
of aeration while an H/C of ca. 1.5 represented a 'low' level of
aeration.

Results and Discussion

Kinetics of Ethyl Acetate Accumulation and Effect of Iron. Candida
utilis accumulated significant levels of ethyl acetate when grown on
glucose in medium limited for iron [Figure 1] whereas the addition
of iron (FeCl₃ at 100 μM) severely inhibited this capability (12).
In both cultures, grown under adequate aeration (H/C = 1.5), ethanol
and cell mass accumulated until glucose was depleted and the yield
of ethanol was similar (about 90% of theoretical yield). In the
culture without iron supplementation, the cells began to utilize
ethanol after a lag period and accumulated acetic acid and cell
mass. Following this, ethyl acetate accumulation began and continued while acetic acid declined. It was demonstrated that both
acetic acid and ethyl acetate resulted from ethanol utilization.
Ethanol is also known to be an intermediate for ethyl acetate
synthesis by H. anomala (3,4). Apart from a severe inhibition of
ethyl acetate accumulation, the addition of iron (100 μM) to C.
utilis cells resulted in a significant prolongation of the lag
before the start of ethanol utilization. A more rapid utilization
of ethanol has been also observed previously in cultures of C.
utilis (13) and S. cerevisiae (14) grown under conditions of iron
limitation.
      In order to study the effect of iron on the utilization of
ethanol more closely, FeCl₃ was added at various levels to glucose-
containing medium. The presence of iron did not affect the levels
to which ethanol accumulated from glucose (24 h) [Figure 2]; however
the subsequent utilization of ethanol was inhibited in proportion to
the level of iron present. It is possible that iron may interfere
with some early stage of ethanol utilization thereby leading to
delayed ethyl acetate accumulation.

Effect of Glucose on Ethanol Uptake by C. utilis. Glucose can
repress metabolism of other carbon sources by yeasts (15). In

Escherichia coli the primary effect of glucose is the inhibition of transport of other carbon sources (16). As indicated above [Figure 1] sequential production of various products (ethanol, acetic acid and ethyl acetate) from glucose could be seen under an appropriate degree of aeration. Also cell mass accumulation (A_{620}) showed a diauxic increase [Figure 1] which reflected utilization of different carbon sources by C. utilis. Ethanol utilization began only after glucose was depleted. This suggests a possibility that glucose inhibits ethanol uptake and thus conversion of ethanol to ethyl acetate. It is of interest, especially from a biotechnological standpoint, to know whether ethyl acetate production is possible in the presence of glucose. Therefore, the effect of glucose on the uptake of ethanol by C. utilis was studied [Figure 3].

Candida utilis was isolated at the three different growth phases indicated [Figure 3] and resuspended in fresh medium with or without glucose (10 g/L). Small volume (450 µL) aliquots were placed in tubes to which [^{14}C] ethanol along with non-labelled ethanol carrier was added. The amount of the ethanol taken up by cells without glucose was divided by the uptake by cells with glucose. A ratio greater than 1.0 indicated inhibition of ethanol uptake by glucose. It would appear that the results are consistent with the general phenomenon of catabolite repression (15,16). Thus, without additional environmental or genetic manipulation, production of ethyl acetate may not occur efficiently in the presence of glucose. A production phase for the ester should be separated from the fermentative conversion of glucose to ethanol.

Effect of Inhibition of Acetyl-CoA Formation on Ethyl Acetate Accumulation.

Ethyl acetate has been shown to be formed from acetic acid and ethanol without any cofactors in S. cerevisiae suggesting an esterase mechanism of biosynthesis (17). Other studies (18) found cell-free synthesis of ethyl acetate with ethanol and acetyl-CoA but not with acetic acid as a substrate. From the latter study it was proposed that ester formation in yeasts was primarily via alcoholysis of acyl-CoA compounds. Others have also suggested acetyl-CoA as an intermediate for ethyl acetate accumulation by S. cerevisiae (19).

Yeasts generally catabolize ethanol as follows: ethanol --→ acetaldehyde --→ acetic acid --→ acetyl-CoA --→ other oxidative pathways [Figure 4]. It has been speculated that C. utilis forms ethyl acetate by the reaction of acetyl-CoA with ethanol (5). Thus if the flow of metabolites into the TCA cycle is inhibited or limited, as might occur under iron-limited conditions, acetyl-CoA would be expected to accumulate thus providing a precursor pool for esterification of ethanol.

Acetyl-CoA synthesis is known to be inhibited by arsenite (20). Therefore, to investigate the significance of acetyl-CoA in the formation of ethyl acetate, arsenite was added to ethanol-adapted C. utilis cells suspended in fresh ethanol-containing medium [Figure 5]. Even at low concentrations of arsenite (0.1 and 0.2 mM) ethyl acetate accumulation was severely inhibited. At an elevated level of arsenite (1.0 mM) no ethyl acetate was detected whereas the levels of acetaldehyde and acetic acid increased significantly.

Many metabolic inhibitors exert secondary effects (20).

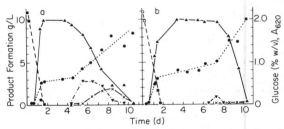

Figure 1. Kinetics of ethyl acetate accumulation by Candida
utilis. Growth on glucose medium without (a) or with iron [100
uM FeCl₃] (b). Glucose (■), ethanol (▲), acetic acid (▼) -
shown 10 X actual level, ethyl acetate (+), A₆₂₀ (●).
(Reproduced with permission from Ref. 12. Copyright 1984, John
Wiley & Sons, Inc.)

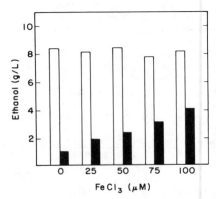

Figure 2. Effect of iron levels on ethanol utilization by
Candida utilis. Ethanol accumulated in cultures from glucose at
24 h (□) and 120 h (■).

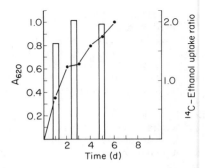

Figure 3. Effect of glucose on ethanol uptake by Candida utilis.
See text for details. A ratio greater than 1.0 indicates
inhibition of ethanol uptake by glucose. Ethanol uptake ratio
(□), A₆₂₀ of original culture (●-●).

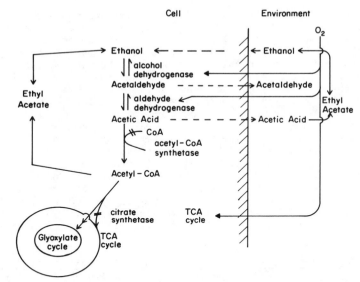

Figure 4.   Pathway of ethanol utilization and ethyl acetate formation (model).

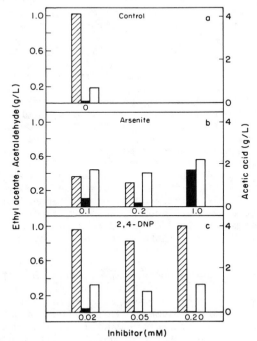

Figure 5.   Effect of selective inhibitors on product accumulation (24 h) by <u>Candida</u> <u>utilis</u>.   Ethyl acetate (▨), acetaldehyde (▮), acetic acid (▢).

Arsenite is known to have some effect on oxidative phosphorylation. However, as can be seen [Figure 5c] the use of an uncoupler of oxidative phosphorylation (2,4-dinitrophenol) did not affect accumulation of ethyl acetate. From this it would appear that the main inhibitory effect of arsenite on ethyl acetate accumulation by C. utilis is at the level of acetyl-CoA formation. Thus acetyl-CoA is implicated as a key precursor for synthesis of ethyl acetate supporting a model presented earlier [Figure 4].

It has been found that in cultures of C. utilis, ethanol concentrations exceeding about 35 g/L causes a progressive shift in product distribution from ethyl acetate to acetaldehyde (21) [Figure 6]. Acetaldehyde is known to inhibit acetyl-CoA synthetase which catalyses the formation of acetyl-CoA from acetic acid (7). The results [Figure 6] suggest that higher ethanol concentrations cause higher levels of acetaldehyde which in turn inhibits acetyl-CoA formation and thus ethyl acetate accumulation is reduced. These latter results along with those involving selective inhibitors implicate acetyl-CoA as a key intermediate in ethyl acetate accumulation in C. utilis.

Ethyl Acetate Accumulation by C. utilis Cells Isolated at Different Phases of Growth. The accumulation of ethyl acetate was previously shown [Figure 1] to be concomitant with the onset of utilization of ethanol. The question to be asked then was whether the accumulation of ethyl acetate is merely dependent upon the availability of the ethanol or is there also a requirement for a modification of the metabolic capacity of the cell? To explore this question, C. utilis was grown under a low degree of aeration (H/C = 1.5) in order to delineate distinct phases of diauxie [Figure 7]. Aliquots of the culture were isolated at different points (indicated by arrows), cells were collected and resuspended in fresh ethanol-containing medium under a high level of aeration (H/C = 4). The capacity of these cells to accumulate ethyl acetate was shown to depend upon the phase of growth during which they were isolated. The results of Table I indicate that cells isolated from the second rise in the diauxic cell mass density curve (stage III) exhibited the greatest efficiency of ethanol conversion to ethyl acetate. It is postulated that the presence of ethanol alone was not sufficient for efficient conversion to the ester, but the cells needed to undergo some adaptation to allow efficient ester formation. However ethanol utilization by C. utilis isolated at the different points were similar. Thus the main difference between the cells isolated at the different stages may be in TCA cycle activity leading to different levels of acetyl-CoA pools rather than the induction of enzymes for ester synthesis.

Effect of Chelating Agents on Product Distribution. In earlier studies, it was shown that the addition of EDTA to a culture of C. utilis, growing on glucose, encouraged a rapid production of ethyl acetate (22). The ester accumulated even though glucose was present. In addition, when EDTA was added to a glucose medium supplemented with a high level of iron, C. utilis produced amounts of the ester comparable to iron-limited cultures. Thus use of EDTA could be beneficial in the rapid fermentation of glucose to ethyl acetate even in the presence of relatively high levels of iron. The

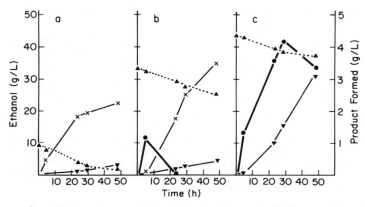

Figure 6. Change in product distribution at different levels of ethanol by <u>Candida</u> <u>utilis</u>. Ethanol (▲), ethyl acetate (x), acetaldehyde (●), acetic acid (▼). (Reproduced with permission from Ref. 22. Copyright 1984, Science and Technology Letters.)

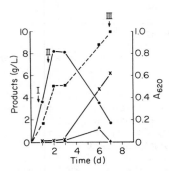

Figure 7. Isolation of <u>Candida</u> <u>utilis</u> at different points of diauxie. See Table 1, footnote 1. Ethanol (●), ethyl acetate (x), acetic acid (▲), A_{620} (■).

Table I. Ethanol Utilization and Ethyl Acetate Accumulation by
          C. utilis Isolated at Different Phases[1]

Time (min)	Percent of Original Ethanol Remaining: Phase			Relative Ethyl Acetate Accumulated per A_{620}: Phase		
	I	II	III	I	II	III
0	100	100	100	0	0	0
30	94	95	92	23	0	31
60	91	91	89	23	23	62
120	78	78	75	31	46	100[2]

[1] Cells isolated at phases indicated (I, II, III) and resuspended in
ethanol (10 g/L) medium. Ethanol utilization and ethyl acetate
accumulation measured at times indicated.

[2] 100 is equal to 0.11 g/L ethyl acetate at A_{620} = 1.0.

following speculations on the effect of EDTA are therefore possible.
In the absence of EDTA, ethanol is readily excreted to an extracellular milieu. When glucose is depleted, uptake of extracellular
ethanol begins and any ethyl acetate intracellularly formed is
excreted to the medium. When ethanol is depleted, utilization of
the extracellular ester is induced. EDTA is known to affect
membrane permeability in various microorganisms. It is therefore
postulated that EDTA removes these membrane regulatory mechanisms by
increasing membrane permeability. Since the ethanol produced
permeates through the membrane into the cell, it is immediately
converted to ethyl acetate.
    In other studies, cells of C. utilis were placed in a medium
containing ethanol as the carbon source. Different chelating agents
(EDTA, NTA or EGTA) were studied for their effects on ethanol
utilization and product accumulation [Table II]. It is interesting
to note [Table II] that the effect of EDTA on product distribution
from this yeast (in the presence of 10 g/L ethanol) was markedly
different than described above when glucose was used as the carbon
source. Although it was demonstrated that exogenously added ethanol
at levels exceeding 35 g/L brought about a shift in product distribution from ethyl acetate to acetaldehyde, the use of EDTA
encouraged this same phenomenon at much lower alcohol levels (10
g/L). This supports the speculation that EDTA is affecting a
permeability barrier. In the presence of EDTA, the ethanol may
diffuse more readily into the cell thereby causing an elevated
intracellular concentration. It is possible that the complete
oxidative metabolism of ethanol is rate-limited at some stage
thereby allowing for an elevated intracellular accumulation of
acetaldehyde. As the activity of acetyl-CoA synthetase may be
adversely affected by acetaldehyde, the subsequent formation of
ethyl acetate would be limited, leading to increased acetaldehyde
accumulation. However, in the presence of glucose, the ethanol
formed via fermentative routes could eventually be oxidized to ethyl
acetate at a rate matching the formation of alcohol from glucose
with the net intracellular levels of ethanol and, therefore,
acetaldehyde never exceeding an inhibitory level.

Table II.   Effect of different chelating compounds on product formation

Chelator added	Ethanol utilized (g/L.d.)	% of theor.[1] yield ethyl acetate	% of theor.[2] yield acetaldehyde	Acetaldehyde formed/A_{620}.d (g/L)/A_{620}.d	Ethyl acetate formed/A_{620}.d (g/L)/A_{620}.d
0	4.3	28	0	0	2.4
NTA	7.9	58	0	0	3.1
EGTA	7.8	71	0	0	4.7
EDTA	1.8	12	87	17	2.4

[1] Theoretical yield (after 24 h incubation) of ethyl acetate based on:
$2C_2H_5OH + O_2 \rightarrow CH_3CO_2C_2H_5 + 2H_2O$.

[2] Theoretical yield (after 24 h incubation) of acetaldehyde based on the conversion of 1 mol ethanol to 1 mol acetaldehyde.

The use of other chelating agents such as NTA or EGTA resulted
in markedly different results than those with EDTA. The addition of
either NTA or EGTA brought about a significant improvement in
ethanol utilization and yield of ethyl acetate compared with
untreated cultures of C. utilis and those with added EDTA. EGTA
allowed for the best overall yield of ethyl acetate and specific
productivity. It is speculated, based on these and other results on
the effect of EDTA (22), that the main action of EGTA and NTA is not
at the level of cell permeabilization since acetaldehyde accumulation does not occur as found with EDTA. An interference with the
utilization and resultant elevation of the acetyl-CoA pool levels
due to inhibition of other biosynthetic, acetyl-CoA consuming,
pathways could explain the present observations.

Conclusions

The selective production of ethyl acetate by C. utilis provides
excellent examples of environmental manipulation of microbial
metabolism. The results have demonstrated that growth under iron
limitation, leading to reduced oxidative metabolism is necessary for
development of the capacity to accumulate significant levels of
ethyl acetate.

Acknowledgments

The author thanks Dr. I.J. McDonald for his constructive comments
and Ms. W. McLorie for skillful technical assistance. The typing/
formatting of the manuscript by Ms. C. Gobey was greatly
appreciated.

Literature Cited

1. Datta, R. Biotechnol. Bioeng. Symp. 1981, 11, 521.
2. Yoshioka, K.; Hashimoto, N. Agric. Biol. Chem. 1981, 45,
    2183.
3. Tabachnick, J.; Joslyn, M. A. J. Bacteriol. 1953, 65, 1.
4. Peel, J. L. Biochem. J. 1951, 49, 62.
5. Thomas, K. C.; Dawson, P. S. S. Can. J. Microbiol. 1978, 24,
    440.
6. Nördstrom, K. J. Inst. Brew. 1962, 68, 188.
7. Prokop, A.; Votruba, J.; Sobotka, M.; Panos, J. Biotechnol.
    Bioeng. 1978, 20, 1523.
8. Yamaguchi, Y.; Komatsu, A.; Moroe, T. J. Agric. Chem. Soc.
    Japan 1977, 51, 61.
9. Anonymous Biotechnology News 1983, 3, 5.
10. Miller, G. L. Anal. Chem. 1959, 31, 426.
11. Herbert, D.; Phipps, P. J.; Strange, R. E. In "Methods in
    Microbiology"; Norris, J. R.; Ribbons, D. W., Eds.; Academic
    Press: New York, 1971; Vol. 5B, pp. 239-242.
12. Armstrong, D. W.; Martin, S. M.; Yamazaki, H. Biotechnol.
    Bioeng. 1984, 26, 1038.
13. Nördstrom, K. Svensk Kemisk Tidskrift 1964, 76, 510.
14. Nagamune, T.; Endo, I.; Ionoue, I. Adv. Biotechnol. 1981, 2,
    219.

15.   Rawn, J. D.   In "Biochemistry"; Harper & Row Publishers: New York, 1983; p. 1092.
16.   Fraser, A. D. E.; Yamazaki, H.   Curr. Microbiol. 1982, 7, 241.
17.   Schermers, F. H.; Duffus, J. H.; MacLeod, A. M.   J. Inst. Brewing 1976, 82, 170.
18.   Howard, D.; Anderson, R. G.   J. Inst. Brew. 1976, 82, 70.
19.   Nördstrom, K.   J. Inst. Brew. 1963, 69, 142.
20.   Hochster, R. M.; Quastel, J. H.   "Metabolic Inhibitors: A Comprehensive Treatise"; Academic Press: New York, 1963; Vol. 1.
21.   Armstrong, D. W.; Martin, S. M.; Yamazaki, H.   Biotechnol. Lett. 1984, 6, 183.
22.   Armstrong, D. W.; Yamazaki, H.   Biotechnol. Lett. 1984, 6, 819.

RECEIVED January 27, 1986

21

Synthesis of 2-Methoxy-3-alkylpyrazines by *Pseudomonas perolens*

R. C. McIver[1] and G. A. Reineccius

Department of Food Science and Nutrition, University of Minnesota, St. Paul, MN 55108

The influence of media components on the production of 2-methoxy-3-isopropyl and 2-methoxy-3-secbutyl pyrazines by *Pseudomonas perolens* (ATCC 10757) and selected mutants of this culture was studied. Pyrazine production was observed only during the stationary growth phase. The parent culture produced a maximum of 42 ng/mL pyrazines while selected mutants were found to produce a maximum of 15,760 ng/mL. The parent and mutant strains were found to exhibit similar responses to nutrient sources. Maximum pyrazine production was observed using pyruvate (1%), lactate (1%) or nutrient broth as carbon source. Nitrogen source had no influence on pyrazine production. Maximum pyrazine formation was observed when phosphate level ranged from 0.4 mM to 1.2 mM in the media.

There are many reports of *Pseudomonas* cultures producing musty, earthy, and potato-like odors (1-6). The work of Morgan *et al*. (7) established 2-methoxy-3-isopropyl pyrazine to be partially responsible for these odors. Subsequently, 2-methyoxy-3-isopropyl pyrazine was found in bell peppers (8), a similar compound 2-methoxy-3-secbutyl pyrazine was identified in galbanum oil (9), and several 2-methoxy-3-alkyl pyrazines were identified in various raw botanicals (10). The odor threshold exhibited by 2-methoxy-3-isobutyl pyrazine (1 part in 10^{12}, 8) indicates flavor significance for these compounds even at the exceptionally low concentrations in which they occur in foods and other natural products. Producing these compounds from microbial fermentations could be an economical source of flavor for the food industry.

The work of MacDonald (11, 12) on the synthesis of aspergillic acid from valine and isoleucine suggests that 2-methoxy-3-alkyl pyrazine, as one of a group of substituted pyrazines and diketopi-

[1]Current address: Sunkist Growers Inc., 760 Sunkist St., Ontario, CA 91761

0097-6156/86/0317-0266$06.00/0
© 1986 American Chemical Society

pyrazines, is formed metabolically by microorganisms. Furthermore, the odor intensity and character of *Pseudomonas perolens* cultures has been shown to be influenced by media composition (4, 5). Because the potential for producing "natural" 2-methoxy-3-alkyl pyrazines via fermentation might be improved by optimizing media composition, we studied the influence of media components on the synthesis 2-methoxy-3-isopropyl pyrazine by cultures of *Pseudomonas perolens* and selected mutant strains.

Materials and Methods

Microbial. *Pseudomonas perolens* subsp gdansk ATCC 10757 was obtained from the American Type Culture Collection (Rockville, MD) and revived according to the procedure recommended by the supplier. The parent strain and subsequently isolated mutants (obtained by treatment with N-methyl-N'-nitro-N-nitrosoguanidine, 13) were maintained by combining equal volumes of an early logarithmic phase nutrient broth culture with an 80% glycerol solution in small vials and storing at -20°C. Cultures were routinely grown in minimal salt media containing per liter: $Na_2HPO_4 \cdot 7H_2O$ 8.2g, KH_2PO_4 2.7g, NH_4Cl 1.0g, $FeSO_4 \cdot 7H_2O$ 2.0mg, $MgSO_4 \cdot 7H_2O$ 0.1g, $Ca(NO_3)_2$ 5mg at pH 7.2. The carbon sources were filter sterilized and added immediately before use. For investigations of different media components, a culture from frozen stock was inoculated into an 0.5% glucose-minimal salts culture and grown overnight. The cells were centrifuged and washed twice with minimal media (no carbon or nitrogen source) then inoculated into fresh media at a level of approximately 10^6 cfu/mL. All cultures were grown at ambient temperature (22°C) with rotary shaking at 180 rpm. At appropriate intervals, samples were taken for cell counts, pH and pyrazine analysis.

Pyrazine Quantitation: Culture samples were analyzed for pyrazine content using a commercial dynamic headspace apparatus (Hewlett Packard, Purge and Trap Headspace Concentrator, model 7675A). This apparatus used Tenax as the adsorbent and desorbed via heating (180°C). Purging of the cultures was done with helium (100 mL/min) for 10 min at ambient temperature. Analysis was performed on a Hewlett Packard model 5880 gas chromatograph using a nitrogen specific detector (NPD). A 15 m x 0.32 mm DB-5 (0.25 μm film thickness) column was used for separation (J & W Scientific, Rancho Cordova, CA). Isobutyl pyrazine was used as the internal standard. Identity of the pyrazines from microbial cultures was confirmed by GC-MS using authentic reference compounds.

Results and Discussion

Effect of Carbon Source. The growth of *Pseudomonas perolens* as well as culture levels of 2-methoxy-3-isopropyl pyrazine in minimal salts plus 1% pyruvate are shown in Figure 1. High cell density was obtained but the presence of 2-methoxy-3-isopropyl pyrazine was not detected until after cell numbers were no longer increasing. At this point, the amount of pyrazine increased rapidly reaching a maximum on the fifth day. The growth and pyrazine synthesis with other carbon sources showed similar behavior. The results are presented in Table I, and plotted for selected carbon sources at 1%

level in Figure 2.   No growth was observed when sucrose or lactose
was the sole carbon source, nor when any carbon source was present
at 5% or greater concentration.

Table I.   The influence of carbon source on the production of
2-methyl-3-isopropyl pyrazine by P.*perolens*.

Carbon Source	% (w/v)	2-Methoxy 3-isoPropyl Pyrazine (ng/mL)	% vs. 1% Lactate
Glucose	0.2	36	75
	0.5	25	52
	1.0	18	38
	2.0	9	19
Fructose	0.2	25	52
	0.5	19	40
	1.0	20	42
	2.0	10	21
Pyruvate	0.2	14	29
	0.5	14	29
	1.0	42	88
	2.0	31	65
Lactate	0.2	15	31
	0.5	14	29
	1.0	48	(100)
	2.0	37	77
Na Acetate	1.0	20	--
Nutrient broth	--	46	96

     With the exception of sodium acetate containing cultures,
viable cell counts were similar in the range of 10^9-10^{10} cfu/mL.   In
all cultures where growth was observed, a strong musty-green potato
odor, characteristic of 2-methoxy-3-isopropyl pyrazine was present.
Although 2-methoxy-3-secbutyl pyrazine was not detected by GC, a
definite green-pea pod like odor was noted at the retention time
corresponding to this compound when the effluent from the gc column
was sniffed, suggesting culture concentrations below the limit of
detection.   The reported threshold of this pyrazine ([14]) is exceptionally low and apparently below the detection limit of the GC.
     As can be seen from Table I, pyruvate and lactate at 1%
resulted in the highest yield of 2-methoxy-3-isopropyl pyrazine.   At
the 2% level these substrates resulted in a decrease of approximately 30% in the amount of pyrazine with a much greater drop at 0.2%
and 0.5% levels.   This is in contrast to glucose and fructose as
carbon source where the amount of pyrazine synthesized was inversely
related to the initial concentration of sugar in the media.   Sodium
acetate yielded about 20 ng/mL which is impressive considering that
the cell numbers were 3-4 orders of magnitude lower than that found

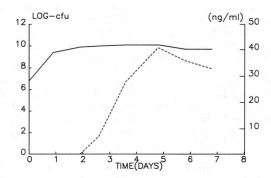

Figure 1. Relationship between the growth of *Pseudomonas perolens* (———) and the production of 2-methoxy-3-isopropyl pyrazine (----).

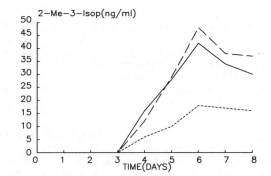

Figure 2. The influence of carbon source on the production of 2-methoxy-3-isopropyl pyrazine (ng/mL). Pyruvate (———), lactate (— ——), Glucose (----).

for the 3- and 6-carbon substrates. The complex growth media supported dense growth and substantial amounts of pyrazine synthesis.

Effect of Nitrogen source. Nitrogen source and concentration exerted a minor influence on the amount of 2-methoxy-3-isopropyl pyrazine synthesized in cultures. It has been reported by a number of investigators that ammonium salts as the source of nitrogen results in good growth but show negative effects on metabolite production (15-19). The influence of increasing concentrations of ammonium chloride as the sole nitrogen source and the effects of differing concentrations of other sources of nitrogen on pyrazine production is shown in Table II. In all cases, the growth curves and pH profiles were similar. Also, appearance of 2-methoxy-3-isopropyl pyrazine coincided with the cessation of the

Table II. The influence of nitrogen source and concentration on the production of 2-methoxy-3-isopropyl pyrazine by *P. perolens*.

| | 2-Methoxy-3-isopropyl pyrazine (ng/mL) | | |
| | N concentration mM | | |
N source	15	50	100
NH$_4$CL	30	25	15
L-valine	31	31	29
L-proline	35	33	36
L-α-amino-n-butyric acid	27	29	30

growth phase. The only instance where the results differed significantly was when ammonium chloride was present at an initial concentration of 100 mM. The fact that L-proline, L-valine and L-α-amino n-butyric acid gave similar results suggests that exogenous addition of valine, the probable precursor of 2-methoxy-3-isopropyl pyrazine, does not increase the amount of the pyrazine made by the cell. Similar results have been reported by Morgan (20).

Effect of Phosphate. In contrast to the limited effects of carbon and nitrogen source and concentration, the amount of 2-methoxy-3-isopropyl pyrazine synthesized by *Pseudomonas perolens* was found to be sensitive to the level of phosphate in the media (Table III). The usual media contained 50 mM phosphate as both a nutrient and to provide buffering capacity. In order to buffer the media at lowered amounts of phosphate, 0.2 M 3-(N-morpholino)-propanesulfonic acid (MOPS) was added (21). Although clear differences were observed in the amounts of 2-methoxy-3-isopropyl pyrazine produced, the data were more striking when population density was considered. When compared on the basis of pyrazine synthesized per 10^{10} cfu/mL, the culture grown in media containing 0.2 mM phosphate produced almost 50 fold more pyrazine than the one grown in 50 mM phosphate.

The important role of phosphate in the control of metabolite synthesis has been recognized for a number of years as reviewed by Weinberg (22). As with the cases of carbon and nitrogen sources, sensitivity to control by these substrates depends on the organism and the metabolite. Recently, Young et al. (19) overcame the

Table III.   The influence of phosphate concentration on the production of 2-methoxy-3-isopropyl pyrazine by *P. perolens*.[1]

initial PO_4 (mM)	2-Methoxy-3-isopropyl pyrazine (ng/mL)	viable cells (cfu/mL)
0.2	74	4×10^8
0.4	130	1×10^9
1.2	133	4×10^9
5.0	78	9×10^9
20.0	46	8×10^9
50.0	37	1×10^{10}
100.0	11	9×10^9

[1]grown in minimal salts + 0.2 M MOPS + 1% pyruvate

inihibitory effects of both phosphate and ammonia on lincomycin synthesis by *Streptomyces lincolnensis*. They added substantial amounts of $MgSO_4$ to the phosphate containing media which resulted in the precipitation of $MgNH_4PO_4$. As ammonia and phosphate were removed from the media by the cells, more of the precipitate would dissolve. The effect was a slow release of the two nutrients. They observed an increase in both the rate of lincomycin synthesis and titer.

Investigation of Mutant Strains.   Murray (14) suggested that the amino acids valine, leucine and isoleucine were precursors of 2-methoxy-3-isopropyl, 3-isobutyl and 3-secbutyl pyrazines, respectively.   The effects of mutations in the branched chain amino acid pathways on the synthesis of these compounds was investigated. During the process of isolating mutants, we noticed that one culture produced a very intense musty-green odor.   Upon isolation and culturing of the strain responsible, it was found to have 2-methoxy-3-isopropyl- and 2-methoxy-3-secbutyl pyrazines at levels of 12,500 and 150 ng/mL, respectively, when grown in nutrient broth culture.   The bacterium isolated was not found to have different growth requirements from the parent strain.

Table IV.   The production of 2-methoxy-3-alkyl pyrazines by a mutant of *P. perolens*.

Carbon Source	Conc. (% w/v)	pyrazine (ng/mL) 2-MeO-3-isoPr	2-MeO-3-secB	ratio isoPr/secB
Glucose	0.5	2030	21	97
	1.0	960	11	87
	2.0	840	9	92
Lactate	0.5	6280	68	92
	1.0	9890	112	88
	2.0	4730	48	99
Pyruvate	0.5	5150	55	94
	1.0	11120	100	111
	2.0	8400	95	88
Glycerol	0.5	3610	86	42
	1.0	3170	69	46
Nutrient Broth	--	12500	145	86

Investigations with different carbon sources (Table IV) showed that although the amounts of the two pyrazines varied according to the carbon source used, the ratio of the isopropyl to secbutyl pyrazine remained relatively constant. This suggested that either availability of precursor followed a set ratio, or else enzyme specificity for the different side chains played a role in determining the relative amounts of each compound synthesized. Accordingly, strains auxotrophic for leucine (leu⁻), isoleucine (ilu⁻), or leucine, valine, isoleucine, and pantothenic acid (leu⁻, val⁻, ilu⁻, pant⁻) were isolated from cultures treated with N-methyl-N'-nitro-N-nitrosoguanidine. Also, a strain that grew well on the leucine analog, 4-azaleucine was isolated.

Metabolic Implications of Auxotrophs. The levels and ratios of the 2-methoxy-3-alkyl pyrazines synthesized by these various mutants are contained in Table V. For the auxotrophs, the required nutrients

Table V.   The production of 2-methoxy-3-alkyl pyrazines by selected mutant strains of *P. perolens*.

Strain	C-source (1%)	pyrazine (ng/mL) 2-MeO-3isoPr	2-MeO-3secB	Ratio isoPr/secB
	glycerol	3170	69	46
	pyruvate	11120	100	111
ilu⁻	pyruvate	12020	<10	>1200
leu⁻	pyruvate	15760	95	165
aza⁻	pyruvate	5180	110	47
val⁻,ilu⁻,leu⁻	pyruvate	<20	<10	--
	pyruvate + t-leu (200 µg/mL)	11330	46	246

were supplied at 20 µg/mL. The ilu⁻ strain had no detectable amounts of 2-methoxy-3-secbutyl pyrazine and the multiple auxotroph possessed neither pyrazine. These results suggest that these amino acids may indeed serve as precursors of the pyrazines in question. The pyrazine that would be derived in part from leucine, 2-methoxy-3-isobutyl pyrazine, was never detected, even when a culture of the multiple auxotroph was incubated with 2 mg/mL leucine added to the media. A parallel culture with valine at 2 mg/mL synthesized 2-methoxy-3-isopropyl pyrazine to a level of 900 ng/mL indicating the ability to synthesize the pyrazines was not lost. This suggests that a ß-methyl group on the amino acid side chain may be required to form the substituted pyrazines.

Even though leucine did not appear to be incorporated into a pyrazine molecule, the diversion of α-keto isovalerate to form leucine decreases the availability of this compound to form valine

and hence 2-methoxy-3-isopropyl pyrazine. Cultures unable to form
leucine (leu⁻) showed increased amounts of 2-methoxy-3-isopropyl
pyrazine and 4-azaleucine resistant cultures less. Jensen (23)
observed that mutants of *Bacillus subtillis* resistant to 4-azaleucine
were leucine excretors. Tert-leucine appeared to act as a negative
effector for the synthesis of 2-methoxy-3-secbutyl pyrazine.
       The mutant which was blocked in the synthesis of branched chain
amino acids produced very low levels of methoxy pyrazines. Cultures
of this mutant did generate a new N peak and produced a strong
butter-like aroma. Two compounds were identified in these cultures
as 2,3,5,6-tetramethyl pyrazine and diacetyl. The synthesis of
tetramethylpyrazine by a *Corynebacterium glutamicum* that was also
metabolically blocked in the branched chain amino acid pathway has
previously been reported (24).
       Further investigation found that synthesis of tetramethyl pyra-
zine was not limited to the *Pseudomonas perolens* with the multiple
auxotrophy. Any one of the high producing strains that was incubated with lactate or pyruvate at 2% concentration was capable of
producing tetramethyl pyrazine. This compound appeared after the
methoxy pyrazine levels had stopped increasing (Fig 3). This would
suggest that a precursor for the methoxy pyrazines must first be
depleted before the excess pyruvate would be directed to tetramethyl
pyrazine synthesis.
       Although this work has shown how various media constituents and
pathway manipulations can influence the synthesis of substituted
pyrazines by *Pseudomonas perolens*, elucidation of the individual
biosynthetic steps is needed to understand fully the controls and
explore the possibility that other substituted pyrazines could be
made.

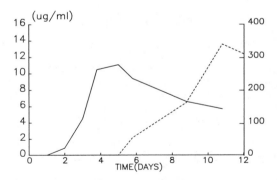

Figure 3.   The relationship between the formation of 2-methoxy-3
  -isopropyl pyrazine (range 0-16 µg/mL, ——) and tetramethyl
  pyrazine (range 0-400 µg/mL, ----).

Acknowledgments

     Published as Paper No. 14642 of the Scientific Journal series
of the Minnesota Experiment Station on research conducted under
Minnesota Agricultural Experiment Station Project No. MN 18-038,
supported by Hatch funds.

Literature Cited

1. Turner, M. D. Aust. J. Expt Biol. and Med. Sci. 1927, 4,
    57-60.
2. Spanswick, M. D. Am. J. Public Health 1930, 20, 73-4.
3. Jensen, L.B. Food Research 1948, 13, 89-93.
4. Castell, C. H.; Greenough, M. F. J. Fish Res. Bd. Canada
    1957, 14, 617-23.
5. Castell, C. H.; Greenough, M. F.; Jenkin, N. L. J. Fish Res.
    Bd. Canada 1957, 14, 775-82.
6. Miller, A III; Scanlan, R. A.; Lee, J. S.; Libbey, L. M.;
    Morgan, M. E. Appl. Microbiol. 1973, 25, 257-61.
7. Morgan, M. E.; Libbey, L. M., Scanlan, R. A. J. Dairy Sci.
    1972, 55, 666.
8. Buttery, R. G.; Seifert, R. M.; Lundin, R. E.; Guadagni, D. G.;
    Ling, L. C. Chem Ind. 1969, 490-91.
9. Bramwell, A. F.; Burrell, J.W.K.; Riezebos, G. Tetrahedron
    Lett 1969, 37, 3215-16.
10. Murray, K. E.; Whitfield, F. B. J. Sci. Fd. Agric. 1975, 26,
    973-86.
11. MacDonald, J. C. J. Biol. Chem. 1962, 237, 1977-81.
12. MacDonald, J. C. Biochem J. 1965, 96, 533-38.
13. Adelberg, E. A.; Mandel, M.; Chen, O. C. Biochem. Biophys.
    Res. Commun. 1965, 18, 788-793.
14. Murray, K. E.; Shipton, J.; Whitfield, F. B. Chem Ind. 1970,
    897-98.
15. Drew, S. W.; Demain, A. L. Ann. Rev. Microbiol. 1977, 31,
    343-56.
16. Aharonowitz, Y.; Dewain, A. L. Can. J. Micro. 1979, 25,
    61-67.
17. Gräfe, U. In. "Overproduction of Microbial Products";
    Kranphanzl, V.; Sikyta, B.; Vanek, Z.; Academic Press, New
    York, N.Y., 1982; pp. 63-75.
18. Tamaski, T.; Tomita, F. Agric Biol Chem. 1982, 46, 1021-26.
19. Young, M. D.; Kempe, L. L.; Bader, F. G. Biotechnol Bioeng.
    1985, 27, 327-33.
20. Morgan, M. E. Biotechnol Bioeng. 1975, 18, 953-65.
21. Aharonowitz, Y.; Demain, A. L. Arch. Microbiol. 1977, 115,
    169-73.
22. Weinberg, E. D. Dev. Ind. Microbiol. 1974, 15, 70-81.
23. Jensen, R. A. J. Biol. Chem. 1969, 244, 2816-23.
24. Demain, A. L.; Jackson, M.; Trenner, N. R. J. Bacteriol 1967,
    94, 323-26.

RECEIVED February 11, 1986

Regulation of Acetaldehyde and Ethanol Accumulation in Citrus Fruit

J. H. Bruemmer

U.S. Citrus & Subtropical Products Laboratory, Agricultural Research Service, U.S. Department of Agriculture, P.O. Box 1909, Winter Haven, FL 33883-1909

Maturation effected the following changes in enzymes and metabolites in orange fruit: ethanol and acetaldehyde accumulated to levels of 10 mM and 0.08 mM, pyruvate decreased about 30%, pyruvate decarboxylase increased over 4 fold, alcohol dehydrogenase increased about 2 fold, the NADH to NAD ratio increased 2 1/2 fold and the terminal oxidase developed CN-insensitivity. The fraction of the total alternative respiratory pathway in actual use increased from 0.46 to 1.08. Induction of the alternative, CN-insensitive oxidase during maturation was interpreted as indicating that membrane function was modified which affected metabolic pathways resulting in the accumulation of ethanol and acetaldehyde.

The "essence" of citrus flavor is a complex mixture of volatile alcohols, aldehydes, esters, hydrocarbons, ketones and oxides. Alcohols are the largest class and ethanol is the main organic constituent of the essence. Esters and aldehydes are considered to contribute most to the characteristic flavor and aroma. In these two classes ethyl butyrate and acetaldehyde were shown to be important components of high quality orange juice (1).

Ethanol is enzymically related to ethyl butyrate and acetaldehyde through reactions catalyzed by alcohol

acetyltransferase (AAT) and alcohol dehydrogenase (ADH)
respectively. In juice from immature Hamlin orange ethanol was
present at 1 mM but accumulated to 10 mM at the mature harvest
stage (2). Acetaldehyde levels increased more slowly than ethanol
from 0.03 mM to a level of 0.08 mM at maturity. Ethanol also
accumulated in juice from mature citrus fruit during hypo-O_2 or
hyper-CO_2 refrigerated storage for 6 weeks (2).

Elevation of the ethanol content occurs during the maturation
stage of fruit development when acidity declines, total solids
increase, and the flavor compounds accumulate to characteristically
ripe levels. Since ethanol is a normal product of anaerobic
respiration in plants, its accumulation signals a change in the
pathway of energy metabolism. This summary of research on the
control of ethanol accumulation in citrus fruit describes an
approach to identify the role of energy metabolism in the
bioregulation of maturation.

Biochemical changes during anaerobic metabolism

Immature grapefruit responded to 16-hr. incubation at $38^{\circ}C$ under
CO_2 by a 15-fold increase in juice ethanol and a much smaller
increase in acetaldehyde compared to air controls (3). Malate, a
precursor of ethanol, was about 25% lower in juice from the
anaerobically stored fruit, but the decline in malate could account
for only 7% of the increase in ethanol. Citrate was also lower and
its metabolism could account for 15% of the increase in ethanol.
The alcohol dehydrogenase activity was about twice as high in juice
from anaerobic-treated fruit, but the levels of malic enzyme (ME),
malate dehydrogenase (MDH), and pyruvate decarboxylase (PDC) were
not greatly affected by the treatment.

Mature oranges responded to 6 weeks storage at $4^{\circ}C$ under 5%
O_2 in N_2 by a 3- to 4-fold increase in juice ethanol and a
consistent decrease in sugar level compared to controls stored in
21% O_2 in N_2 (4). Levels of ME, PDC, ADH, citrate and malate
were similar in treated and control groups. The ratio of oxidized
to reduced forms of NAD was several fold higher in juice from the
low oxygen group than from the 21% O_2 control.

Ethylene stimulated respiration in citrus (5,6) and enhanced
maturation of fruit (7). Mature grapefruit stored in air
containing 20 ppm ethylene contained 7 to 10 times more ethanol
after 4, 8 or 12 weeks at $15^{\circ}C$ than control fruit stored without
ethylene (8). Acetaldehyde in ethylene treated fruit was about 3
times higher. Levels of ME, PDC, and ADH were similar in treated
and control groups. Ethylene treatment decreased juice malate
about 60% in the 12-week period compared to controls. Malate
values were lowest in samples with highest ethanol, which suggested
that ethylene promoted the metabolism of malate to ethanol. In
pome fruit metabolism of malate through the decarboxylating system
is considered part of the ethylene-promoted ripening process (7).

Decarboxylation of malate to ethanol would explain changes in
these metabolites after short term incubation in CO_2 and after
storage with 10 ppm ethylene in air. Two of the enzymes (PDC and
ADH) in this decarboxylation pathway (Figure 1) were isolated and
purified from oranges and their reaction properties examined.

PDC and ADH reactions

PDC was isolated from orange juice sections and purified about 65-fold for kinetic measurements of the reaction (9):

$$\text{pyruvate + NADH} \xrightarrow{\text{PDC}} \text{acetaldehyde + NAD}^+ + CO_2$$

The optimum pH was 4.7, which was lower than that (6.1) reported for yeast PDC (10). The Hill plot was typical of a mono-catalytic site enzyme and, although activity was dependent on reduced sulfhydryl groups, substrate binding was not. Binding kinetics of Mg^{++} and thiamine pyrophosphate (TPP) and formation of an active cyclic ternary complex in the presence of pyruvate are properties similar to yeast PDC. About 15% of orange PDC was in the active cyclic form without added pyruvate (9). Activation of PDC by pyruvate was demonstrated in orange juice (11). Addition of 10 mM pyruvate to juice increased 5-fold the rate of acetaldehyde formation, suggesting the presence of about 20% in the active form. Activation of PDC by pyruvate could explain the stimulation of ethanol and acetaldehyde accumulation in citrus by anaerobic treatment (3) and by storage in hypo-oxygen atmospheres (4) without noticeable increase in PDC level. Anaerobiosis in the fruit would suppress the oxidative pathway and increase availability of pyruvate for activation of PDC.

ADH was isolated and partially purified from orange juice vesicles and examined for substrate specificity, maximum relative velocity (Vr) and affinity (1/Km) (12). Ethanol is the preferred saturated alcohol for reduction to the aldehyde based on Vr and 1/Km. Unsaturated alcohols, 2-propenol, 2-butenol and 2-hexenol, had comparable to or higher Vr's and 1/Km's than ethanol. ADH had 5- to 30-fold greater affinity for saturated aldehydes than the corresponding saturated alcohols, whereas affinities of the unsaturated alcohols and aldehydes were similar. The apparent equilibrium constants (Kapp = 0.003 for ethanol - acetaldehyde pair) favor alcohol formation in the saturated series. Other aldehydes compete with acetaldehyde for the enzyme but the concentration of acetaldehyde is much higher than other aldehydes in juice vesicles and the 1/Km for acetaldehyde is 10 X higher than for other aldehydes found in the juice vesicles.

Regulation of ADH and malate dehydrogenase (MDH)

The redox ratios of NAD and NADP in mature citrus fruit are comparable to ratios reported for other fruit (13). The ratio NADH/NAD was more than six times higher in mature, very ripe oranges (°Brix-acid ratio of 20) than in immature oranges (°Brix-acid ratio of 6). In contrast the NADPH-NADP ratio for immature and mature fruit were comparable. This shifting toward higher redox ratio of NAD has been associated with maturation and transition toward anaerobiosis in plant tissue (14). When oranges were incubated under N_2 at 34°C for 18 hr the NADH-NAD ratio in juice was twice as high as in juice from control fruit incubated under air (15). The redox ratio of NAD affects MDH and ADH activities in extracts from orange juice vesicles (15).

Dehydrogenase activity of MDH was suppressed by NADH at 5% of the
NAD concentration. Oxidative activity of ADH was suppressed when
NADH approached the concentration of NAD. The reductase activities
were not suppressed by NAD at even 10 times the concentration of
NADH. The increase in the redox ratio in senescing and anaerobic
citrus fruit could inhibit the oxidation of malate to oxalacetate
and increase the flux via the decarboxylation pathway through
pyruvate to ethanol (Figure 1).

Pyruvate metabolism during maturation

Juice ethanol and acetaldehyde increased in Hamlin oranges during
maturation (Figure 2) (16). The ethanol-acetaldehyde ratios for
October through February were 44, 55, 100, 109, 145. The series of
values showed a noticeable difference before and after December
when the fruit reached ripe maturity. The monthly series of values
for pyruvate also showed a marked decline between November and
December (Figure 3), but the other acids exhibited variable trends.
    PDC and ADH increased in juice vesicles of Hamlin orange during
maturation from October to December (Figure 4). The largest
increase occurred from November to January. Pyruvate dehydrogenase
(PDH) increased slightly from October (2.2 U) to November (2.9 U)
in the juice vesicles and then plateaued through February (2.9 U)
(Figure 5). The higher PDC activity would increase competition
with PDH for pyruvate and thereby increase decarboxylation to
acetaldehyde (Figure 1). The higher PDC activity could thus
explain the decline in pyruvate and increase in acetaldehyde and
ethanol in December. The higher ADH level in December could
explain the higher concentration ratio of ethanol to acetaldehyde.
ME increased steadily (Figure 5) which probably influenced the
equilibrium between malate and pyruvate in the cytoplasm; malate
increased from 4.5 mM to 7.0 mM and pyruvate decreased from 68 μM
to 48 μM over the season. Phosphoenolpyruvate carboxylase (PEPC)
increased during the early part of the season (Figure 5). However,
because MDH was much more active than PEPC (100-fold), the increase
in PEPC probably had little effect on the equilibrium concentration
of oxalacetate, which actually declined (Figure 3).
    Increase in ethanol and decrease in pyruvate level during
ripening could result from stimulation of the pyruvate
decarboxylase reaction promoted by the higher enzyme level.
However, activity of the competing enzyme for pyruvate, PDH, is
controlled by ratios of NADH to NAD and ATP to ADP in plant
mitochondria (18). During maturation of Hamlin orange the ratio of
NADH to NAD in juice vesicles increased from 0.09 in October to
0.24 in March, while the phosphorylated ratio (NADPH/NADP) was
constant (17). The PDH from broccoli was very sensitive to
increases in the mole fraction of NADH (19). A 10 to 15% increase
in ratio in whole tissue decreased PDH activity 15 to 25%. The
ratio of ATP to ADP in juice vesicles increased initially from 0.7
in October but plateaued at 1.0 after December. ATP inactivated
PDH by enzymic phosphorylation in mitochondria from pea leaf (20).
The phosphorylated PDH was activated by a Mg^{++}-dependent
phosphatase. Both reactions were inhibited by ADP which suggests

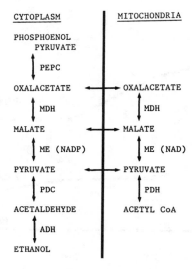

Figure 1.   Pyruvate metabolism in citrus fruit (16).

ANAEROBIC METABOLITES

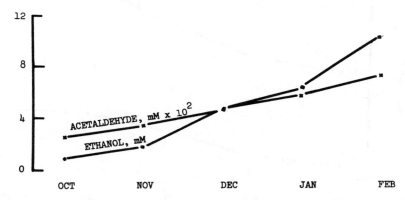

Figure 2.   Ethanol and acetaldehyde during fruit maturation (16).

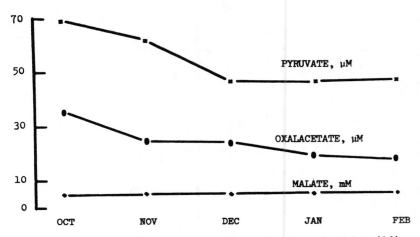

Figure 3.   Aerobic metabolites during fruit maturation (16).

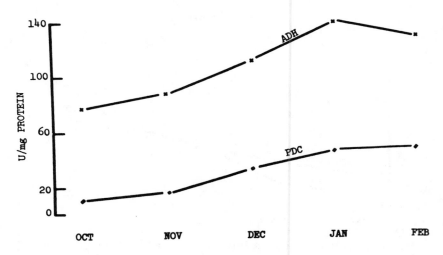

Figure 4.   ADH and PDC during fruit maturation (16).

the potential of the ATP-ADP couple for PDH regulation. The ubiquinol to ubiquinone ratio, (UQH/UQ) increased during the season from 0.4 in October to 0.7 in March. As an indicator of the redox state of the respiratory pathway, increase in this ratio indicates a change in equilibrium between substrate availability and oxidative function of the pathway. Thus, increases in UQH/UQ and NADH/NAD ratios suggest that the oxidative capacity of the pathway is inadequate to maintain redox equilibrium.

Terminal oxidases

The NADH oxidase of the mitochondrial fraction from Hamlin orange juice vesicles became less sensitive to KCN as the fruit matured (Table I).

Table I. NADH oxidation (nmoles O_2, mg protein^{-1}, min^{-1}, mean ± SE, n=4) (21)

	NADH	NADH + 1mM KCN	NADH + 1mM KCN + 1mM SHAM
Sept	281 ± 2.8	9 ± 1.4	0
Oct	294 ± 3.1	12 ± 0.9	0
Nov	303 ± 4.5	8 ± 0.3	6 ± 0.7
Dec	366 ± 4.7	116 ± 3.1	36 ± 0.5
Jan	379 ± 4.7	129 ± 3.2	45 ± 1.0

Fractions prepared from September, October, and November fruit were almost completely inhibited by 1 mM KCN, but preparations from December and January fruit were inhibited 65 to 70%. The KCN-insensitive respiration was inhibited 65 to 70% by salicylhydroxamic acid (SHAM). The SHAM-sensitive oxidase, or alternative oxidase, accounted for about 22% of the total, and residual oxidase (activity in presence of KCN + SHAM) about 10% in mitochondria from mature tissue.

Induction of the alternative respiratory pathway has been observed in maturing and aging organs, tissues and cells (22). Aged sweet potato root slices contained two types of mitochondrial membranes (23). The denser type was deficient in phospholipids and possessed alternative oxidase activity. The lighter type was identical to membranes from fresh slices and had no alternative oxidase. Submitochondrial particles from aged sweet potato root tissue were low in phospholipid and possessed alternative oxidase activity (24). Addition of phospholipid to the particles increased the lipid content and eliminated KCN-insensitive respiration.

Submitochondrial particles (membranes from washed sonicated mitochondria) prepared from juice vesicles of Hamlin oranges harvested in September contained KCN-insensitive respiratory activity (46% of total) using a substrate mixture containing 0.05 M malate, 0.05 M succinate, 0.01 M glutamate and 0.01 M TPP (21).

This activity contrasts to the small (3%) KCN-insensitive
respiratory activity observed when NADH was used as substrate with
the mitochondria (Table I). To calculate the contribution of each
oxidase to the total oxygen uptake, the titration method of Bahr
and Bonner (25) was used. O_2-uptake of submitochondrial
particles (membranes) prepared from juice vesicles of Hamlin
oranges harvested in September and January was measured in the
presence and absence of 1mM KCN titrated with a series of SHAM
concentrations using the malate, succinate, glutamate substrate
mixture (Table II).

Table II. Activity of alternative pathway (21)

	SHAM	O_2-uptake ± 1 mM KCN	
		-KCN	+KCN
	mM	% of Control	
Sept	0	100	100
	0.25	85	83
	0.5	77	67
	1.0	75	46
Jan	0	100	100
	0.25	75	87
	0.5	67	78
	1.0	63	66

The set of values (as % of control) obtained in the absence of KCN
was plotted against the set in the presence of KCN (Figure 6).
The direct linear relationship between the sets of values is
described by the equation

$$V_T = \rho \cdot g(i) + Vcyt. \quad (25)$$

Where V_T is the total respiration rate, Vcyt is the CN-sensitive
cytochrome mediated respiration, and g(i) is the maximal
contribution of the CN-insensitive alternative respiration at
given concentrations of the alternative path inhibitor, SHAM. The
slope of the line, ρ is the fraction of the alternative path which
is operating or in actual use and $\rho \cdot g(i)$ represents the actual
contribution of the alternative path to the total respiration.
The slope for the September preparation was 0.46, and the slope
for the January preparation was 1.08. The slope of 1.0 indicates
that the maximal capacity of the alternative path was in actual
operation in the mature January fruit. These data suggest that
during maturation, membrane function was altered which increased
the contribution of the alternative pathway to the total
respiration.

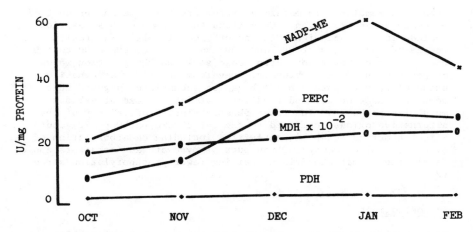

Figure 5.   Enzymes of aerobic metabolism during fruit maturation (16, 17).

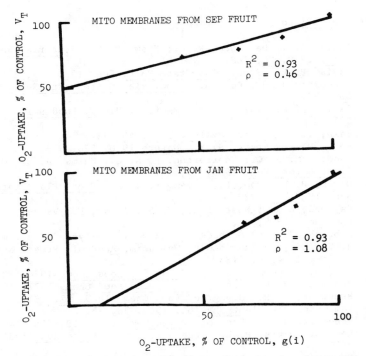

O_2-UPTAKE, % OF CONTROL, g(i)

Figure 6.   Estimation of the contribution of the alternative pathway to total respiration of mitochondrial membranes from September and January fruit (21).

Conclusion

Ethanol accumulated in maturing citrus fruit as the end product of
pyruvate decarboxylation. Conditions that promote this reaction
include low O_2, and high CO_2, and ethylene levels. Maturation
increased the levels of PDC and ADH and increased the NADH to NAD
ratio. The higher redox ratio could slow the PDH reaction which
competes with PDC for pyruvate. Development of the alternative
oxidase activity when ethanol began to accumulate suggests that
membrane function was modified which affected rates of various
metabolic pathways. The lower phosphorylation efficiency of the
alternative oxidase compared to the cytochrome pathway (22) could
affect numerous metabolic activities including decarboxylation of
pyruvate. Also, membrane transport of pyruvate and cofactors could
be altered in mitochondria containing fewer phosphorylation sites
(26).

Acknowledgment

Mention of a trademark or proprietary product is for identification
only and does not imply a warranty or guarantee of the product by
the U.S. Department of Agriculture over other products that may also
be suitable.

Literature Cited

1.   Ahmed, A.M.; Dennison, R. A.; Shaw, P. E. J. Agric. Food
     Chem. 1978, 26, 368.
2.   Davis, P. L. Proc. Fla. State Hort. Soc. 1970, 83, 294.
3.   Bruemmer, J. H.; Roe, B. Proc. Fla. State Hort. Soc. 1970,
     83, 290.
4.   Davis, P. L.; Roe, B.; Bruemmer, J. H. J. Food Sci. 1973, 38,
     225.
5.   Eaks, I. L. Plant Physiol. 1970, 45, 334.
6.   Reid, M. S.; Pratt, H. K. Nature 1970, 226, 976.
7.   Rhodes, M. J. C.; Wooltorton, L. S. C.; Hulme, A. C. Qual.
     Plant Mater. Veg. 1969, 19, 167.
8.   Davis, P. L.; Roe, B.; Bruemmer, J. H. Proc. Fla. State Hort.
     Soc. 1970, 87, 222.
9.   Raymond, W. R.; Hostettler, J. B.; Assar, K., Varsel, C. J.
     Food Sci. 1979, 44, 777.
10. Ulbrich, J. In "Methods of Enzymatic Analysis"; Bergmeyer,
     H. V., Ed.; Verlay Chemic, Weinheim; Academic Press: New York,
     1974; Vol. 4, p. 2186.
11. Roe, B.; Bruemmer, J. H. J. Agric. Food Chem. 1974, 22, 285.
12. Bruemmer, J. H.; Roe. B. J. Agric. Food Chem. 1971, 19, 266.
13. Bruemmer, J. H. J. Agric. Food Chem. 1969, 17, 1312.
14. Yamamato, Y. Plant Physiol. 1963, 38, 45.
15. Bruemmer, J. H.; Roe. B. Phytochem. 1971, 10, 255.
16. Roe, B.; Davis, P. L.; Bruemmer, J. H. Phytochem. 1984, 23,
     713.
17. Bruemmer, J. H.; Roe B. Phytochem. 1985, 24, 2105.
18. Randall, D. D.; Rubin, P. M. Plant Physiol. 1977, 59, 1.
19. Rubin, P. M.; Randall, D. D. Arch. Biochem. Biophys. 1977,
     178, 342.

20. Randall, D. D.; Williams, M.; Rapp, B. J. Arch. Biochem.
     Biophys. 1981, 207, 437.
21. Bruemmer, J. H.; Roe, B. Phytochem. 1986, in press.
22. Henry, M.; Nyns, E. Sub-Cell. Biochem. 1975, 4, 1.
23. Nakamura, K.; Asahi, T. Arch. Biochem. Biophys. 1976, 174,
     393.
24. Maeshima, M.; Li, H.; Asahi, T. Plant & Cell Physiol. 1984,
     25, 999.
25. Bahr, J. T.; Bonner, W. D., Jr. J. Biol. Chem. 1973, 248,
     3441.
26. Robertson, R. N. "The Lively Membranes"; Cambridge
     University: Cambridge, 1983; p. 136.

RECEIVED February 23, 1986

23

Enzymic Generation of Methanethiol To Assist in the Flavor Development of Cheddar Cheese and Other Foods

R. C. Lindsay and J. K. Rippe

Department of Food Science, University of Wisconsin–Madison, Madison, WI 53706

Quantitative studies on the enzymatic generation of methanethiol from methionine showed that methioninase obtained from Pseudomonas putida could be used for the development of flavors. Reactions carried out under anaerobic conditions yielded only methanethiol while aerobic conditions favored conversion of substantial amounts of methanethiol to dimethyl disulfide. Incorporation of free or fat-encapsulated methionine/methioninase systems into Cheddar cheeses resulted in the formation of volatile sulfur compounds, including carbon disulfide, and accelerated rates of development of aged Cheddar-like flavors. Methanethiol, when present alone, was observed not to cause the true, Cheddar-like flavor note in experimental cheeses.

Methanethiol has been implicated as an influential aroma and flavor compound in a variety of foods (1, 2), but its occurrence is frequently associated with overall aromas that are distinctly putrid, fecal-like, and sulfurous. Such putrid-type aromas are found, for example, in spoiling meats, poultry and fish (3, 4, 5, 6) where microbiologically generated methanethiol is generally accepted as a strong contributor to their occurrence. On the other hand, similar sulfurous aromas are considered desirable and appropriate in some surface-ripened cheeses, including Limburger (7, 8), Trappist (9, 10), and Muenster (11), and again methanethiol is an accepted contributor to the characteristic aroma and flavor of each. Likewise, some of the characterizing sulfurous notes of asparagus and the crucifer vegetables, such as cabbage, broccoli, and cauliflower, can be attributed to methanethiol generation upon tissue disruption or cooking (12).

Although methanethiol is most noted for its role in distinctive sulfurous aromas and flavors, other more subtle involvements of methanethiol in flavors occur which deserve attention in developing concepts for controlled enzymic generation of flavors. The flavors and aromas of some fruits and fruit

juices appear to be influenced by methanethiol where its presence
causes subtle flavor modifications. Methanethiol has been found
in orange and grapefruit juices (13), pineapple (1) and
strawberries (1), and it is believed to enhance the perception of
ripeness of some fruit and berry flavors.

The development of many thermally-induced reaction flavors
involves the participation of methanethiol that is derived from
either enzymic or nonenzymic reactions. Meat flavors are noted
for their dependence on volatile sulfur compounds which include
methanethiol (14-21), and a variety of naturally occurring
precursors have been suggested as contributors to
nonenzymically-formed methanethiol (22). However, methionine is
generally accepted as the most important precursor for
methanethiol via nonenzymic mechanisms that involve either
Strecker degradations or cation-catalyzed mechanisms. Strecker
degradation of methionine involves an initial reaction with a
dicarbonyl compound to yield methional, and this is followed by a
secondary formation of methanethiol and acrolein (23, 24).
Methionine is also converted to methional and a variety of other
volatiles, including methanethiol, ethylene, dimethyl disulfide,
and formic acid, when it is reacted with iron or manganese ions
and sulfite (25, 26, 27). Reactions involving methanethiol and
either saturated or unsaturated aldehydes along with hydrogen
sulfide yield a variety of methylthioalkanes and other sulfur
compounds that exhibit vegetable, citrus-, and meat-like aromas
(28). Additionally, methylthioalkanes have been found in Gouda
cheese by Sloot and Harkes (29), and could serve as flavor
contributors to cheeses.

Aside from the distinctively-flavored, washed, surface-ripened
cheeses mentioned earlier (9, 11, 30), methanethiol has been
recognized as a contributor also to the flavor of mature mold
surface-ripened cheeses, including Camembert and Brie (31, 32,
33). In these cheeses Brevibacterium linens or related
coryneforms (7, 34, 35) are responsible for the formation of
methanethiol. Perhaps the most significant but least understood
occurrence of methanethiol in cheeses is that of Cheddar cheese
where it appears to be associated with the development of
distinctive, true Cheddar-type flavors.

Methanethiol was first isolated from aged Cheddar cheese by
Libbey and Day (36) who believed that it was a significant factor
in aged Cheddar flavors. Subsequently, Manning and coworkers
(37-41) investigated the volatile sulfur compounds in Cheddar
cheese, and have concluded that methanethiol is somehow important
in the development of distinctive Cheddar-like flavors. It is
unlikely that lactic starter bacteria could be directly involved
in the formation of methanethiol in Cheddar cheese, but certain
nonstarter bacteria that occur adventitiously in cheese could
contribute this compound through enzymic mechanisms (42-46).
Regardless of the usual means for the development of distinct
Cheddar-like flavors in aging cheese, a great deal of interest has
been directed towards the development of means to accelerate the
development of Cheddar-like flavors in cheeses and cheese slurries
(47, 48). Because of the correlation of methanethiol
concentrations to Cheddar-like flavor intensities found in Cheddar

cheeses by Manning and coworkers (39, 40), means to utilize
methanethiol in enhancing Cheddar cheese flavors have been widely
sought.
    Methanethiol is a very volatile (b.p. 6.2°C) compound
possessing a intensely putrid, fecal-like aroma even at low
concentrations. The detection threshold value for methanethiol
has been reported as 0.02 ppb in air (49, 50), and it readily
undergoes oxidative condensation with itself in the presence of
oxygen to yield dimethyl disulfide (51) which also exhibits
pronounced aroma properties (12 ppb detection threshold in air;
49, 50). Besides the difficulties in handling and encapsulating
methanethiol for flavor applications, its propensity to adsorb to
surfaces and react with other organics makes the use of this
compound in flavor concentrates very troublesome indeed.

Enzymic Generation of Methanethiol Via Methioninase

A number of microorganisms have been reported to produce
methanethiol via the degradation of methionine. Included are a
variety of bacteria, actinomycetes, and filamentous fungi:
Brevibacterium linens (7, 35), Pseudomonas putida (52, 53),
Escherichia coli (54), Clostridium sporogenes (55), Proteus
vulgaris, Sarcina lutea, Bacillus subtilis, Streptomyces griseus,
Rhizopus nigricans, Fusarium culmorum, and Aspergillus oryzae (56,
57). Methioninase is most well-known for its degradation of
methionine to yield methanethiol, alpha-ketobutyric acid, and
ammonia (Figure 1; 52, 53) where it functions as
an L-methionine-alpha-deamino-gamma-mercaptomethane-lyase.
However, the enzyme can also effect gamma-eliminations with a
variety of substrates, including homocysteine and other
alkyl-S-substituted analogs (53). The specificity for the dual
elimination processes extends to include cysteine and its
alkyl-S-substituted analogs (Figure 1) where the reaction leads to
the liberation of hydrogen sulfide, pyruvate, and ammonia (58)
through alpha-,beta-eliminations. Furthermore, methioninase
catalyzes the exchange of either alkylthiols or hydrogen sulfide
for the terminal sulfur moieties in compounds of either the
homocysteine or cysteine series (Figure 1; 59, 60, 61) through
gamma- or beta-exchanges, respectively. This combination of
reactions by methioninase provides unique possibilities for
manipulating both methanethiol and hydrogen sulfide concentrations
and occurrences in flavor systems, especially in those of Cheddar
cheese and meat-like reaction flavor ingredients.
    The relative reactivity of methioninase with substrates other
than methionine is sometimes less than that observed for
methionine (K_m = 1.3 - 1.6 mM; 52, 53), including that for
homocysteine (K_m = 7.2 mM; 53) and methionine sulfone (K_m =
2.6 mM; 53). The enzyme is pyridoxyl-5'-phosphate dependent, and
Ito et al. (53) have noted that it was activated by heating to 52°
and 62°C. The critical temperature at which a change-over from
higher to lower activation energy occurred was at 40°C (E_a =
15.5 and 2.97 kcal/mole, respectively), and the data indicated
that there were two transitional conformations of the enzyme each
with different kinetic properties.

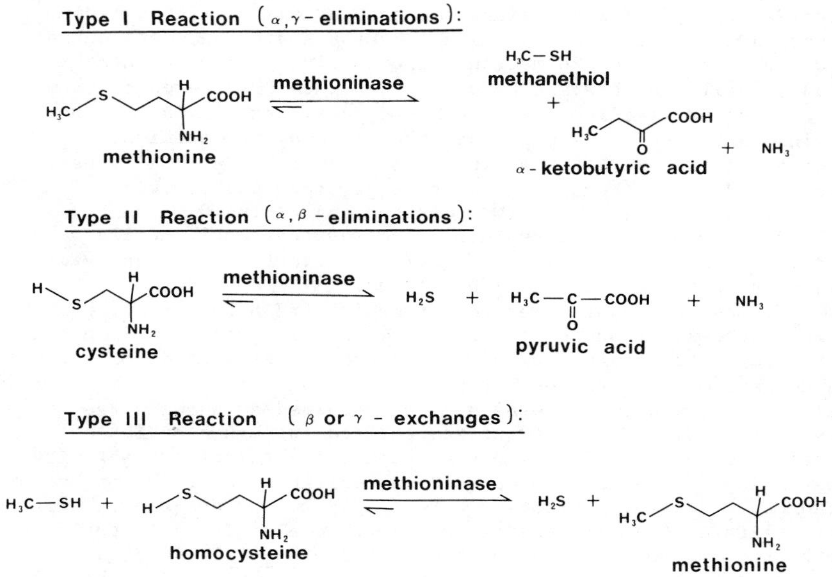

Type I Reaction (α,γ - eliminations):

Type II Reaction (α,β - eliminations):

Type III Reaction (β or γ - exchanges):

Figure 1. Some reactions mediated by P. putida methioninase.

Pseudomonas putida as a Source of Methioninase

As a portion of studies on the mechanisms involved in Cheddar
cheese flavor development, initial investigations on methanethiol
production have been focused on methioninase from P. putida. This
organism was chosen as a model for the production of methanethiol
and hydrogen sulfide because its methioninase has been more
thoroughly characterized than for other bacteria. Cultures of P.
putida ATCC 17453 were maintained on agar slants with a medium
containing glycerol (2 g/l) as a carbon source and methionine
(0.2 g/l) to induce formation and/or activity of methioninase
(53). Cells for harvesting were grown in a 20 l batch culture
fermentor in a medium that contained nutrients similar to the
maintenance medium, except that the glycerol and methionine
concentrations were 4.0 and 0.4 g/l, respectively. Cells were
collected with a continuous centrifuge, then washed twice with 100
mM phosphate buffer (pH 7.0) solution containing 0.02 mM
pyridoxal-5'-phosphate and 0.01% 2-mercaptoethanol. Cells were
harvested at the end of the logarithmic growth phase, and yields
were about 8 g cell paste/l of medium.

Washed cell pastes were diluted with fresh buffer using a
ratio of 3 ml of buffer for each 1 g of cell paste. A French Cell
Press operated at 16,000 lb/in^2 was used to disrupt cells. The
crude enzyme preparation was then centrifuged at 10,444 X g for 30
min at 0°C, and the supernatant was retained as a cell-free
extract (CFE). This extract was carried through a series of
activation and purification steps and the methioninase activity
was assayed after each step using a modification of the method of
Tanaka et al. (52). Protein was assayed by the dye-binding method
of Bradford (62), and enzyme solutions were dialyzed in tubing
prepared by procedures described by Brewers et al. (63).
As noted by Ito et al. (53), methionase was activated by heating
at 60°C for 10 min. The greatest increase in activity for the
methioninase was observed for a 50% saturated ammonium sulfate
precipitate (4.6-fold), but this was substantially less than the
20-fold increase obtained by Tanaka et al. (52, 59) when a
DEAE-cellulose column fractionation step was also employed before
salt precipitations. A DEAE-cellulose step was not included in
the current study because emphasis was directed towards provision
of a more stable medium than maximum activity. Loss of activity
noted during dialysis steps was later found to be retarded by
incorporation of 2-mercaptoethanol into the dialysis buffer.
However, loss of activity of the P. putida methioninase when in
solution occurred readily, and the data in Figure 2 illustrate
this behavior. Freeze-dried methioninase powder lost little
activity when assayed after holding at -20°C for periods up to one
year, and when rehydrated was more stable than the initial cell
free extract (64).

Loss of P. putida methioninase activity has been noted also by
Ito et al. (53), and Kreis and Hession (55) have reported similar
data for C. sporogenes methioninase. However, Tanaka et al. (52)
and Ohigashi et al. (54) did not comment about loss of activity.
Loss of activity could be caused by the dissociation of the

cofactor, pyridoxal-5'-phosphate, from methioninase, but addition
of 0.02 mM pyridoxal-5'-phosphate to enzyme solutions did not
prevent loss of activity as measured by alpha-ketobutyric acid
production. Trials were also conducted where reducing agents were
added to minimize oxidizing environments, and possibly stabilize
the enzyme system. However, addition of 1.3 mM of either
dithiothreitol, glutathione, or 2-mercaptoethanol into rehydrated
freeze-dried methioninase preparations had little, if any, effect
on the rate of loss of activity as compared to a control, and as
determined by alpha-ketobutyric acid formation. These
observations suggest initially that the enzyme preparation may
have contained a protease that eventually destroyed activity.
Yet, as will be discussed later, measurement of methanethiol and
dimethyl disulfide in model methioninase systems provided data
that indicate that the enzyme was not inactivated to the same
extent during incubation that alpha-ketobutyric acid measurements
indicate.

Because of the heterogeneous nature of methioninase catalysis
(Figure 1), measurement of alpha-ketobutyric acid (as well as
other carbonyls that may be present) may not provide accurate
indications of the activity of the enzyme. Comparison of the data
in Figure 2 for cell-free extracts and semi-purified freeze dried
methioninase solutions suggests that either protease activity was
involved or that complex interactions of potential substrates and
products occurred. Use of alpha-ketobutyric acid assays for
methioninase activity in samples incubated for prolonged periods
seems potentially vulnerable to such analytical pitfalls.
However, if protease activity were involved, such preparations of
methioninase from P. putida could provide self-destructing enzyme
systems for encapsulated, flavor-producing systems (65) that would
prevent over-production of methanethiol and hydrogen sulfide.
Alternatively, it might be possible to use other sources of
methioninase, such as Brevibacterium linens (7, 35) or Aspergillus
oryzae (56, 57), as systems with potentially less protease
activity. In any event further studies will be required to
determine if alpha-ketobutyric acid assays provide suitable data
for indexing methioninase activity in flavor systems.
Additionally, more clear definition of causes of apparent loss of
methioninase activity needs to be pursued.

Methanethiol Production by Crude P. putida Methioninase Preparations in Model Systems

Since methioninase has potential applications in both aerobic and
anaerobic food systems, a number of experiments were conducted to
characterize and evaluate the performance of methioninase under a
variety of these conditions. In these studies freeze-dried
methioninase preparations (either 0.779 units/mg protein or 2.45
units/mg protein) were employed for generation of methanethiol,
and selective flame photometric detection gas chromatography was
use for quantifying methanethiol. The column was a 3 m x 4 mm id
glass column packed with Carbopak BHT-100, and it was operated at
60°C isothermally for methanethiol analysis. When higher
molecular weight sulfur compounds were measured also (e.g.,

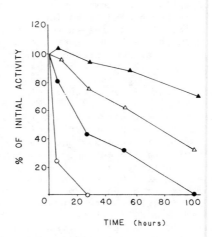

Figure 2. Methioninase stability as determined by
alpha-ketobutyric acid assay in a cell-free
extract and in hydrated freeze-dried enzyme
preparations at 8° and 30°C.

● – Cell free extract, incubated at 8°C;
    Initial specific activity = 0.0097.

○ – Cell-free extract, incubated at 30°C;
    Initial specific activity = 0.0110.

▲ – Rehydrated freeze-dried enzyme, incubated
    at 8°C;
    Initial specific activity = 0.0076.

△ – Rehydrated freeze-dried enzyme, incubated
    at 30°C;
    Initial specific activity = 0.0072.

dimethyl disulfide), the oven was programmed from 60° to 130°C at
25°/min (Figure 3). Purified nitrogen flow through the column was
40 ml/min, and air, oxygen, and hydrogen flow rates were adjusted
for each operating period to a high-sensitivity, reducing flame.
For most studies, the hydrogen, oxygen, and compressed air flow
rates were 126, 11, and 53 ml/min, respectively, and the
oxygen:hydrogen ratio was 0.18. Under these conditions, 0.06 ng
of methanethiol could be detected. The flame was set for each
analysis session according to the flows noted, and the detector
response was then fine-tuned with a standard dilution of
propanethiol to obtain a standard response which replaced the
requirement for an internal standard in the analyses.

Samples were contained in glass flasks of either the
configuration of a serum bottle (100 ml) capped with a Mininert
valve or of a specially-designed Erlenmeyer flask (125 ml)
constructed with a screwtop and fitted with a sidearm and a Teflon
stopcock. The latter flask was capped with a silicone septum held
with an open-topped screw-cap, and allowed syringe-sampling as
well as evacuation. Samples for headspace analysis were
equilibrated at 30°C. Calibration curves were calculated as log
(area) versus log (mass of methanethiol or dimethyl disulfide).
The methanethiol amounts shown in graphs refer to the mass of
methanethiol that was present in a 10 ul injection, and not the
total mass present in a given headspace bottle. To obtain the
amount of methanethiol present in a bottle headspace, the mass
obtained from plots must be multiplied by 9500 (i.e., 99 ml
headspace/10 ul injection). Further discussions about employing
this technique for measuring thiols can be found in reports by
Rippe (64), Jansen *et al*. (66), and Banwart and Bremner (67). It
was assumed that the differences in partial pressures of
methanethiol and dimethyl disulfide were insignificant between
calibration and model systems. However, over time methanethiol
oxidizes to dimethyl disulfide in the aerobic environment of
sample bottles.

Methioninase activity under aerobic environments was evaluated
in a series of trials where reactants were combined into a serum
vial, the vial capped, and reaction time commenced upon capping.
The reaction mixture buffer contained 0.02 mM
pyridoxal-5'-phosphate and 100 mM potassium phosphate (pH 8.0),
and the methioninase preparation had 0.779 units of activity per
mg protein. Samples were immediately placed in a waterbath at
appropriate intervals during the incubation period. When only
methanethiol was measured (Figure 4A), the accumulation rate of
this methioninase product diminished after about 50 hr of
incubation. However, when total methanethiol (sum of methanethiol
and dimethyl disulfide) was measured (Figure 4B), the accumulation
rate remained linear through at least 200 hr of incubation at 30°
which indicates that the methioninase was not nearly as unstable
as earlier experiments using alpha-ketobutyric acid measurements
had indicated (Figure 2). For the total methanethiol data, the
mass of dimethyl disulfide was converted to micromoles, and then
was multiplied times 2 to obtain the equivalent micromoles of
methanethiol from which it was initially derived. Additionally,
when trials were conducted with a constant methioninase

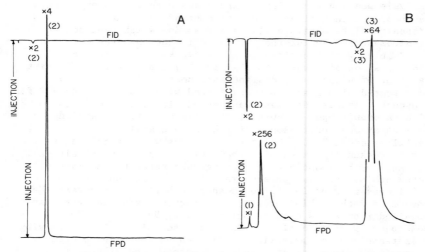

Figure 3. Headspace FPD/FID gas chromatographic profiles of
products of methioninase action on a methionine
substrate at 30°C after A) 1 hour, and B) 100
hours of incubation. Temperature programmed 60°
to 130°C at 25°C/min; Peaks are 1) hydrogen
sulfide, 2) methanethiol, and 3) dimethyl
disulfide eluted from a packed Carbopack BHT-100
column.

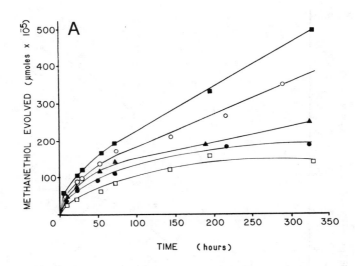

Figure 4A.   Accumulation of methanethiol from
             methionase activity on methionine
             (100mM) in an aerobic environment
             during extended incubation at 30°C;
             FDE=freeze-dried enzyme preparation.

      □- 1.0 mg FDE/ml (5.0 mg total)

      ●- 2.5 mg FDE/ml (12.5 mg total)

      ▲- 5.0 mg FDE/ml (25.0 mg total)

      ○- 7.5 mg FDE/ml (37.5 mg total)

      ■- 10.0 mg FDE/ml (50.0 mg total)

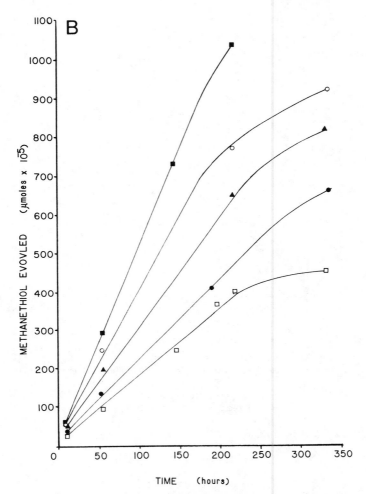

Figure 4B. Accumulation of total methanethiol (as the
            sum of methanethiol plus dimethyl disulfide)
            from methioninase activity on methionine
            (100mM) in an arobic environment during
            extended incubation at 30°C; FDE=freeze-
            dried enzyme preparation.

            □ -1.0 mg FDE/mL (5.0 mg total)

            ● -2.5 mg FDE/mL (12.5 mg total)

            ▲ -5.0 mg FDE/mL (25.1 mg total)

            ○ -7.5 mg FDE/mL (37.5 mg total)

            ■ -10.0 mg FDE/mL (50.0 mg total)

concentration and various methionine concentrations (Figure 5), similar continued production of total methanethiol occurred over a period of 200 hr at 30°C, especially for concentrations above 3 mM methionine.

Experiments conducted under reduced-oxygen conditions (sonicated reaction mixtures under vacuum with nitrogen gas replacement) gave results for methanethiol production similar to that obtained for total methanethiol (methanethiol plus dimethyl disulfide) production under aerobic conditions. Very little dimethyl disulfide was produced in reaction flasks when near anaerobic conditions were obtained. The atmospheric oxidation of methanethiol to dimethyl disulfide is well recognized (6, 50, 68, 69), and the redox of foods influences the relative concentrations of each. In the presence of hydrogen sulfide, and perhaps elevated temperatures, methanethiol is first oxidized to the sulfenic acid and then two molecules of the sulfenic acid react with a molecule of hydrogen sulfide to yield dimethyl trisulfide (69). Thus, overall production of methanethiol appears to be only one of the parameters for consideration in developing flavor systems employing methioninase, and redox-established equilibria between methanethiol and dimethyl disulfide may be of even greater significance to final flavors.

Butterfat-Encapsulated P. putida Methioninase for Methanethiol Production in Cheddar Cheese

Methanethiol has been found to be correlated with the development of Cheddar cheese flavor by Manning and coworkers (37, 39, 40), and both nonenzymic (22) and enzymic generation of methanethiol have been proposed as the source of this compound in Cheddar cheese (46). Although the correlation of methanethiol to Cheddar flavor appears statistically valid, difficulties have been encountered in explaining the nature of its flavor-conferring properties in cheese. In addition, uniform production of methanethiol is difficult to achieve commercially, and the rate of its natural formation in accelerated-ripening may not be suitable for achieving typical Cheddar flavors (47, 48).

A recently-developed technique for fat-encapsulation of enzymic flavor generating systems (65, 70) was used to produce capsules containing various P. putida methioninase for incorporation into Cheddar cheese during the cheesemaking process. In these studies, freeze dried methioninase extracts (0.779 units activity/mg protein) were prepared in buffer (100 mM potassium phosphate, pH 8.0; and 0.02 mM pyridoxal-5'-phosphate), and were encapsulated by the procedure outlined by Magee et al. (65). The butterfat encapsulant was heated to 60°C and Glycomul TS and Span 60 were each added at a rate of 1.5% before cooling the mixture to 54°C with stirring. The carrier system that was encapsulated contained buffer, methionine substrate (100 mM), reduced glutathione (1.3 mM), and pyridoxal-5'-phosphate (0.02 mM). Capsules were prepared by injecting butterfat-carrier (58 g and 15 ml, respectively) emulsions at 53°-54° into a dispersion fluid of skim milk (822 g; 13°-16°C) using an airless paint sprayer. About 80% efficiency in the encapsulation of the

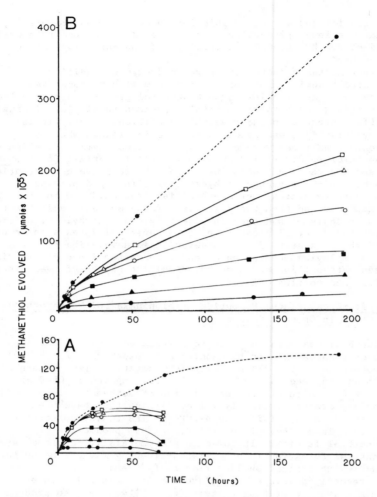

Figure 5. Accumulation of A) methanethiol and B) total
          methanethiol (as the sum of methanethiol plus
          dimethyl disulfide) from methioninase activity
          (constant concentration, 2.5 mg freeze-dried
          extract/ml) on varying concentrations of
          methionine in an aerobic environment during
          extended incubation at 30°C.

●- 100 mM L-Methionine   ■- 2.0 mM L-Methionine

□- 5.0 mM L-Methionine   ▲- 1.0 mM L-Methionine

△- 4.0 mM L-Methionine   ●- 0.5 mM L-Methionine

○- 3.0 mM L-Methionine

enzyme-containing carrier system was obtained, and capsules were prepared immediately before addition to cheesemaking milk at the time of traditional renneting.

Traditional Cheddar cheese making procedures were used except that the temperature was reduced to 30°C (86°F) before renneting, and the cooking temperature was only increased to 37.8°C (100°F) to assist in maintaining the integrity of the butterfat capsules. A higher rate of renneting (0.11 ml/lb milk; single-strength) than usual was employed to compensate for curd-softening that occurred with the reduced temperatures. Frozen, concentrated starter cultures SG-1 (containing Streptococcus cremoris and Streptococcus lactis) and FCJB (containing S. cremoris, S. lactis, and Leuconostoc sp.) (Marshall, Miles Laboratories, Madison, WI) were used at a rate of 0.8% inoculum. Cheese was manufactured in experimental miniature vats (80 lbs milk each) in the University of Wisconsin-Madison Dairy Products Laboratory. Fresh raw whole milk was pasteurized at 62.8°C for 30 min, cooled to 20°C, and then stored at 4°C for use the next day. After pressing, cheeses were packaged under reduced pressure (10-14 mm Hg) in barrier pouches using a heat-sealing machine.

The cheese from each mini-vat provided 7 to 8 lbs of sample, and after pressing each lot was split before barrier-packaging under vacuum so that two ripening temperature regimes could be evaluated (i.e., continuous 10°C, and 21°C for 21 days followed by continuous 10°C, through 4 months). The Cheddar cheeses were typical in average moisture (37.5%), salt (1.7%), fat (31.5%), and pH (5.10) levels even though they were manufactured on a miniature scale compared to commercial-scale batch operations.

Samples of cheeses for analysis of volatile sulfur compounds by FPD gas chromatography were taken from the interior of cheese blocks, and each (12 x 8 x 2 cm) was placed in a closed polyethylene bag and held at -96°C until analysis within one month. For analysis, 25 g of either thawed or fresh cheese which had been shredded (3 mm slots) was added to a specially-fabricated, screw-capped (with a silicone septum), 125 ml Erlenmeyer flask fitted with a side-arm and stopcock. The assembled sampling flask was rapidly evacuated (10 sec to about 25 mm Hg), and then the headspace was brought back to atmospheric pressure with purified nitrogen. Flasks were then held at 30°C in a waterbath for 1.5 hr before sampling for volatile sulfur compounds, and analysis on a Carbopack BHT-100 column. Identities of peaks were assigned on the basis of retention times, and confirmation by GC-MS using equipment described earlier (71). The aroma and flavor properties of each sample were evaluated by three individuals experienced in dairy products evaluation to provide an indication of the overall sensory qualities of the cheeses.

Each trial consisted of a vat of control cheese (butterfat capsules contained only phosphate buffer), a vat of cheese with the encapsulated enzymic flavor generating system, and a vat of cheese containing both butterfat capsules with only phosphate buffer and 15 ml of the unencapsulated reaction mixture (i.e., enzyme, substrate, and cofactor). Such an experimental design allowed assessment of influences of both the addition of free methioninase systems and encapsulated methioninase systems on the

production of methanethiol in Cheddar cheese. Only methanethiol
and dimethyl disulfide were quantified because the anaerobic
sample preparation resulted in variable loss of hydrogen sulfide,
but over 95% of methanethiol was found retained through the
procedure. A designated flask was employed for all analyses where
comparative uses of the results were anticipated because some
variability in headspace volumes existed between flasks.

While the relationship of headspace concentrations of
methanethiol and other volatiles to their actual concentrations in
cheese remains to be established, headspace concentrations of
volatile sulfur compounds appear appropriate to assess relative
effects of various treatments (64). A baseline of information on
methanethiol concentrations found in cheeses manufactured in the
University of Wisconsin Dairy Plant was initially developed to
serve as a magnitude guide for the current trials as well as for
comparative purposes with those published for Cheddar cheese by
others (37, 40, 48). Concentrations of methanethiol are shown in

Table I. Calculated concentrations of methanethiol in the
headspace of selected Cheddar-type cheeses.

Cheese sample	Calibrated peak area (mm^2)	Concentration of methanethiol in headspace (ng/ml)
Cheddar, aged, full flavor	251	0.84
Cheddar, mild, clean	103	0.50
Colby, mild	58	0.37

Table I, and reflect the Cheddar flavor intensity correlation
reported by Manning and coworkers (37). In addition to
methanethiol, peaks for dimethyl disulfide were quite large in the
aged Cheddar and the Colby cheese headspace profiles, but very
little of this compound was found in the mild Cheddar sample.
Based on the concentration of methanethiol in the aged sample,
conditions were established for the methioninase system which
would approximate what might be encountered naturally, i.e., about
2 orders of magnitude greater than the target figure that was
defined as the aged, full-flavored Cheddar cheese.

Concentrations of methanethiol measured in headspace samples
of the experimental cheeses are summarized in Table II for the
analysis times of 1 day, 21 days and 4 months for each of the two
ripening conditions employed. Notably, the cheese made with only
encapsulated buffer did not contain methanethiol after 1 day at
either temperature. However, the encapsulated methioninase system
yielded significant amounts of methanethiol at 1 day, and
continued to increase through 4 months. Generally, the final
concentration of methanethiol in the encapsulated-buffer control

Table 2. Summary of concentrations of methanethiol in the headspace of samples of experimental Cheddar cheeses containing methioninase.

Age of Cheese	Encapsulated Buffer Control		Unencapsulated enzyme & substrate		Encapsulated enzyme & substrate	
	10°C	Ripening temperature 21°C for 21d, then 10°C	10°C	Ripening temperature 21°C for 21d, then 10°C	10°C	Ripening temperature 21°C for 21d, then 10°C
	(——————————————— ng methanethiol/ml headspace ———————————)					
1 Day	0	0	0	0.35	0.24	0.96
21 Days	0.29	1.09	0.48	1.15	0.58	2.13
4 Months	1.89	N.D.[1]	2.75	33.00	N.D.	60.50

[1] N.D. = Not determined because of insufficient sample.

sample of cheese was in relative agreement with the concentration found in the well-aged Cheddar sample analyzed initially (Table I). In each of the samples which employed unencapsulated enzyme and substrate, the production of methanethiol was enhanced compared to the control, and this indicates that encapsulation would not be required to effect the production of methanethiol in Cheddar cheese using the conditions of these experiments. However, encapsulation provided more rapid and extensive generation of methanethiol than was observed for the unencapsulated system. Thus, the advantages of encapsulation which include close proximity of enzyme and substrate, stability of microenvironments within capsules, and better retention of enzyme in cheese appear to be borne out by the current experiments.

Flavor evaluation of cheeses indicated that the methanethiol found in 1 day old cheese containing the encapsulated methioninase did not cause a Cheddar-like flavor in the otherwise bland cheese flavor background. Thus, more than the presence of methanethiol is required to evoke Cheddar flavor. At the lower ripening temperature of 10°C, flavor differences between the encapsulated buffer control cheese and the two methioninase cheeses were modest after 21 days, but the enzyme-containing cheeses generally exhibited slightly more sulfury flavors that seemed to give more sharp tastes than the control. Cheeses ripened initially at an elevated temperature (21°C) responded to the accelerated ripening conditions in terms of both flavor intensity and methanethiol production. A pasty, atypical body which developed during the initial ripening of these cheeses became less noticeable after the final ripening for 3 months at 10°C. The flavor of both of these methioninase-containing cheeses became noticeably sulfury, cooked-vegetable-like, and toasted-cheese-like in character soon after manufacture, and the cheese containing the encapsulated methioninase exhibited the most pronounced flavor of this type. This flavor was not particularly offensive or unclean, but it was decidedly unbalanced and not characteristic of Cheddar cheese. However, as ripening at 10°C progressed the flavors of these cheeses became more intensely Cheddar-like, but the quality of flavor still suffered from unbalanced, sulfur-like flavors. This was true to some extent for the cheeses ripened at 10°C also after 4 months, but the intensity of sulfury flavors were less.

After 4 months aging, the samples containing the unencapsulated methioninase system exhibited distinctly the most favorable Cheddar flavor quality and intensity among those made with enzyme. Analysis of cheese headspaces indicated that hydrogen sulfide levels were apparently greater in these cheeses than those made with encapsulated methioninase, and methanethiol levels were substantially less also. It seems likely that this improved flavor quality in cheeses with added unencapsulated methioninase resulted from the action of the enzyme on cysteine or its analogs with resulting release of hydrogen sulfide (Figure 1). Thus, the inclusion of either capsules containing both methionine and cysteine or separate capsules containing each of the two substrates seems to be worthy of further investigation. Based on the results of these trials, it can be concluded that potential applications of methioninase in cheese flavor development seem worthy of pursuit, particularly where accelerated rates of flavor development are sought.

Overall, the results agree with earlier literature reports regarding the apparent necessity of methanethiol and hydrogen sulfide (39) for the development of Cheddar flavor and aroma. However, evidence is accumulating that the association of methanethiol and other volatiles with Cheddar cheese flavor intensity is not a direct combination effect of the volatiles known to date. Evidence in the current study strongly supports the view that methanethiol alone in the early stages of ripening does not convey a Cheddar flavor to cheese. Rather than the "component balance theory" advanced some time ago to account for the flavor of Cheddar cheese (72), evidence which has been accumulated could just as readily be interpreted that the small molecular weight compounds seen in Cheddar cheese headspace profiles are the result of a degradation of an unstable, sulfur-containing compound which possesses a distinct Cheddar-like aroma. At least six low molecular weight sulfur compounds occur regularly in the headspace profiles of distinctly-Cheddar flavored aged cheeses (Figure 6), and several non-sulfur, low molecular weight compounds are also seen regularly (41, 48). Although these were not investigated in the current study, they include diacetyl, ethanol, pentan-2-one, acetone, butanone, and methanol (40, 48). Of these, Manning (40) has reported that pentan-2-one, acetone, methanol along with methanethiol correlate best with Cheddar flavor intensity and Cheddar quality.

In the current study, special attention was given to confirming the presence of carbon disulfide in the headspace of Cheddar cheese, and conclusive mass spectral evidence was developed for this compound. The concentration of carbon disulfide varied widely in the samples analyzed for the methanethiol studies, and became quite abundant in the most-aged sample of cheese that contained the encapsulated methioninase system (Figure 6). This sample also contained a lower molecular weight sulfur compound which remains unidentified, but it was observed only in the cheese prepared with encapsulated methioninase.

The extremely elusive nature of the chemical basis of Cheddar-like flavors supports the hypothesis that an unstable aroma compound is involved in the flavor of Cheddar cheese, and this view deserves thorough investigation. The circumstantial evidence in favor of the existence of a Cheddar-like aroma also includes considerations relating to the sensitivity of Cheddar flavor to heat and oxygen, and the fact that the redox of cheese is quite low. Additionally, sulfury or sulfide-like defects as well as brothy flavor-like defects are often encountered in Cheddar cheeses of various compositions and origin. These flavors could reflect either production of excessive amounts of certain sulfur compounds or the absence of certain essential compounds that are initially required to allow formation of a Cheddar compound. While attempts to date have not resulted in the isolation of such a compound, this could reflect the very unstable nature of the proposed compound. Other similar circumstances appear to occur in freshly roasted coffee and nuts where transient

aroma compounds provide characterizing, desirable flavors and
aromas. Based on the low molecular weight compounds found to be
correlated to Cheddar flavor (40), it is difficult to reconstruct
a compound accommodating all of the circumstantial evidence.
However, a compound based on the combined structures (Figure 7) of
certain sulfur compounds, such as lenthionine (73, 74) and
trithiahexanes (75), and certain sulfur-oxygen compounds,
including 2-methyl-4-propyl-1,3-oxathiane (76) and thiophenones
(77), would appear to provide properties that would account for
the fragments observed. Of interest also is the reported
instability of 2,5-dimethyl-2,4-dihydroxy-3(2H)-thiophenone on
Carbowax 20M column (77). If a Cheddar aroma compound can be
confirmed, perhaps the role for methioninase in the development of
cheese flavors can be more clearly positioned in terms of
technology development.

Additional Means for Manipulation of Methanethiol and Other Sulfur
Volatile Compounds in Foods

Besides controlling the usual enzymic reaction parameters for
manipulating methioninase activity in systems discussed to this
point, some other avenues exist for potentially influencing
concentrations of methanethiol and some of the other volatile
sulfur compounds commonly encountered in foods. As has been
demonstrated by Manning (22), methanethiol and hydrogen sulfide
can be generated in Cheddar cheese via chemical reactions
involving products of added methionine, sodium caseinate, and
cysteine substrates along with a reducing agent, such as
glutathione and dithiothreitol. Using chemical acidification with
delta-gluconolactone, methanethiol was not produced, but when
precursors were added the compound was formed. This was
interpreted that starter organisms probably assisted in Cheddar
cheese flavor development by liberating amino acid precursors, and
that methanethiol and hydrogen sulfide formation followed because
of chemical reaction under the low redox conditions of ripening
cheese. Thus, production of methanethiol might be possible
through controlled chemical reactions as well as through enzymic
mechanisms.

The extremely potent odor properties of low molecular weight
sulfur compounds, including methanethiol, make over-production or
imbalanced-production of these compounds somewhat of a troublesome
problem in applications of enzymes as well as in usual
processing. One of the most researched flavor problems associated
with volatile sulfur compounds is that related to the cooked or
heated flavor of pasteurized or ultra-high-temperature (UHT)
processed milk. Suppression or removal of this flavor through the
use of sulfhydryl oxidase, an enzyme normally present in milk, is
now well-documented (78, 79). Sulfhydryl oxidase is a
metalloglycoprotein which catalyzes oxidation of thiols to their
corresponding disulfides using molecular oxygen as an electron
acceptor. Cysteine, peptides, and proteins all serve as
substrates for this oxidative activity which results in binding of
volatile thiols, such as hydrogen sulfide and methanethiol that
are present in heated milk (78). Thus, sulfhydryl oxidase

FID
3
X64
4
2
X2
5
1
7
6
FPD
60
130
TEMP (°C)
0 1 2 3 4 5 6 7
TIME (MIN)
1 - H2S
2 - COS
3 - H3C-SH
4 - H3C-S-CH3
5 - CS2
6 - UNKNOWN
7 - H3C-S-S-CH3

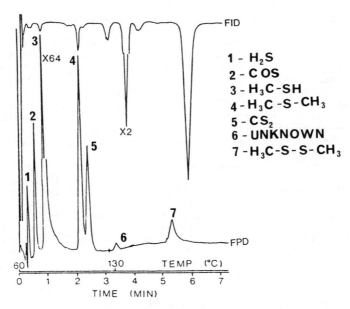

Figure 6. Gas chromatogram of sulfur volatiles from Cheddar cheese containing encapsulated methioninase and ripened at 21°C for 21 days, then 3 months at 10°C (Carbopak BHT-100 column).

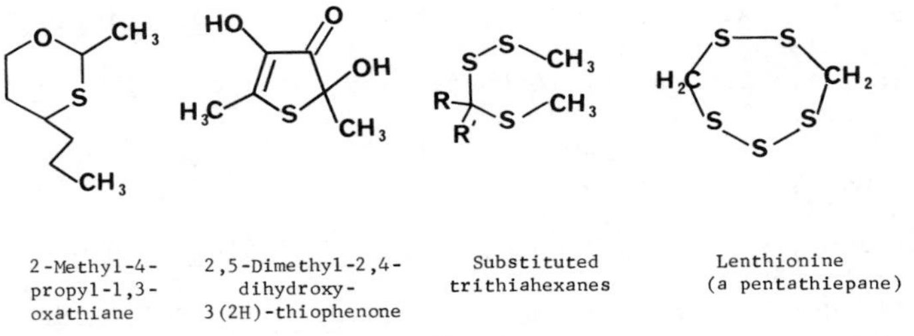

2-Methyl-4-propyl-1,3-oxathiane
2,5-Dimethyl-2,4-dihydroxy-3(2H)-thiophenone
Substituted trithiahexanes
Lenthionine (a pentathiepane)

Figure 7. Sulfur aroma compounds showing structural features that could lead to low molecular weight compounds similar to those found in the headspace of Cheddar cheese.

possesses properties that could be useful in regulating
concentrations of free thiols in enzymic preparations for flavor
applications. Additionally, xanthine oxidase is an oxidative
enzyme found widely which has the primary function of controlling
the last stages of purine catabolism by converting hypoxanthine
and xanthine to uric acid. This enzyme also occurs abundantly in
milk (80), and significant concentrations of xanthine oxidase
frequently survive usual milk processing conditions (81).
Although xanthine oxidase does not mediate sulfhydryl oxidations
per se, it is activated by the binding of hydrogen sulfide and
perhaps other simple thiols (82). Thus, the effects of this
enzyme also might be useful in fabricating enzymic
flavor-generating systems where it could serve as an assist in
limiting free thiols in the medium.

In conclusion, these investigations have shown that
methanethiol generation by methioninase has potential applications
in the development of cheese flavors as well as other for other
foods. The use of fat encapsulated enzyme systems functioned well
in experimental cheeses, and their use should provide assistance
in controlled delivery of methanethiol into food systems during
further efforts to elucidate the complex nature of Cheddar cheese
flavor.

Acknowledgments

This research was supported by the College of Agricultural and
Life Sciences, University of Wisconsin-Madison.

Literature Cited

1. van Straten, S.; de Vrijer, F.; de Beauveser, J. C.
   "Volatile Compounds in Food"; Central Institute for Nutrition
   and Food Research: Zeist, The Netherlands, 1977.
2. Maga, J. A. CRC Crit. Rev. Food Sci. Nutr. 1976, 7(2), 147.
3. Lee, M.; Smith, D. L.; Freeman, L. R. Appl. Environ.
   Microbiol. 1979, 37, 85.
4. Thomas, C. J.; McMeekin, T. A. J. Appl. Bacteriol. 1981, 51,
   529.
5. Herbert, R. A.; Ellis, J. R.; Shewan, J. M. J. Science Food
   Agric. 1975, 26, 1187.
6. Miller, A.; Lee, J. S.; Libbey, L. M.; Morgan, M. E. Appl.
   Microbiol. 1973, 25, 257.
7. Boyaval, P.; Desmazeaud, M. J. Lait 1983, 63, 187.
8. Parliment, T. H.; Kolar, M. G.; Rizzo, D. J. J. Agric. Food
   Chem. 1982, 30, 1006.
9. Grill, H.; Patton, S.; Cone, F. J. J. Agric. Food Chem.
   1967, 15, 392.
10. Grill, H.; Patton, S.; Cone, F. J. J. Dairy Sci. 1966, 49,
    409.
11. Cuer, A.; Dauphin, G.; Gergomard, A.; Dumont, J. P.; Adda, J.
    J. Agric. Biol. Chem. 1979, 43, 1783.
12. Shankaranarayana, M. L.; Raghaven, B.; Abraham, K. O.;
    Natarajan, C.P. In "Food Flavours, Part A., Introduction";
    Morton, I. J.; Macleod, A. J., Eds.; Elsevier Scientific:
    Amsterdam, 1982; p. 169.
13. Shaw, P. E.; Ammons, J. M.; Braman, R. S. J. Agric. Food
    Chem. 1980, 28, 778.
14. Farbood, M. T. Ph.D. Thesis, Pennsylvania State University,
    Pennsylvania, 1977.

15. Boelens, M.; Van der Linds, L. M.; de Valois, P. J.; Van Dort, H. M.; Takken, H. J. J. Agric. Food Chem. 1974, 22, 1071.
16. Baines, D.A.; Mlotkiewicz, J.A. In "Recent Advances in the Chemistry of Meat"; Bailey, A.J., Ed.; Royal Society of Chemistry, London, 1984; p. 119.
17. Schutte, L.; Koenders, E. B. J. Agric. Food Chem. 1972, 20, 181.
18. Schutte, L. CRC Crit. Rev. in Food Technol. 1974, 4, 457.
19. Takken, H. J.; van der Linde, L. M.; de Valois, P. J.; van Dort, H. M.; Boelens, M. In "Phenolic, Sulfur, and Nitrogen Compounds in Food Flavors"; Charalambous, G.; Katz, I., Eds.; ACS Symposium Series No. 26, American Chemical Society: Washington, D. C., 1976; p. 114.
20. Wilson, R. A.; Katz, I. J. Agric. Food Chem. 1972, 20, 741.
21. Wilson, R. A.; Mussinan, C. J.; Katz, I.; Sanderson, A. J. Agric. Food Chem. 1973, 21, 875.
22. Manning, D. J. J. Dairy Res. 1979, 46:531-537.
23. Ballance, P. E. J. Sci. Food Agric. 1961, 12, 532.
24. Visser, M. K.; Lindsay, R. C. Proceed. Annual Meet. Amer. Soc. Brew. Chem; American Society of Brewing Chemists: St. Paul, MN, 1971; p. 230.
25. Mazelis, M. Nature Lond. 1961, 189, 305.
26. Lieberman, M.; Kunishi, A. T.; Mapson, L. W.; Wardale, D. A. Biochem. 1965, 97, 449.
27. Wainwright, T.; McMahon, J. F.; McDowell, J. J. Sci. Food Agric. 1972, 23, 911.
28. Zanno, P.; Parliment, T. H. Perfum. Flavor. 1982, 7(2), 48.
29. Sloot, D.; Harkes, P. D. J. Agric. Food Chem. 1975, 23, 356.
30. Ades, G. L.; Cone, J. F. J. Dairy Sci. 1969, 52, 957.
31. Moinas, M.; Groux, M.; Horman, I. Lait 1975, 55, 414.
32. Dumont, J. P.; Roger, S.; Cerf, P.; Adda, J. Lait 1974, 54(538), 501.
33. Dumont, J. P.; Roger, S.; Adda, J. Lait 1976, 56(559), 559.
34. Cuer, A.; Dauphin, G.; Kergomard, A.; Dumont, J. P.; Adda, J. J. Appl. Environ. Microbiol. 1979, 38, 332.
35. Hemme, D.; Bouillanne, C.; Metro, F.; Desmazeaud, M. J. Sci. Aliments 1982, 2, 113.
36. Libbey, L. M.; Day, E. A. J. Dairy Sci. 1963, 46, 859.
37. Manning, D. J.; Robinson, H. M. J. Dairy Res., 1973, 40, 63.
38. Manning, D. J. J. Dairy Res. 1974, 41, 81.
39. Manning, D. J.; Price, J. C. J. Dairy Res. 1977, 44, 357.
40. Manning, D. J. J. Dairy Res. 1979, 46, 523.
41. Manning, D. J.; Moore, C. J. Dairy Res. 1979, 46, 539.
42. Tokita, F.; Hosono, A. Jap. J. Zootechn. Sci. 1968, 39, 127.
43. Law, B. A.; Castanon, M.; Sharpe, M. E. J. Dairy Res. 1976, 43, 117.
44. Sharpe, E.; Law, B. A.; Phillips, B. A.; Pitcher, D. G. J. Gen. Microbiol. 1977, 101, 345.
45. Law, B. A.; Sharpe, E. J. Dairy Res. 1978, 45, 267.
46. Green, M. L.; Manning, D. J. J. Dairy Res. 1982, 49, 737.
47. Law, B.A. Dairy Ind. Int. 1980, 45(5), 15.
48. Aston, J. W.; Douglas, K. Aust. J. Dairy Technol. 1983, 38, 66.
49. Fazzalari, F. A. (Ed.) "Compilation of Odor and Taste Threshold Values Data"; American Society for Testing Materials: Philadelphia, 1978.

50. Buttery, R. G.; Guadaghi, D. G.; Ling, L. C.; Seifert, R. M.;
    Lipton, W. J. Agric. Food Chem. 1976, 24, 829.
51. Schaefer, J. Agric. Environ. 1977, 3, 121.
52. Tanaka, H.; Esaki, N.; Yamamoto, T.; Soda, K. FEBS Letters
    1976, 66(2), 307.
53. Ito, S.; Nakamura, T.; Eguchi, Y. J. Biochem. 1976, 79, 1263.
54. Ohigashi, K.; Tsunetoshi, A.; Ichihara, K. Medical J. of
    Osaka Univ. 1951, 2(2), 111.
55. Kreis, W.; Hession, C. Cancer Res. 1973, 33, 1862.
56. Segal, W.; Starkey, R. L. J. Bacteriol. 1969, 98(3), 908.
57. Ruiz-Herrera, J.; Starkey, R. L. J. Bacteriol. 1969, 99(3),
    764.
58. Miwatani, T.; Omukai, Y.; Nakada, D. Medical J. of Osaka
    Univ. 1951, 5(2-3), 347.
59. Tanaka, H.; Esaki, N.; Yamamoto, T.; Soda, K. Biochem. 1977,
    16, 100.10.
60. Ito, S.; Nakamura, T.; Eguchi, Y. J. Biochem. 1975, 78, 1105.
61. Ito, S.; Nakamura, T.; Eguchi, Y. J. Biochem. 1976, 80, 1327.
62. Bradford, M. M. Anal. Biochem. 1976, 77, 248.
63. Brewer, J. M.; Rose, A. J.; Ashworth, R. B. "Experimental
    Techniques in Biochemistry"; Prentice Hall, NY, 1974; p. 3.
64. Rippe, J. K. M.S. Thesis, University of Wisconsin-Madison,
    Wisconsin, 1982.
65. Magee, E. L.; Olson, N. F.; Lindsay, R. C. J. Dairy Sci.
    1982, 64, 616.
66. Jansen, H. E.; Strating, J.; Westra, W. M. J. Inst. Brew.,
    1971, 77, 154.
67. Banwart, W. L.; Bremmer, J. M. Soil Biol. Biochem. 1974, 6,
    113.
68. Kadota, H.; Ishida, Y. Ann. Rev. Microbiol. 1972, 26, 127.
69. Maruyama, F. T. J. Food Sci. 1970, 35, 540.
70. Braun, S. D.; Olson, N. F.; Lindsay, R. C. J. Food Biochem.
    1983, 7, 23.
71. Josephson, D. B.; Lindsay, R. C.; Stuiber, D. A. J. Agric.
    Food Chem. 1983, 31, 326.
72. Kosikowski, F. V. In "Chemistry of Natural Food Flavors";
    Mitchell, J. H.; Leinen, N. J.; Mrak, E. M.; Bailey, S. D.,
    Eds.; Quartermaster Research & Engineering: Washington, D. C.,
    1957; p. 133.
73. Morita, K.; Kobayashi, S. Tetrahedron Lett. 1966, 6, 573.
74. Yasumoto, K.; Iwami, K.; Baba, Y.; Mitsuda, H. J. Jap. Soc.
    Food Nutr. 1971, 24, 467.
75. Dubs, P.; Stussi, R. Helv. Chim. Acta 1978, 71, 2351.
76. Mosandl, A.; Heusinger, G. Lebensmittel Chem. Gerichtl.
    Chem. 1985, 39, 29.
77. Shu, C. K.; Hagedorn, M. L.; Mookherjee, B. D.; Ho, C. T. J.
    Agric. Food Chem. 1985, 33, 638.
78. Swaisgood, H. E. Enzyme Microb. Technol. 1980, 2, 265.
79. Swaisgood, H. E.; Abraham, P. J. Dairy Sci. 1980, 63, 1205.
80. McPherson, A. V.; Kitchen, B. J. J. Dairy Res. 1983, 50, 107.
81. Zikakis, J. P.; Wooters, C. J. Dairy Sci. 1980, 63, 893.
82. Walsh, C. "Enzymatic Reaction Mechanisms"; W. H. Freeman:
    San Francisco, 1979; p. 440.

RECEIVED May 28, 1986

BIOGENERATION OF SELECTED AROMAS

24

Generation of Flavor and Aroma Components by Microbial Fermentation and Enzyme Engineering Technology

Ian L. Gatfield

Haarmann & Reimer GmbH, 3450 Holzminden, West Germany

The flavor and aroma of many foodstuffs
are formed as a result of the action of
enzymes and microorganisms. This knowledge
can be exploited to obtain both complex
flavor mixtures and individual flavor
compounds via microbial fermentation
and/or enzyme technology. Many different
classes of compound can be obtained this
way including esters, lactones, aldehydes,
ketones and acids.

Microorganisms and enzymes play a very important role in the
production of flavor compounds in a wide variety of foodstuffs
and man has unwittingly relied upon them for centuries to
improve the quality of his food. Long before microorganisms
and enzymes were known to exist, man began to learn how to
encourage and exploit their fermentative action. Initially,
progress was made on an empirical basis without knowing
what was happening biologically. Today, with microbial activity
fairly well understood, fermented foods and drinks account for
a large and important sector of the food industry.
    Probably the most important reason for ancient man to
continue making fermented foods, after their chance discovery,
was the increase in shelf-life usually bestowed upon them by
the fermentation process. This can be due to a number of factors,
such as the drop in the pH value in converting for example milk
to yogurt, and this acidification protects the latter from subsequent deterioration. In the case of fermented fruit juices,
these are rendered more stable as a result of the conversion
of sugar to ethanol, the general antimicrobial properties of
which are well-known. The fact that this type of "technology"
provided ancient man with foodstuffs with greater shelf-lives
was very important and meant that not so much time and effort
was necessary to obtain his daily food and that stockpiling
was possible. It was only after a certain, unknown length of
time that these types of products became popular in their own

0097–6156/86/0317–0310$06.00/0

right. In the case of alcoholic drinks, one could imagine that
these were soon consumed, mainly as a result of the physiological
effect of the alcohol present. But soon after one became partial
to the flavors of these fermented foodstuffs which were distinctly
different from those of the corresponding starting materials.
This resulted in a very welcome flavor variety in days gone by.
The increased shelf-life of fermented foods completely lost importance with the onset of such reliable and cheap alternative
procedures such as canning and freezing. The reason why this type
of traditional food processing technology has persisted to the
present day is partly due to the flavor generated by these processes.

Microorganisms are living systems and some produce metabolites
which are flavor-active. The metabolic reactions are catalyzed by
the enzymes present in the microorganisms. Flavor-active metabolites can belong to most classes of chemical compounds and
include acids, alcohols, lactones, esters, aldehydes, ketones,
etc. If these materials accumulate in the foodstuff, then they
influence its final flavor. It is conceivable that this knowledge could be exploited for the manufacture of flavors. Two
distinct strategies are possible involving the use of enzymes
or microorganisms to produce either complex, multicomponent
flavor systems or individual flavor compounds.

Multicomponent Flavor Systems

Milk was probably one of the first agricultural products and is
rapidly infected by bacteria which sour it by converting lactose
to lactic acid. As a result, one obtains such products as yogurt
and also cheese. During the ripening of cheese, a number of reactions proceed simultaneously, representing an extremely dynamic
system. Some of the reactions are caused by microbial enzymes
and others by those enzymes present in the milk. Milk fat,
protin and carbohydrate are degraded to varying degrees thereby
yielding a very complex mixture of compounds, some of which
contribute to the characteristic cheese flavor.

Fatty Acids. One of the very first fermentation flavors,
which was produced on a large scale was a product known as
lipolyzed milk fat (1) in which cultured cream was subjected
to a controlled enzymatic hydrolysis using lipases. This type
of enzyme hydrolyzes triglycerides and liberates individual
acids. The composition of these fatty acids depends upon the
specificity of the lipases used. Some show a high degree of
specificity towards short chain fatty acids (Figure 1) (2,3),
some towards the long chain variety, and some display no particular preference. Another type of specificity is displayed
towards the position of the fatty acid residue in the trigly-
ceride molecule. The liberation of the fatty acids has to be carefully controlled since at higher concentrations an unpleasant,
soapy flavor profile is obtained mainly from the long chain acids.
The importance of free fatty acids in the characteristic
flavors of various dairy products is an undisputed fact and the

different fatty acids display considerably different flavor characteristics. For example, butyric acid is of major importance in determining both the intensity and characteristic flavor of Romano and Provolone cheeses (3). The flavors of other types of cheeses, especially the hard type of Swiss cheese, rely to a considerable extent upon the presence of propionic acid which does not originate from the milk fat but rather is produced fermentatively by Propionibacteria present in the typical Swiss cheese starter culture. These bacteria not only give the cheese a characteristic flavor but also give rise to the holes typical of such cheeses. Such lipolyzed milk fat products can be used for the enhancement of butter-like flavors and flavor development in milk chocolate (5, 6, 7).

An extension of this technology lead to the development of enzyme modified cheeses or EMC which are a very important product. The basic function of the process was to shorten the ripening time of a mature cheese without losing flavor. The potential financial gains are obvious. Young cheeses are subjected to a controlled lipolysis and proteolysis which is brought about by adding suitable microbial enzymes (8). After thermal inactivation of the enzymes, a pasty product is obtained which can have a flavor intensity of up to 20 times that of the mature cheese.

Methyl Ketones. A very popular type of cheese is the blue-type of which there are many different varieties. The most important factor they have in common is the deliberate inoculation of the young cheese with a mold, usually species of a Penicillium. Penicillium roqueforti is the principal species used and it imparts to the cheese its typical flavor. Proteolytic and lipolytic enzymes of the mold play a major role in the development of the final flavor. The enzymes are excreted into the milk where the protease hydrolyzes the casein present thereby producing both flavor precursors and substrates for mold growth. The lipase produces free fatty acids which are flavor active. However, fatty acids are also toxic to P. roqueforti, and the degree of toxicity depends on the concentration of the acids, their chain length and the pH of the medium (9). It seems that this mold has developed a detoxifying mechanism and converts the fatty acids into methyl ketones which are generally considered the key flavor components of blue cheese (10).

A great deal of research has been directed towards the elucidation of this oxidation mechanism which is summarized in Figure 2. The free fatty acids liberated by the Penicillium roqueforti lipase are oxidized to β-keto acids which undergo decarboxylation to form the methyl ketones. Under certain circumstances, the methyl ketones are reduced to the corresponding secondary alcohols (11). Both the spores and the mycelium seem capable of producing methyl ketones from fatty acids (12, 13). Furthermore, both short chain and long chain fatty acids are metabolized, thereby giving rise to an homologous series of methyl ketones, the main ones being 2-pentanone, 2-heptanone and 2-nonanone (14). A number of processes have been developed and patented for producing blue cheese flavor via the fermentation of milk fat (15, 16). Usually the thermally inactivated fermentation broth is spray dried thereby yielding the final multicomponent flavor. The enzyme modified cheeses, lipolyzed milk fats and the blue cheese flavor produced by fermentation, are examples of complex multicomponent flavors which are being produced using enzymes and/or via microbial fermentation.

Single Flavor Components

The other strategy mentioned above is the biosynthesis, isolation and purification of individual flavor-active species. This approach involves exploiting specific bio-conversions, such as oxidation and reduction or de novo syntheses by either microbial fermentation or by using specific enzyme systems.

Low Molecular Weight Carbonyl Compounds. In the dairy field, a major product made this way is starter distillate. The main component is diaceyl which is a very important aroma compound responsible for the characteristic buttery flavor of fermented dairy products such as sour cream or buttermilk. The dairy industry relies upon fermentation by lactic streptococci for the production of diacetyl in cultured products. Starter distillate is a natural product rich in diacetyl which is produced by distilling such lactic cultures. The key intermediate in the biosynthesis of diacetyl is &-acetolactic acid which is decarboxylated to form diacetyl (Figure 3). The starting material of the biosynthetic pathway is citrate which is a natural component of milk.

Another important, low molecular weight flavor compound is acetaldehyde. This material plays a significant role in the flavor of yogurt and certain fruit flavors, such as orange. A procedure was patented in 1984 for the conversion of ethanol to acetaldehyde (17) using the enzyme alcohol dehydrogenase (ADH). This process for the production of natural acetaldehyde is complex (Figure 4) and suffers from the same drawbacks as all enzymatic oxidation reactions which rely upon cofactor regeneration. Specifically, the process involves the use of alcohol dehydrogenase and its cofactor nicrotinamide adenine dinucleotide, which during the course of the reaction is reduced and then regenerated by light-catalyzed oxidation with flavin mononucleotide (FMN). The reduced flavin mononucleotide ($FMNH_2$) is reconverted to FMN by oxidation with molecular oxygen. The by-product of this reconversion is hydrogen peroxide which is in turn decomposed, by the action of the enzyme catalase, to oxygen and water. Conversion rates in the range of 10 - 20% are typically obtained, and in a continuous batch reactor system, the concentration of acetaldehyde achieved is some 2.5 g/l after a period of 9 hours.

Lactones. Lactones are generally very pleasant, potent flavor materials which are very widely distributed in Nature (18). Lactones have been isolated from all the major classes of food including fruits, vegetables, nuts, meat, milk products and baked products.

Relatively old publications indicate that some microorganisms can biosynthesize lactones. Thus it was shown (19) that various

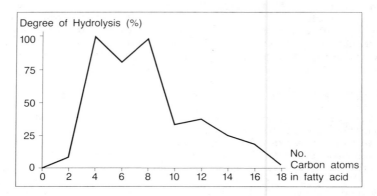

Figure 1. Effect of chain length of fatty acid upon degree of hydrolysis of synthetic triglycerides by the esterase-lipase from Mucor miehei.

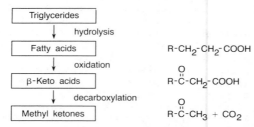

Figure 2. Schematic oxidation mechanism of fatty acids by Penicillium roqueforti.

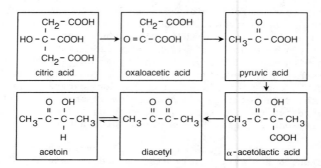

Figure 3. Biosynthesis of diacetyl from citric acid.

Candida strains convert ricinoleic acid into γ-decalactone, which
displays the fatty, fruity aroma typical of peaches. Ricinoleic
acid (12-hydroxy octadec-9-enoic acid) is the major fatty acid
in castor oil (approx. 80 %). The yeast can lipolyze castor oil
glycerides and the liberated ricinoleic acid is subsequently
metabolized via β-oxidation and eventually converted to 4-hydroxy-
decanoic acid (Figure 5). Recently a European patent has been
filed (20) essentially covering the same procedure. Shake culture
fermentations were carried out on 100 ml scale for one week. The
4-hydroxydecanoic acid formed was converted to γ-decalactone by
boiling the crude, acidified (pH 1.5) fermentation broth for a
period of 10 minutes. The lactone was isolated via solvent extraction and a yield of some 5 g/l was obtained. The same lactone
was detected as the major volatile component formed when the yeast,
Sporobolomyces odorus was grown in standard culture medium (21).
Although the culture medium displayed an intense fruity, typical
peach-like odor, the concentration of γ-decalactone amounted to no
more than 0.5 mg/l.

Related studies have shown that other microorganisms are
capable of producing lactones when grown on standard culture
media devoid of special substrates or precursors. Thus, Trichoderma viride, a soil fungus, produces a strong, coconut-like
aroma when grown on potato dextrose medium (22). The volatile
oil obtained via steam distillation of the culture medium, contained over 90 % 6-pentyl-α-pyrone which was responsible for
the characteristic aroma and which has also been detected in
peaches. The yield of the lactone amounted to some 170 mg/l of
the culture medium; low when compared to the production of
γ-decalactone by similar means. However, the chemical synthesis
of the pyrone is complex and requires seven distinct steps (23).
Recently it was reported that the wood-rotting fungus, Polyporus
durus, when grown on synthetic medium produced an intense fruity,
coconut-like odor (24). This aroma was due to the presence of
a number of γ-lactones, of which γ-octalactone was the main
component at a concentration of 7 mg/l culture medium. These
low concentrations indicate one of the main problems encountered
when considering the feasibility of the fermentative production
of individual aroma components. The low threshold levels
intrinsic to many aroma chemicals mean that cultures with very
promising odors more often than not, do not produce the required
materials at anything like acceptable levels.

Esters. Esters are extremely important aroma compounds and there
are many reports that esters are biosynthetic products of bacterial action. Thus, the fruity flavor defect sometimes found
in cheddar cheese is due to the presence of esters, principally
ethyl butyrate and ethyl caproate (25). Similar esters can be
found in beer in which both fusel alcohols and the short chain
fatty acids, acetic and butyric, are also present. These materials can undergo esterification, which in this case is mediated
by the enzyme alcohol acetyltransferase present in the yeast used
for beer fermentation (26). There are a number of esters present
in wine which are metabolically produced by the yeast. Of these,

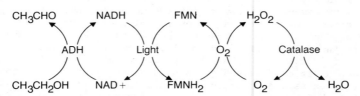

Figure 4.   Enzymatic oxidation of ethanol to acetaldehyde according to U.S. Patent 4 481 292.

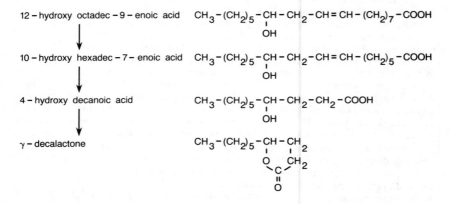

Figure 5.   Oxidative degradation of ricinoleic acid by Candida lipolytica.

ethyl acetate seems to play an important role as far as taste and
aroma are concerned.

The fungus Ceratocystis moniliformis has been described as
having potential as a source of flavor concentrates (27). Selected cultures when grown on synthetic media, gave rise to fruity
aromas, the composition of which varied depending on the carbon
source. All extracts, however, contained acetates of the fusel
alcohols which are commonly found in fruit. Unfortunately, no
estimation of product yields was made. Significant amounts of
ethyl acetate can be produced by the yeast Hansenula anomala
from glucose or ethanol, but only after an extensive fermentation
of several weeks (28). A more promising procedure would appear
to be that mediated by the yeast Candida utilis which has been
published recently (29). The rates of production and yields
of ethyl acetate from ethanol were superior to other known ester-
producing microorganisms. Under optimal conditions, at pH 7.0
in phosphate buffer and with an initial ethanol concentration
of 10 g/l, an ethyl acetate concentration of approximately 4.5 g/l
was obtained after a 5 hour incubation period.

An entirely different strategy for ester synthesis involves
the use of isolated enzymes. A number of lipases are able to synthesize esters from alcohols and acids under certain conditions.
Thus, Iwai and co-workers (30, 31) were able to show that certain
microbial lipases, for example that from Aspergillus niger, could
synthesize terpene esters such as the propionates, butyrates and
caproates of the primary terpene alcohols, geraniol and citronellol.
The best results were obtained when a mixture of geraniol and butyric acid, at a 2:1 molar ratio, was stirred at 30° C for 18 hours
with the enzyme solution. The final system contained some 70 %
water, and the yield of geranyl butyrate amounted to 40 % of the
theoretical yield. Secondary and tertiary alcohols were not
esterified by these enzymes. This strategy has been extended to
other areas of the food industry and has been described for the
inter-esterification of fats using both immobilized (32) and
soluble (33) lipase preparations. It is thus possible to alter
the fatty acid composition of naturally occurring triglycerides
to obtain fat with different melting characteristics. Thus,
cocoa butter-like fat can be produced from palm oil via enzymatic
interesterification.

We have investigated the ester synthesizing capabilities of
the esterase-lipase enzyme preparation derived from Mucor miehei
(34) which is used in cheese manufacture. The performance of the
enzyme in non-aqueous systems was of major interest since, if
successful, this could provide a way of drastically improving
not only the ester yield but also of simplifying the isolation
and purification procedures when compared with aqueous fermentation systems.

The replacement of water by an organic solvent is almost
invariably accompanied by a dramatic decrease in the catalytic
activity of an enzyme and a decline in its substrate specificity.
Thus, significant activities in non-aqueous media are the exception rather than the rule (35, 36). Surprisingly, the esterase-
lipase preparation showed very good esterification properties
and was tested extensively in a model system consisting of oleic
acid and ethanol. Equimolar quantities of oleic acid and ethanol
undergo rapid esterification when stirred at room temperature

with 3 % by weight, relative to the acidic component, of the
Mucor miehei enzyme (Figure 6). Increasing the ethanol concentration of the heterogeneous system causes only slight inhibition
of the enzyme such that a yield of some 75 % is obtained after
three days when a sevenfold molar excess of ethanol is employed.
The enzyme is remarkably stable and can be reused a number of
times. Thus, the same enzyme was used for 12 cycles over a 36
day period in the model system, without any apparent loss of
activity. The esterification yield varied in a random fashion
between 83 % and 87 % during the 12 cycles.

The enzyme exhibits pronounced substrate specificity.
The degree of esterification achieved depends, among other things,
upon the chain length of the carboxylic acid (Figure 7). Propionic
and acetic acids are very poor substrates and do not undergo
esterification to any considerable extent under these conditions.
The addition of a higher molecular weight acid such as capric
acid, is beneficial, and under these conditions high yields of
propionates are obtained, presumably via a transesterification
mechanism. Using this enzymatic technique it is possible to
carry out relatively large scale preparative syntheses of natural
flavor-active esters, using natural raw materials as substrates.
This enzyme system is also capable of producing certain terpene
esters in high yield, for example geranyl butyrate. Furthermore,
incubation of toluene solutions of hydroxy acids, such as 4-hydroxy-
butyric acid and ω-hydroxypentadecanoic acid, with the enzyme
system, gives rise to the corresponding γ-butyrolactone and the
macrocyclic lactone pentadecanolide.

The synthetic capability of the Mucor miehei enzyme was
examined at low temperatures where the model system was diluted
with an equal weight of n-hexane in order to permit stirring.
The rate and degree of esterification determined at -22° C were
only nominally poorer than those determined at room temperature
(Figure 8) (37). Recently it was demonstrated that dry porcine
pancreatic lipase can not only withstand heating at 100° C for
many hours, but it also exhibits a high catalytic activity at
that temperature (38). From these results, it would appear
that certain enzymes can be used under surprisingly severe
conditions to bring about novel types of reactions. It is
expected that this type of technology, using immobilized enzymes
for example and possibly in organic solvents, will be developed
further and used more extensively in the aroma field. A related
example is the recently published enzymatic synthesis of a mixture of flavor-active ethyl esters from butter fat using the
lipase from Candida cylindracea (39). Both hydrolyzed and non-
hydrolyzed butter fat underwent considerable esterification with
ethanol in the presence of this lipase. The ethyl ester profiles
indicated that the enzyme showed a high specificity towards butyric
and caproic acids and that caproic acid exhibited the slowest rate
of esterification.

Pyrazines. Pyrazines are heterocyclic, nitrogen-containing compounds which exhibit very interesting flavor characteristics.
It has been claimed that no other class of flavoring compounds
is as important in flavoring foods (40). Pyrazines have been
detected in many processed foodstuffs and are often associated
with "roasted" and "nutty" flavors. Some raw vegetables owe
their typical flavor to the presence of certain pyrazines,

possibly the best example being 2-methoxy-3-isobutyl pyrazine
in green, bell peppers.

   Early research by Demain showed that a mutant of Corynebacterium glutamicum was able to convert certain amino acids, including leucine, isoleucine and valine into tetramethyl pyrazine
(41) which accumulated in the culture medium at a concentration
of 3 g/l after 5 days. Scattered reports indicate that certain
microorganisms can produce pyrazines thereby giving rise to
off-flavors in food. For example 2-methoxy-3-isopropyl pyrazine
was identified as the compound responsible for a musty potato
flavor defect in eggs and milk (42). This compound had been produced by Pseudomonas taetrolens. Although, the production of
pyrazines in these systems gives rise to flavor defects, the
responsible organisms clearly do have potential for producing
pyrazines destined for application in flavors.

Green Components. The major volatile flavor constituents of
many fruits and vegetables arise from the action of endogenous
enzymes upon flavor precursors present in the tissue. Thus,
the action of lipoxygenase on the unsaturated fatty acids,
linoleic and linolenic, (Figure 9) followed by the action of
hydroperoxide lyase gives rise to hexenal and 2-hexenal in
apples, 2,6-nonadienal in cucumbers and 2- or 3-hexenal in
tomatoes (43). Enzymatic systems for the preparative scale production of such "green" flavor components via the oxidation of fatty
acids have not been described. The likely reason for this is
the lack of availability of the lyase enzyme. Efforts to obtain
these types of flavor compounds seem to be restricted to the
use of endogenous enzymes present in the plant tissues. Thus
Guadagni has shown that aroma formation in apples is more intense
in the skin than in the flesh, and that a strong apple aroma is
developed within 1 or 2 days if apple peels are stored in a
closed vessel at room temperature (44).

Mustard Oils. The characteristic pungent taste of mustard,
cress and horseradish is due to the formation of mustard oils
or isothiocyanates from the corresponding, odorless precursors
known as glucosinolates by the action of the endogenuos enzyme
myrosinase (Figure 10) (45). A system has been described (46)
for the use of immobilized myrosinase in the continuous production of horseradish aroma. Thus, horseradish extract is either
pumped through a plug-flow reactor containing the immobilized
myrosinase or through a membrane reactor which has the myrosinase
immobilized on the membranes. The liberation of the isothiocyanates is extremely rapid, whereby the turnover and breakdown of
the glucosinolates depends upon the enzyme-substrate contact time.
The half-life of immobilized myrosinase is in excess of 100 days
when determined under operational conditions.

Future Prospects

One area where biotechnology could make an impact is the flavor
and fragrance field, and some of the flavor and fragrance firms
are active in this area (48). As described in this review there
are a number of enzymatic and/or fermentative processes already
being used to produce materials of interest to the industry.

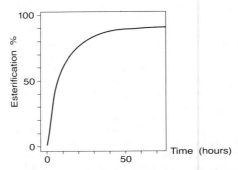

Figure 6. Esterification profile for the model system oleic
acid-ethanol using the esterase-lipase from Mucor miehei.

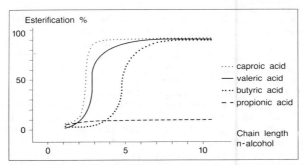

Figure 7. Effect of chain length of carboxylic acid upon the
degree of esterification obtained with n-alcohols.

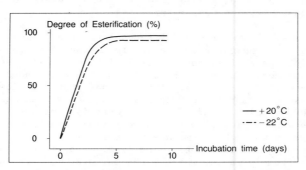

Figure 8. Effect of temperature upon the esterification profile for oleic acid-ethanol model system.

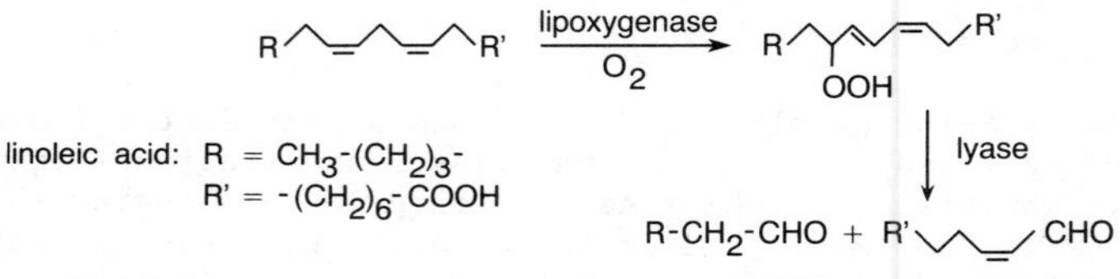

linoleic acid: R = CH_3-$(CH_2)_3$-
                R' = -$(CH_2)_6$- COOH

Figure 9. Enzymatic formation of aldehydes from unsaturated
fatty acids.

Furthermore, there are very many reports in the literature of the biosynthesis of volatile compounds by microorganisms, some of which could be produced commercially by fermentation. One of the best reviews in this field is that of Schindler and Schmid (49).

It is conceivable that this type of technology could be used to develop perfumery materials as well. There are already examples of this approach including the formation of *w*-hydroxy fatty acids (50) useful as raw materials for synthetic musks, and the potential is great for utilizing the excellent synthetic capabilities of microorganisms to obtain materials, difficult to synthesize chemically. This strategy has been industrialized in the manufacture of steroids and attempts to produce new perfumery materials in this way, are beginning to appear in the literature (51, 52).

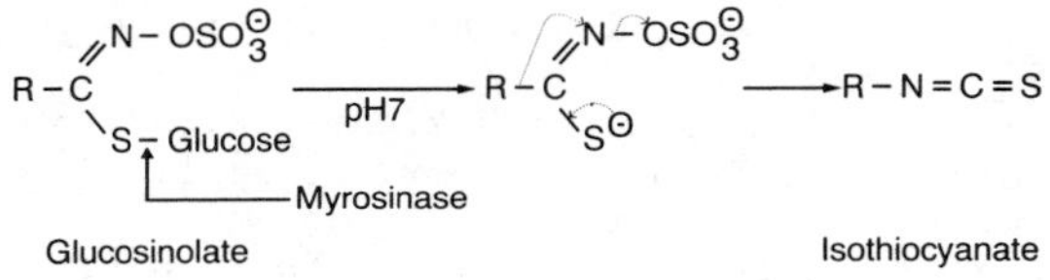

Figure 10. Action of myrosinase upon glucosinolates.

Literature Cited

1.   Pangier, D.J.; U.S. Patent 3 469 993, 1969.
2.   Moskowitz, G.J.; Cassaigne, R.; West, I.R.; Shen, T.; Feldman, L.I. J. Agric. Food Chem., 1977, 25, 1146.
3.   Kanisawa, T.; Yamaguchi, Y.; Hattori, S., Nippon Shokuhin Kogyo Gakkaishi 1982, 29, 693.
4.   Arnold, R.G.; Shahani, K.M.; Dwivedi, B.K.; J. Dairy Sci. 1975, 58, 1127.
5.   Anti, A.W.; Food Proc. Dev. 1969, 3, 17.
6.   Higashi, T.; Kobunshi 1967, 16, 1220.
7.   Tanabe Seiyaku Ltd. Product Information Bulletin on Lipase RH, Osaka, Japan.
8.   Sood, V.K.; Kosikowski, F.V.; J. Dairy Sci. 1979, 62, 1865.
9.   Pressmann, B.C.; Lardy, H.A.; Biochem. Biophys. Acta 1956, 21, 458.
10. Dwivedi, B.K.; CRC Crit. Rev. in Food Techn. 1973, 457.
11. Hawke, J.C.; J. Dairy Res. 1966, 33, 225.
12. Gehrig., R.F.; Knight, S.G.; Nature 1958, 182, 1237.
13. Lawrence, R.C.; Hawke, J.C.; J. Gen. Microbiol. 1968, 51, 289.
14. Moskowitz, G.J.; La Belle, G.G.; in "Flavor of Foods and Beverages"; Charalambous G.; Inglett, G., Ed.; Academic: New York, 1978; p. 21.
15. Knight, S.G.; U.S. Patent 3 100 153, 1963.
16. Watts, J.C.; Nelson, J.H.; U.S. Patent 3 072 488, 1963.
17. Raymond, W.R.; U.S. Patent 4 481 292, 1984.
18. Maga, J.A.; CRC Crit. Rev. in Food Sci. Nutr. 1976, 10, 1.
19. Okui, S.; Uchiyama, M.; Mizigaki, M.; J. Biochem. 1963, 54, 536.

20. Farbood, H.; Willis, B.; Europ. Patent PCT 1072, 1983.
21. Tahara, S.; Fujiwara, K.; Ishizaka, H.; Mizutani J.; Obata, Y.; Agr. Biol. Chem. 1972, 36, 2585.
22. Collins, R.P.; Halim, A.F.; J. Agric. Food Chem. 1972, 20, 437.
23. Nobuhara, A.; Agr. Biol. Chem. 1969, 33, 1264.
24. Drawert, F.; Berger, R.G.; Neuhauser, K.; Chem. Mikrobiol. Technol. Lebensm. 1983, 8, 91.
25. Hosono, A.; Elliott, J.A.; J. Dairy Sci. 1975, 57, 1432.
26. Yoshioka, K.; Hashimoto, N.; Agr. Biol. Chem. 1981, 45, 2183.
27. Lanza, E.; Ko, K.W.; Palmer, J.K.; J. Agric. Food Chem. 1976 24, 1247.
28. Davies, R.; Falkiner, E.A.; Wilkinson, J.F.; Peel, J.L. Biochem J. 1951, 49, 58.
29. Armstrong, D.W.; Martin, S.M.; Yamayaki, H.; Biotechn. Bioeng. 1984, 26, 1038.
30. Iwai, M.; Akumura, S.; Tsujusaka, Y.; Agr. Biol. Chem. 1980, 44, 2731.
31. Akumura, S.; Iwai, M.; Tsujisaka, Y.; Biochem. Biophys. Acta 1979, 575, 156.
32. Yokozeki, K.; Yamanaka, S.; Takinami, K.; Hirose, Y.; Tanaka, A.; Sonomoto, K.; Fukui, S.; Eur. J. Appl. Microb. Biotechn. 1982, 14, 1.
33. Stevenson, R.W.; Luddy, F. E.; Rothbart, H.L.; J. Amer. Oil, Chem. Soc. 1979, 56, 676.
34. Gatfield, I.L.; Sand, T.; German Patent 3 108 927, 1981.
35. Butler, L.G.; Enzyme Microb. Technol. 1979, 1, 253.
36. Klibanov, A.M.; Samokhin, G.P., Martinek, K.; Berezin, I.V.; Biotechn. Bioeng. 1977, 19, 1351.
37. Gatfield, I.L.; unpublished data.
38. Zaks, A.; Klibanov, A.M.; Science 1984, 224, 1249.
39. Kanisawa, T.; Nippon Shokuhin Kogyo Gakkaishi 1983, 30, 572.
40. Maga, J.A.; CRC Crit. Rev. Food Sci. Nutr. 1982, 16, 1.
41. Demain, A.L.; Jackson M.; Tanner, N.R.; J. Bact. 1967, 94, 323.
42. Morgan, M.E.; Libbey, L.M.; Scanlan, R.A.; J. Dairy Sci. 1972, 55, 666.
43. Ruttloff, H.; Nahrung 1982, 26, 575.
44. Guadagni, D.G.; Bomben, J.L.; Hudson, J.S.; J. Sci., Fd. Agric. 1971, 22, 110.
45. Ettlinger, M.E.; Dateo, G.P.; Harrison, B.W.; Mabry, T.J.; Thompson, C.P.; Proc. Nat. Acad. Sci. (USA) 1961, 47, 1975.
46. Gatfield, I.L.; Schmidt-Kastner, G.; Sand, T.; German Patent 3 312 214, 1983.
47. Schwimmer, S.; Friedmann, M.; Flavour Ind. 1972, 137.
48. van Brunt, J.; Biotechnology 1985, 3, 525.
49. Schindler, J.; Schmid, R.D.; Process Biochem. 1982, 17, (5), 2.
50. Mikami, Y.; Jap. Patent 56-17075, 1981.
51. Krasnobajew, V.; Europ. Patent 12246, 1979.
52. Mikami, Y.; Fukumaga, Y.; Arita, M.; Kisaki, T.; Appl. Environ. Microb. 1981, 41, 610.

RECEIVED March 10, 1986

Generation of Flavor and Odor Compounds through Fermentation Processes

L. G. Scharpf, Jr., E. W. Seitz, J. A. Morris, and M. I. Farbood

Research and Development Department, International Flavors & Fragrances, Union Beach, NJ 07735

Since Neolithic times, microorganisms have been used to achieve desired flavors in foods and beverages. In recent years, a commercial demand has developed for flavoring materials derived from natural sources. Microbiologists and flavorists are exploiting the fermentative action of microorganisms to improve the flavor of alcoholic beverages, cheese, yogurt, bread, fruit, and vegetable products. Many chemicals produced by microbial processes are odorants or tastants. Categories of molecules will be reviewed as a function of substrates and microbial strains. That microorganisms are also capable of de novo synthesis of tastants and odorants from suitable substrates will be illustrated by several examples.

Since the beginning of time, flavors and fragrances have played an important role in providing man with happiness, beauty and satisfaction.

Up to this century many natural flavor materials were obtained from animals and higher plants. Supplies of many of these materials have dwindled due to social, economic, and political factors, conservation, wildlife protection and industrial growth.

The use of microorganisms may offer an alternative method for producing natural flavor and fragrance materials. The release of odors by microbial cultures is well known by microbiologists. Many of the volatiles are produced in only trace quantities but, due to their strength, sufficient for taste and smell.

Harnessing microorganisms to produce volatile flavor and odor materials in large quantity is a real technical challenge, for reasons which will be presented. Nevertheless, fermentation technology offers the potential of producing flavor and aroma substances of interest to today's consumer.

0097–6156/86/0317–0323$07.00/0

Sensory Properties of Microbial Metabolites

Many microbial metabolites are volatile compounds and in terms of
their sensory properties can be broken into two broad categories;
odorants and tastants (Table 1).   Tastants include salty, sour,
sweet, and bitter compounds such as amino acids, peptides, and sugars.   Primary odorants typically are quite volatile and include
carbonyl compounds, esters, and terpenes.   There is considerable
overlap between the two categories; lactones, for example, have both
taste and odor properties.   In keeping with the theme of this symposium, volatile aroma substances will be the primary focus.

Microbial Sources of Flavor and Fragrance Materials

A.   Diversity of Metabolic Requirements

Many microorganisms require oxygen for their growth, reproduction, and maintenance of cells.   These requirements vary widely,
depending on the genus of the organism.   The strict anaerobes
metabolize and grow in the absence of free oxygen.   Examples are
members of the bacterial genus Clostridium, which produce the
volatile chemicals, ethanol, butanol, acetone, and acetate(1).
Strict aerobes metabolize and grow only in the presence of molecular
oxygen.   Among the aerobes are the filamentous fungi which produce
many flavor/odor substances such as aldehydes, ketones and esters.
Facultative microorganisms are capable of switching their metabolic
processes from an aerobic mode to an anaerobic mode, depending on
the environment in which they find themselves.   Industrial yeasts
such as Saccharomyces, which produce bread and wine flavors, are
members of this class.

Besides oxygen, microorganisms also have other requirements
for their growth:
1. Carbon   Source   -   Carbon   compounds   are   principle
   nutritional   components   in   a   fermentation,   providing
   both   energy   and   building   blocks   for   microbial
   metabolism.   The most common sources of carbon are the
   carbohydrates such as starch and sugars.   Microorganisms
   are versatile, however, and in some cases can be forced
   to grow on other organic molecules (even hydrocarbons)
   as their sole carbon source.   For example, amino acids
   can be produced from n-alkanes by amino acid auxotrophs
   derived from Corynebacterium(2).
2. Nitrogen and Phosphorus - Both of these elements are
   incorporated into the structural and functional cell
   components.   They can also become part of product molecules.
3. Other Nutrients - Minerals, vitamins, and trace minerals
   are also required in most fermentations.

B.   Primary vs. Secondary Metabolites

The biosynthetic pathways of microorganisms can be manip-
ulated to produce certain molecules that are normally limited to
very small amounts by cellular regulatory mechanisms.   Industrial

Table 1. Sensory Properties of Microbial Metabolites

PRIMARY ODORANTS	TASTANTS & ODORANTS	PRIMARY TASTANTS
Aldehydes . acetaldehyde . phenylacetaldehyde	Amines	Amino Acids
Ketones . diacetyl . acetophenone	Fatty Acids	Peptides
Esters . ethyl butyrate	Pyrazines	Sugars
Alcohols . butanol	Lactones	Polyols
Terpenes . citronellal		

microbiologists have developed techniques to select mutant strains
in which the regulatory process is altered in a way that certain
metabolites can be over produced. Primary metabolites are
substances such as sugars and amino acids which are essential for
cell growth(3). Some of these substances are taste and odor active.
Certain amino acids and peptides are sweet while others are bitter
or salty.

Antibiotics are good examples of secondary metabolites
produced by fermentation on an industrial scale. Secondary
metabolites are substances which are not required by the cell for
biosynthesis. Most volatile and odor-active materials fall into
this category of metabolites, including alcohols, aldehydes,
ketones, terpennoids, and lactones. The highly odorous simple
esters formed by fungi are secondary metabolites. They are thought
to provide a mechanism for removing both acid and alcohol precursors
from cells and the media; if allowed to accumulate these chemicals
could become toxic to the organism(5). Secondary metabolites
presumably contribute to the microorganisms' survival by possibly
inhibiting competitive species that could otherwise occupy the same
environmental niche,(3). Collins (4) suggests that with fungi:

1. The release of volatiles into their surrounding environment helps to regulate their competitors. This phenomenon known in higher plants, is called alleopathy.
2. Some volatiles stimulate spore germination.
3. Some volatiles serve as attractants.

Microorganisms that produce secondary metabolites generally
undergo a period of logarithmic growth in which the synthesis of the
secondary metabolite is negligible. When the culture enters the
stationary phase, secondary metabolite production is often trig-
gered.

C.   Traditional Whole Food Fermentations

The most ancient example of the production of volatiles by
microorganisms are the traditional food fermentations. Fermented
foods are technological products which have been converted by microorganisms to foods which are more appealing than the raw material
used. One significant outcome is the production of volatile aroma
and taste substances. Some fermented foods are produced using pure
cultures of microorganisms. Others make use of organisms
indigenous to the raw material. Texts by Margalith(6) and Rose(7)
review the flavor compounds produced by microorganisms in fermented
foods.

1. Milk Substrates   Cheese is perhaps the oldest of the
   fermented foods. (Fig. 1) The basic underlying microbial transformation in all cheese manufacture is the conversion of lactose of milk into lactic acid. The microorganisms in the starter culture contribute
   significantly to the flavor of the cheese. The
   secondary microbial flora of the cheese also elaborate
   taste and odor active substances. These organisms may
   be present as chance contaminants or introduced
   intentionally(8), and result in distinctive types of
   cheeses such as cheddar, blue veined and swiss.

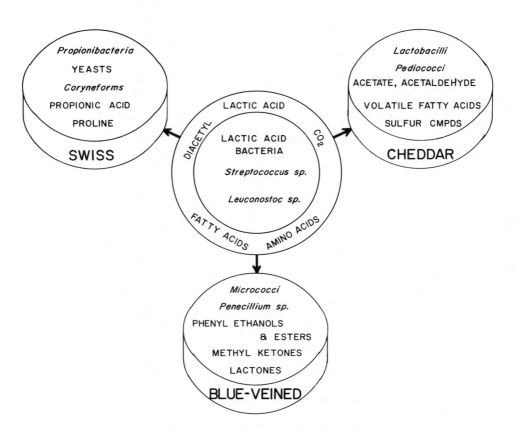

Figure 1.   Cheese odorants and tastants.

Unripened cheeses are made using simple lactic acid
bacterial cultures or streptococcal strains which are
facultative anaerobes.   Cottage cheese is a good
example in which there is bacterial production of
diacetyl the principal flavor component, derived as a
by-product of pyruvate metabolism.   Some cottage cheese
starter cultures also produce acetaldehyde, which may
impart a harsh flavor.   The starter cultures also contribute to the ripening process by degrading proteins
to primary metabolites such as peptides, free amino
acids, and sulfur compounds, and degrading lipids to
free fatty acids, all of which are odor or taste active.
These transformations can be accomplished by living
cells, enzymes released from dead cells, or residues of
rennin left from the milk-clotting step.

Cheese ripening consists of a series of biochemical
reactions, the most predominant ones being lactic acid
metabolism,   lipolysis,   proteolysis,   and   oxida-
tion/reduction.   The basic taste and textural sensation
of cheese is thought to be due to fat, amino acids,
proteins, peptides, lactic acid, and salt(9).   Upon this
flavor base are superimposed the free fatty acids and
alpha-amino acids, which in turn are converted to the
important flavor volatiles, alpha-keto acids, amines,
amides, esters, and ketones.   For example, Ney, et
al(10) report C_2-C_{10} fatty acids as the main volatile
constituents of cheddar.   Small amounts of branched C_4
and C_5 acids alpha-keto acids, 2-alkenones, 2-aldehydes
and amines are also present.

Cheddar cheese is made with the same streptoccocal
bacterial strains used in unripened cheese but
subsequent bacterial growth of lactobacilli and
pediococci can also contribute to flavor and aroma
development.   Volatile compounds from carbohydrate
metabolism of the starter streptococci include diacetyl,
acetic acid, and acetaldehyde.   The starter bacteria
may also be indirectly involved in key flavor/aroma
compound formation by creating environmental conditions
favorable for non-enzymatic reactions.   For example,
the low redox potential resulting from starter culture
activity may be instrumental in the production of sulfur
compounds such as hydrogen sulfide and methanethiol,
which some investigators believe are very important in
cheddar flavor/aroma(8).   Volatile (C2-C10) fatty acids
which play a role in the general background aroma of
cheddar, may be produced by lactic acid starters, and
lactobacilli.

Mold-ripened cheeses are inoculated with mold
spores which germinate and, via metabolic
transformation, produce additional characteristic
flavor compounds.   Blue-vein cheeses are good examples.
In these cheeses, surface molds, yeasts, and bacteria
(micrococci) become dominant as the cheese pH drops due
to the lactic flora early in maturation.   The main

flavor notes in blue vein cheeses are due to fatty acids and methyl ketones produced by <u>Penicillium roqueforti</u>.(8) Methyl ketones are secondary metabolites formed from fatty acids by beta-oxidation and subsequent decarboxylation. Gamma-lactones may also contribute to blue-vein cheese flavor. Yeasts and molds have been reported to produce lactones as secondary metabolites from gamma-hydroxy acids or esterified gamma-keto acids of milk glycerides by several biochemical mechanisms.(8). In well-ripened cheeses, a floral note can often be detected which is apparently due to phenylethanol and phenylethyl esters(12).

Swiss cheeses are distinguished from other varieties by different starter cultures used and the subsequent growth of propionibacteria; with gruyere cheeses, yeasts and coryneforms. Fermentation of lactic acid and residual sugars by propionic bacteria to propionic acid is vital in flavor development, and follows initial lactic acid fermentation by the starters. The propionibacteria also apparently contain peptidases which release the sweet-tasting amino acid proline, according to some investigators(13), an important swiss cheese tastant.

The other two traditional fermentations of great economic importance are alcoholic beverages and bread production. (Figure 2) These two fermentations are based on yeast and utilize sugar as the substrate. The principal function of the fermentations is the conversion of sugar to ethanol and carbon dioxide. The Finnish workers Suomalainen and Nykanen(14) have shown that the aroma components produced by yeast in a nitrogen free sugar solution are very similar to those produced in beer, wine, whiskey, and cognac, and during the fermentation of bread dough. Several of the higher alcohols, which are secondary metabolites from yeast fermentation, are present in both breads and alcoholic beverages.

2. <u>Grain-Derived Substrates</u>. The manufacture of bread exploits the production of CO_2 as a leavening agent, with only a small amount of ethanol remaining in the bread. Alcoholic and a combination of homo- and hetero-lactic fermentations are required for production of high quality breads. The microorganisms used in these fermentations are yeasts and bacteria. <u>Saccharomyces cerevisiae</u> and <u>Candida</u> <u>utilis</u> are involved in flavor development in bread, the latter being used for rye breads. Certain bacteria also contribute to flavor development of bread by producing organic acids. These bacteria occur as contaminants of commercial baker's yeast and belong to the genera <u>Lactobacillus</u>, <u>Streptococcus</u>, <u>Pediococcus</u>, and <u>Leuconostoc</u>.

During bread fermentation lactic and acetic acids, are formed, along with ethanol, esters, aldehydes, ke-

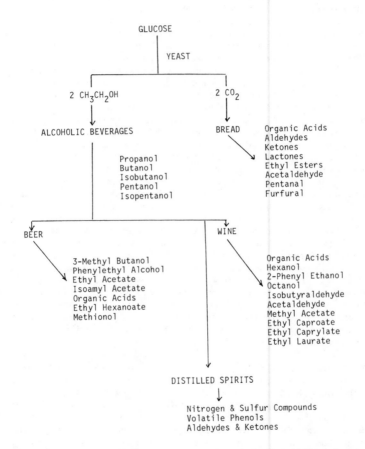

Figure 2. Volatile chemicals produced in traditional yeast
fermentations.

tones, and lactones.(15)   Sour dough breads utilize
Saccharomyces exiguus, S. inusitatus, and Lactobacillus
fermentations(15, 16).   Important flavor acids in this
bread are produced by the lactobacilli from maltose via
the maltose phosphorylase pathway(17).   Carboxylic
acids identified in sour dough breads include lactic and
acetic as the major ones, and trace amounts of the
unusual 2-hydroxypropanoic, 2-hydroxy, 1,2,3-propanetri-
carboxylic, and dihydroxy butanedioic acids (15).
atel, et al.,(18) suggested that the conversion of
alcohols to 2-methyl ketones by the yeast NAD-dependent
alcohol dehydrogenase is important in bread flavor
formation.

Additional aroma components common to yeast-
leavened bread are lower alcohols, acetaldehyde,
propanal, pentanal, and furfural, and ethyl esters, such
as ethyl acetate(19).

Only a few of the volatiles found in fresh beer
exceed their flavor threshold values.   These include
ethanol, 3-methylbutanol, ethyl acetate, isoamyl
acetate, and ethyl hexanoate(20).   Although the origin
of these compounds is unknown, the metabolic
capabilities of Saccharomyces strains could allow
formation of these products via pathways previously outlined.   Sulfur-containing compounds derived from yeast
are also important in beer flavor.   Dimethyl sulfide
increases during the course of fermentation by either
enzymatic or non-enzymatic reactions(21).   Methionol is
one of the main sulfur-containing aroma compounds in
beer and wine.   It is formed from methionine by the
yeast through deamination and decarboxylation to the
aldehyde and reduction to the corresponding
alcohol(21).   Other yeast derived sulfur containing
compounds include methionyl acetate, ethyl-3-methyl
thiopropionate, and 2-methyltetrahydrothiophene-3-
one(21).

One of the more interesting metabolic sequences
thought to occur in beer leads to lactone formation.
Levulinic acid is a degradation product of glucose and
4-oxononanoic acid is derived from linoleic acid during
malt preparation and wort boiling.   It is thought that
these oxo acids are reduced by the yeast to the corresponding hydroxy acids, which form 4-pentanolide and 4-
nonanolide, respectively(22).

3. Fruit Substrates.   Wines are generally described as
acidic alcoholic beverages which contain a special bouquet of volatile constituents.   Factors affecting the
taste-aroma bouquet include strain of yeast used, the
flora of the must conditions of fermentation, residual
sugar, and the type of raw materials.   Although some of
the flavor compounds of wine are indigenous to the
grape itself, much of the flavor of the finished product
arises from the biochemical actions of the yeast.

The products of yeast autolysis, such as amino acids may also play a role in wine flavor. The autolysate by-products may also serve as substrates for secondary fermentations. Malic and tartaric acids, the principle organic acids in grape must and wine, can also serve as substrates for secondary fermentation.

Lactobacillus and Leuconostoc sp. are involved in the conversion of malic acid to lactic acid and CO_2 which improves flavor and mellowness of wines(16). Important wine flavor compounds derived from yeast metabolism during fermentation are: Alcohols – Although there are strain specific variations, the principle higher alcohols involved in wine aroma are: 1-propanol,   2-methyl-1-propanol,   1-butanol,   2-phenylethanol, 2-methyl-1-butanol, 3-methyl-1-butanol, 1-hexanol, 1-octanol(23).   These are likely to be derived in part, from amino acids via the Erhlich pathway. Organic Acids present are: acetic, isobutyric, isovaleric, hexanoic, and decanoic(24).   Aldehydes present are: acetaldehyde, isobutyraldehyde, isovaleraldehyde, furfural(20).   Esters – It is known that ester formation in wine is mediated by yeast. Only diethyl malonate and ethyl-2-methylbutyrate were significant for differentiation of wine varieties. Fermentation derived esters also reported to contribute to wine aroma are: methyl and ethyl acetates, ethyl caproate, ethyl caprylate, and ethyl laurate(21).

Numerous reviews are in the literature on the fermentation-derived components of distilled spirits. Over 280 components have been identified in whiskies(25) and over 400 components have been found in rum(26). The conditions of fermentation and distillation determine the concentration of volatiles. For example the concentration and composition of trace sulfur compounds in grain spirits reflects the manner in which the distillation was carried out. No attempt will be made to compile the complete list of compounds identified to date. Rather, several newly identified components made possible by the recent advances in gas chromatography/mass spectrometry will be reported.

Nitrogen and sulfur containing compounds may be formed during yeast fermentation in the production of distilled spirits(26). Newly found sulfur-containing compounds in rum include methanethiol, alkyl disulfides, and dimethyl sulfoxide. Ethyl nicotinate, a highly odorous compound thought to be formed during fermentation(25), has also been found in rum. A new sulfur containing compound, 2-formylthiophene, is found in scotch whiskey(25).

Volatile phenols can be produced by microorganisms from the decarboxylation of phenolic acids such as ferulic and coumaric, and tyrosine(27). Phenols may contribute to the dryness characteristics in scotch whiskey(25). Cresols and guiacols have been reported to have an effect on the aroma of whiskeys(28).

Aldehydes and vicinal diketones produced during fermentation also contribute to the flavor profile of distilled beverages. Butyr- and valeraldehydes along with alcohols and esters appear to be involved in the "core" of whiskey aroma(29).

Pure Culture Transformations on Partially Defined Media

The release of odor-active materials by microbial cultures is well known by microbiologists. In fact, odor properties have been used as classification markers by microbial taxonomists for years. Omelianski(30) was among the first investigators to define microorganisms on the basis of the odors they produced.

Since that time, microbiologists have learned that the production of volatiles varies as a function of many additional chemical, physical, and biological factors including the following: The pH of the media must be monitored and, in many cases, held constant. Temperatures must be carefully controlled during fermentation to maximize yields. The nature of the assimilable carbon and nitrogen sources directly determines the quality of the odor released and the range of secondary metabolites produced. Optimization of the production of volatiles must also take into account the physiological state of the microorganisms. For bacteria, appropriate inocula may be several days old. For some of the higher fungi the use of 20 day inocula and 2 week cultures to seed the production media are common(31, 32).

With the expansion of the fields of microbial taxonomy and biochemistry, additional microorganisms have been identified which elaborate flavor and odor compounds. Table 2 compiles a few examples from the literature of microorganisms which produce odors in defined media. Most of these fermentations have been carried out only on small laboratory scale. In many studies the volatiles produced are not well characterized and little is known about the mechanism and pathways of biosynthesis.

A.   Bacteria & Actinomycetes

Aroma/taste producing bacteria tend to utilize amino acids as their principal carbon source. When grown on coarse wheat as a substrate, volatile fatty acids of C_2-C_5 are major products after 72 hours(33). With bacteria even slight changes in the oxidation state of the media may affect the number and type of odoriferous metabolites. In general, odors can be detected when bacterial counts reach 10^7-10^{11} cells per gram of media(33). In mixed cultures such as those present in dairy products, bacteria modify flavor substances that are produced by molds and yeasts. However, the odorants produced by bacteria are often overpowered by those produced by yeasts and molds due to the longer biological cycles of the latter species.

Perhaps the most important bacteria in terms of flavor and odor production are the lactic streptococci discussed earlier. In addition to acetaldehyde, these organisms produce a wide variety of neutral and acidic carbonyl compounds which result in the sharp, buttery, fresh taste of dairy products. Some bacteria including

Table 2. Chemicals Produced by Pure Cultures
of Microorganisms and Their Sensory Descriptions

	ORGANISM	SENSORY DESCRIPTOR	VOLATILES PRODUCED
BACTERIA			
	Lactic Acid Streptococci, Lactobacilli, Leuconostoc sp	Sharp, buttery, fresh	Acetaldehyde, Diacetyl, Acetoin, Lactic Acid
	Propionibacterium	Sour, sharp	Acetoin, dienals, Aldehydes
	Pseudomonads Bacillus sp. Corynebacterium sp.	Malty, milky granary	Acetoin, 3-Methyl-1-butanal, 2-Methyl-2-hydroxy-3-keto butanal, pyrazines
ACTINOMYCETES			
	Streptomyces sp	Damp forest soil odor	2-MeO-3-isopropyl pyrazine, Geosmin
YEASTS			
	Saccharomyces sp.	Aromas associated with bread & alcohol fermentation	Higher alcohols (amyl. isoamyl, phenylethyl, esters, lactones, thio-compounds
	Kluyveromyces sp.	Fruity, rose	Phenylethanol and esters, terpene alcohols (linalool, citronellol, geraniol), short chain alcohols & esters
	Geotrichum sp.	Fruity, melon	Ethyl esters, higher alcohol esters
	Hanensula sp.	Floral soil odor	Ethyl esters, higher alcohol esters
	Dipodascus sp.	Apple, pineapple	Higher alcohol esters
	Sporobolomyces sp.	Peach	Lactones
MOLDS			
	Aspergillus sp.	Fungal, musty mushroom	Unsat. alcohols
	Penicillium sp.	Mushroom, blue cheese, rosey	1-Octene-3-ol, methyl ketones, 2-phenylethanol, thujopsene, nerolidol
	Ceratocystis sp.	Banana, pear, peach, plum	Alcohols, esters, monoterpene alcohols, lactones
	Trichoderma sp.	Coconut, anise, cinnamon	6-pentyl-a-pyrone Sesquiterpenes. cinnamate derivatives
	Phellinus sp.	Fruity, rose, wintergreen	Methyl benzoates & salicylates, benzyl alcohol
	Septoria sp.	Anise or cinnamon	Alkyl & alkoxyl pyrazines Cinnamate derivatives
	Lentinus sp.	Aromatic, fruity	Higher alcohols, sesquiterpenes

Bacillus, Corynebacterium and Pseudomanas species are capable of
producing dairy, milky, granary, and vegetable-type odors when cultured on appropriate media(33) probably due in part to the presence
of pyrazines(34).

Pyrazines are heterocyclic nitrogen containing compounds with
unique flavor properties. Although the pathways for synthesis of
pyrazines are not known, it has been suggested that the mechanism
for synthesis in pseudomonads may be similar to that involved in raw
vegetables. In vegetables it has been hypothesized that pyrazine
synthesis involves condensation between alpha-amino acids and 1,2-
dicarbonyls(35). Corynebacterium mutants have been reported to
accumulate up to 3 g of tetramethylpyrazine per liter of culture
media after 5 days of growth(36).

Actinomycetes, primarily streptomyces, are capable of
producing highly odorous volatile metabolites in low yields in
submerged culture. This subject has been reviewed in detail(37).
Among the more important volatiles identified are: geosmin, the
earthy odor, methyl isoborneol, having a camphor or menthol odor; 2-
methoxy-3-isopropyl pyrazine, with a musty vegetable odor; and
miscellaneous compounds such as sesquiterpenoids and lactones.

B.   Yeasts

Perhaps more is known about the production of volatile components in baker's yeast (Saccharomyces cerevisiae) than in any other
microbial species. The type and quantity of esters formed during
yeast fermentations appear to be influenced by the yeast strain,
fermentation temperature, pH, and alcohol concentration(38).

With baker's yeast, higher amounts of esters were found to be
transferred from the cells into the media at higher temperatures.
Using ethyl caprylate as a substrate, Soumalainen(38) found that the
equilibrium between synthesis and hydrolysis depended not only on
the ester concentration but also on pH. A higher amount of ester
remains in solutions of lower pH.

The production of the higher alcohols, the acetates of isoamyl
alcohol and phenylethyl alcohol, and the ethyl esters of the C6-C10
fatty acids has been studied in semiaerobic sugar fermentations by
strains of S. cerevisiae and S. uvarum. S. cerevisiae generally
produced more esters than S. uvarum. Isoamyl acetate was the main
ester produced by S. cerevisiae, and others, in decreasing order,
were ethyl caprylate, ethyl caproate, ethyl caprate and phenylethyl
acetate(39). Several unusual thio compounds have been produced by
Saccharomyces in model anaerobic fermentations using amino acids
such as methionine as the sole carbon source(21). These model
fermentations produce methylthiopropanal and traces of other sulfur
containing compounds, such as methionyl acetate and 2-
methyltetrahydrothiophene-3-one.

Another important class of taste/aroma chemicals produced by
S. cerevisiae are lactones. Four and five oxo acids are transformed
by this yeast into optically active lactones, oxoacid ethyl esters,
and corresponding p-hydroxyacid ethyl esters(22).

Volatile product accumulation kinetics in cultures of
Kluyveromyces strains includes short chain alcohols and esters such
as 2 phenylethyl acetate. These cultures typically have a fruity,

rosey odor(40).   Although the range of compounds was qualitatively
comparable within the three strains investigated, quantitative differences were significant and strain-dependent responses to the
culture media were observed.   For example, the addition of
phenylalanine favored the formation of 2-phenylethyl derivatives in
two of the strains, whereas the addition of tyrosine led to lower
amounts of 2-phenylethyl derivatives in all strains.
     Kluyveromyces lactis is capable of producing mono-terpenes
such as citronellol and linalool, although at very low yields of 50
ug/L.(41).   Changing culture conditions altered the yield and type
of product produced.   Increasing temperature and concentrations of
the nitrogen source asparagine increased the yield of citronellol
substantially.
     Yeasts from the Geotrichum and Oospora genera, have been iso-
lated from ripe strawberries and used in the fermentation of heated
mashed fruits and synthetic media to produce aromas that were described as being close to those from fresh, non-heated fruits(42).
The amino acids present in the mashed fruit were thought to be the
precursors of the aroma substances.   Another Geotrichum strain has
been reported to release a melon odor(43).   Hansenula strains produce a rosey floral aroma, thought to be due to the presence of
ethyl esters and higher alcohol esters(43).
     Diplodascus strains of yeast release an intense fruity aroma
reminiscent of apple and pineapple in liquid glucose media(44).   The
peachy aroma emanating from Sporobolomyces is due to the presence of
two lactones, gamma-deca, and gamma-cis-6-dodecene lactone(45, 46).

C.   Molds and Higher Fungi

     A number of molds and higher fungi secrete a wide variety of
odor-active substances.   Within this category, basidiomycetes
(which include large fleshy fungi such as mushrooms) are reported to
be among the most interesting microbial sources of odoriferous sub-
stances(4).   In general, volatile odor producing molds and yeasts
are predominantly aerobic and utilize carbohydrates and organic
acids and rarely alcohol as principal carbon sources.   Much of the
culture work on odor producing fungi has been carried out on natural
media such as potato dextrose broth, nutrient broth, or malt extract broth(47).   When grains and cereals are used as substrates,
the volatile fractions of the higher fungi are similar.   The higher
alcohols, butanol, methyl butanols, hexanols, and octenols, are
common secondary metabolites.   The principal components of the
volatile fraction appear to be unsaturated octanols containing one
double bond.   These alcohols originate in part from the metabolism
of amino acids such as leucine, isoleucine, valine, and
phenylalanine via the Erhlich pathway.   Among the molds,
Aspergillus oryzae grown on a coarse wheat meal substrate produced
primarily 1-octene-3-ol, the characteristic powerful mushroom
aroma(33).
     In many cases, molds such as Penicillium and Trichoderma pro-
duce volatiles only in the spore producing stage.   Methyl ketone and
blue cheese aroma generation by certain Penicillium species is a
good example.   Another example is Penicillium decumbens which
produces characteristic odor only on media which allow sporulation

to occur. This organism produces as major odor compounds 3-octanone, phenylethyl alcohol, 1-octen-3-ol and nerolidol(4).

Ceratocystis species have been studied in some detail for their ability to produce fruit volatiles(48). These organisms grow readily in submerged culture and most members of this genus produce yeast-like cells. Moreover, they do not require unusual culture conditions(4). A variety of distinctive fruit like aromas ranging from banana to pear are produced by Ceratocystis moniliformis depending on the time course and carbon and nitrogen source in the medium(49). For example (Table 3) banana-like odor was produced when glucose and urea were used as carbon and nitrogen sources, respectively. Canned peach aroma was observed using glycerol and urea. Citrus aromas were observed when galactose and urea were used as substrates, and analysis revealed the presence of monoterpenes such as geraniol and citronellal. All cultures showed the presence of ethanol and short-chain esters such as isoamyl acetate, which was suggested to be derived directly from the amino acid leucine. Further analysis, suggested that this organism began to produce volatiles by day 3 or 4, reached a production peak at about day 6, and then stopped production. Additional studies using various volatiles as sole carbon sources, indicated that the esters were lost by evaporation, but the ethanol was neutralized by the organism(49). Monoterpene production by this species has been found to follow the mevalonate pathway after depletion of the nitrogen source(50). This is typical of secondary metabolite production. Geraniol is the first compound to appear, decreasing after two days. Citronellol, nerol, geraniol, and neral appear after the second day, and linalool and alpha-terpineol do not appear until the fourth day.

Some volatile products accumulate at significant levels only at specific stages in the life cycles of microorganisms. This may be due in part to toxicity of metabolites which, when they reach threshold levels in the culture media, inhibit the organism. For example, Schindler and Bruns(51) found that Ceratocystis variospora is inhibited by higher concentrations of its own terpene metabolites. They were able to significantly improve yields of these products by trapping the end products on resins.

A fungus widely distributed in the soil, Trichoderma viride, produces an interesting compound with a coconut, peachy aroma in a non-agitated liquid medium containing potato extract and salts(52). After three to four days cultivation, the culture sporulates releasing the metabolite 6-pentyl-alpha-pyrone. The maximum level reported, however, is 0.17 gram/L.

Fungi are very adaptable to extreme culture conditions and by changing the culture conditions the production of secondary metabolite can be stimulated or repressed. For example, Lentinus lepidus (Fr.) when cultured on asparagine as the sole nitrogen source produces primarily cinnamic acid derivatives. After three weeks, additional compounds such as the higher alcohols, 1-octanol and 1-decanol appear. However, when cultured on phenylalanine as the sole nitrogen source, it produced a high level of sesquiterpenes, in addition to cinnamates, which resulted in an aromatic-fruity odor(53).

Table 3. Volatiles Produced by <u>Ceratocystis</u> <u>Moniliformis</u>

CARBON SOURCE	NITROGEN SOURCE	AROMA	CHEMICALS IDENTIFIED
Dextrose	Urea	Fruity, banana	Acetate Esters, ethanol
Dextrose	Leucine	Fruity, over-ripe, banana	Isoamyl acetate, ethanol
Galactose	Urea	Citrus, grapefruit, lemon	Monoterpenes, ethanol
Corn Starch	Urea	Cantaloupe, tropical flower banana	
Dextrose	Glycine	Pineapple, lemon, sweet	
Dextrose	Methionine	Weak potato	
Glycerol	Urea	Canned pear, peach	Decalactones, ethanol

From Lanza <u>et al.</u> (49)

The production of benzenoid compounds such as methyl benzoates and salicylates has been noted in species of fungi belonging to the genus Phellinus(54). Pyrazines have been reported to be present in cultures of Septoria species(55).

Specific Microbial Transformations on Defined Media. Volatiles may be produced from precursors which represent the sole source of carbon. The microorganisms selected to accomplish the specific biotransformation may not grow on such a substrate. A sufficient mass of the cells may have to be obtained and then the precursor introduced into the culture in small quantities. In this way the microorganism is "forced" to carry out the "targeted" bioconversion.

In general, microbial reactions offer the following supplements to chemical syntheses:

1. Attack on molecular positions not affected by chemical methods (because the positions cannot be sufficiently activated or they require a number of intermediate steps before they will react)(56).

2. Stereospecific introduction of oxygen functions or other substituents (or alteration of such functions with the possible formation of optically active centers)(56).

3. Combination of multiple reactions into one fermentation step, (can be programmed to occur in a specific sequence with a suitable microorganism)(56).

4. Conditions for microbial reactions are mild (since many volatiles are sensitive to heat, this is a real advantage)(56).

5. Large cell numbers and high rates of growth may allow certain microorganisms to be forced to sustain unique reactions.

6. Low reaction rates can be overcome by the sheer number of individual cells and the ease of probing metabolic pathways by environmental manipulation.

The list of known microbial transformations of defined chemical substrates has grown considerably since 1960. Kieslich (56) has developed an extensive compilation of these reactions which has become a valuable reference for fermentation technologists. The transformations have been classified by structural class by Kieslich ranging from alicylics to peptides. Many of the chemical classes are relevant to the objective of producing volatiles by microbial conversions.

Microbial conversions from Kieslich's compilation which could be utilized in the production of volatile odorants and tastants are(56):

1. Oxidation. Oxidation of aromatic alcohols to corresponding aldehydes; Penicillium, Pseudomonas and enzymes from Caldariomyces sp(57). have the ability to convert benzyl, cinnamyl and other alcohols to the corresponding aldehydes.

2. Reduction. Aromatic aldehydes can be effectively reduced to their corresponding alcohols by microorganisms.

3. Hydrolytic Reactions. Microorganisms can selectively hydrolyze various esters. For example, certain bacterial strains can hydrolyze selected l-menthol esters such as the formates, acetates and caproates.

4.   Dehydration Reactions. Alpha-terpineol can be formed by the
      dehydration of cis-terpin hydrate by Brevibacterium strains.
5.   Degradation Reactions. Oxidative decarboxylations of amino
      acids serve as a good example. Schizophyllum commune is capa-
      ble of degrading phenylalanine to phenylacetic acid as well as
      other acids.
6.   Formation of New CC Bond. These reactions, such as the forma-
      tion of cyclohexane rings from terpenes, can be accomplished
      by microorganisms. Pseudomonas and Penicillium sp. are
      capable of forming menthol from citronellal.
      Over the past 20 years, research has uncovered many microbial
transformations of the terpenoids. Terpenes are important constituents of flavors and fragrances, and can be the center of a wide
variety of microbial hydroxylations, oxidations, reductions,
degradation and rearrangement reactions. See Ciegler(58),
Wood(59), Collins(47), and Schindler and Schmid(60) for separate
reviews of the biological transformation of non-steroidal terpenes.

Scale-Up of Microbial Processes for Volatile Production

Microbial Sources. In many cases, screening for the microbial production of volatiles can be achieved by selecting pure cultures of
microorganisms from public collections such as the American Type
Culture Collection (ATCC). The chances of obtaining an effective
culture can often be improved by selecting organisms known to
perform the desired types of reactions.
      Another source for microorganisms is the isolation of
appropriate strains from the environment. Here one selects
organisms by enrichment techniques from air, soils, and water
sources such as lakes and streams and the soils and surroundings of
plants which produce the substrate to be transformed. The
environmental sample is incubated in a growth medium containing the
substrate which serves as the sole carbon source. The microorganisms isolated are then "screened" by allowing them to grow in or
on appropriate media using standard microbiological techniques.

Screening Program. Small scale fermentations or biotransformations
typically ranging from 10-100 ml are initially carried out in small
flasks called "shake flasks". Incubations are carried out in incubator shakers which provide suboptimum aeration and precise temperature control.
      Once an effective microorganism has been identified from
screening studies, scale-up can be undertaken first in laboratory
fermentors which have efficient stirring and aeration capabilities.
The capacity of these fermentors typically range from 3-10 liters.
Here one studies additional operating parameters which can affect
microbial metabolism and the production of volatile chemicals.
      The pH of the media must be monitored and, in many cases, held
constant. Temperatures must be carefully controlled during fermentation to maximize yields. The nature of the media, such as carbon
and nitrogen sources, directly determines the quality of the odor
released and the range of secondary metabolites produced. Optimization of the production of volatiles must also take into account the
physiological state of the microorganisms.

For pilot scale production of volatile compounds, pilot plant fermentations are usually carried out at the 20-150 liter scale.

Production. Production scale fermenters for volatile compounds would normally range from 1,000 to 10,000 liters as compared to tens or even hundreds of thousands of gallons for high volume fermentation products such as antibiotics or food proteins.

Submerged culture fermentations replaced solid state fermentations in western countries after World War II and are now the most common method of producing primary and secondary metabolites. The main component of the media in these fermentations is water.

When pure cultures are used, fermentations must be carried out aseptically. Thus, the media must be sterilized and the fermenter operated under aseptic conditions. If foreign organisms contaminate the culture, they can over-run the desired culture or destroy the product.

Selection of Feasible Production Routes

At the present time, the generation of volatiles by microorganisms is more of an academic curiosity than a commercial reality. Although the metabolic events leading to volatile production have been described in many microorganisms, few processes have been scaled-up to commercially practiced processes. This is due to a variety of technical difficulties:

1. Low product yields. Most volatiles are produced in only trace amounts by microorganisms due to unfavorable energetics or inherently difficult metabolic routes. For specific chemicals, the literature typically reports only milligrams/liter of culture media. In some cases the substrate can be completely degraded to CO_2 and biomass, with no accumulation of product.

2. Substrate/Product Toxicity. Many volatile organic substrates such as terpenoids can be toxic to the organism and cannot exceed threshold concentration. Many end-products are secondary metabolites with intrinsic toxicological or inhibitory effects on the producing organisms. This can in some cases be minimized by the continuous stripping of finished product from the fermentation broth.

3. Long Fermentation Times. Many volatile producing organisms are the slower growing higher fungi. The longer the fermentation time, the greater the cost and increased possibility of contamination by unwanted organisms.

4. Organism Morphology. Some microorganisms (such as molds) can affect the rheological and oxygen transfer properties of the fermentation broth. Highly viscous broths may require more energy for agitation and aeration to maintain growth and product yields. This increases cost.

5. Product Recovery. The production of volatiles by fermentation often results in a low concentration of a water soluble product in a very large quantity of water. Components present in the fermentation broths often present problems for extraction and recovery. Due to their volatility, some products can be lost in fermentor off-gasses.

6.   **Product Mixes**. Recovered volatiles may be mixtures of
volatiles some of which may have deleterious effects on the
sensory properties of the finished product and must be
removed.

Economics of Producing Volatile Chemicals by Fermentation

The ultimate criterion for the production of any specialty chemical
is economics.
     Although fermentation may appear to be an attractive route to
the production of volatiles, several factors can limit the commercialization of this approach:
1.   Intrinsically low product yields.
2.   Relatively low market volumes for products which create unfa-
      vorable economies of scale.
3.   High capital costs associated with critical control & asepsis
      requirements.
     IFF has been evaluating the production of several flavor mate-
rials by fermentation.   Table 4 summarizes yield, market projection,
and costs for two volatile materials, A & B.   Based on a 5,000 L.
fermentation scale, the lowest projected cost of volatile A is
$800/kg.   This is due to relatively low yields (3g/L) and projec-
tions for low volumes.   Although the economics for volatile B are
more favorable due to high yield and greater market demand, they
fall far short of the economics achieved for two other food
additives and flavoring materials, MSG and citric acid, which by
contrast, are obtained at considerably higher yields and enjoy much
greater use(60).   Although Penicillin G has yields comparable to
those of the volatiles, production volumes result in greater
economies of scale and more favorable costs.
     Although the economics for fermentation derived volatiles may
appear to be discouraging, certain factors may still make the exercise attractive in some cases.   These include:
1.   Some fermentation-derived products may not be readily
      available from other biological or chemical sources.
2.   Some chemicals derived from fermentation may have high flavor
      or odor impact with sufficiently low use levels to justify the
      cost.
     Ultimately the final decision to commercialize will be the
result of a trade-off between economics, functionality, uniqueness
of odor and taste, and consumer demands.   A higher price can be
carried by a chemical which has a unique effect on a finished flavor
or fragrance.

Large Scale Production of Volatile Substances by Fermentation

In spite of unfavorable economics and technical problems, several
flavoring materials and intermediates have been produced on a large
scale by fermentation.
     Fermentation derived natural carboxylic acids are important to
the flavor industry for use in dairy and sweet flavors and as substrates for the production of natural esters.   A process has been
developed for the production of butyric acid which utilizes the
anaerobic conversion of dextrose to butyric acid by the bacterium

Clostridium butyricum(62).   Maximum yields obtained were 1.2% in
the medium representing a 40% sugar conversion.

Blue cheese flavors have been prepared via submerged culture
fermentations in a sterile milk-based medium using Penicillium
roqueforti(63). The fermentations are conducted under pressure with
low aeration rates with optimal flavor production occurring from 24-
72 hours. Similarly, Kosikowski and Jolly(64) prepared blue cheese
flavors from the fermentation of mixtures of whey, food fat, salt
and water by P. roqueforti. Dwivedi and Kinsella(65) developed a
continuous submerged fermentation of P. roqueforti for production
of blue cheese flavor.

Among the amino acids, L-glutamic acid can enhance or improve
the flavor of foods. Glutamic acid is produced via fermentation
directly from sugar using organisms such as Corynebacterium
glutamicum and Brevibacterium flavum.(66).

The so-called starter distillates used by the dairy industry
are now produced on a commercial scale from lactic acid cultures.
These distillates in which 70% of the substrate is converted to
diacetyl have been patented(67) and are used to impart a buttery
taste to edible oils. They are manufactured by the steam distillation of cultures of bacteria grown on a medium of skim milk fortified with 0.1% citric acid. Organisms used are Streptococcus
lactis, S. cremoris, S. lactis subsp. diacetylactis, Leuconostoc
citrovorum and L. dextranicum. Diacetyl comprises 80-90% of the
flavor compounds in the aqueous distillate but is present at only
10-100 ppm.

Farbood and Willis(68) in a recent patent application
disclosed a process for production of optically active alpha-
hydroxy decanoic acid (gamma-decalactone) by growing Yarrowia
lipolytica on castor oil as a sole source of carbon. This is a good
example of a commercial application of a volatile chemical produced
by a microorganism. Yields of up to 6 grams per liter culture media
were obtained making this a promising industrial fermentation.

Table 4. Economics of Fermentation Products

PRODUCT	YIELD (gm/l)	MARKET VOL. (kg/yr)	PRICE ($/kg)
Volatile A	3	200	800
Volatile B	20	2,000	100
MSG[1]	80-100	22,000	2
Citric Acid[1]	120-150	180,000,000	1-2
Penicillin G[1]	10-15	1,250,000	42

[1] From Bartholomew & Reisman (61)

Conclusion

The agricultural production of flavor and fragrance materials has
several disadvantages, including variation in consistency and quality, and dependency on climatic, seasonal, geographic, and even
political factors.   The microbial production of flavor and
fragrance materials may compliment or offer an alternative to
traditional sources of these materials.   Fermentation may be
particularly suited to the production of unique, highly intense
character impact components, i.e., substances that can potentiate
the aroma and flavor of fruits, dairy and other flavors at low
levels (100 ppm in the finished flavor).
        However, technical improvements in fermentation processes
will have to be made to improve product costs and stimulate use.
These include:
1.   Better understanding of underlying microbial metabolism.
2.   Increase in product yields.
3.   Development of more efficient product recovery methods.
4.   Search for a wider variety of microorganisms and low-cost
        substrates.
        The progress that is being made in applying fermentation tech-
niques to the production of volatile materials is encouraging.
Several products are being actively pursued by industry and it is
hoped new results will be available by the time the next symposium
is held.

Literature Cited

1.   Sani, B.K., K. Das, and Ghose, T.K.,   Biotech. Letters 1982 4,
        19-22 .
2.   Tokoro, Y., Oshima, K., Okii, M., Yamaguichi, K., Tanaka, K.
        and Kinoshita, S.,   Agr. Biol. Chem. 1970, 34, 1516.
3.   Phaff, H.I., Sci. Am. 1981, 345(3), 77-89.
4.   Collins, R.P., "The Production of Volatile Compounds by
        Filamentous Fungi" in Developments in Industrial Microbiology
        Series Vol. 20. Amer. Society of Industrial Microbiolgy
        1979, p. 239.
5.   Collins, R.P. and Morgan, M.E.,   Phytopath. 1962, 52, 407.
6.   Margalith, P.Z., "Flavor Microbiology" Chas. C. Thomas,
        Springfield, Ill. 1981.
7.   Rose, A.H., In "Fermented Foods;" Rose, A.H. Ed.; Economic
        Microbiology Series Vol. 7, Academic Press London, 1982, p. 1.
8.   Law, B.A., Ibid. pp. 149-198.
9.   Ney, K.H., In "The Quality of Foods and Beverages"
        Charalambous, G., and Inglett, G., Academic Press, New York,
        1981, 389.
10.   Ney, K.H., Wirotama, I. and W. Freytag, U.S. Pat. No.
        3,922,365, 1977.
11.   Ney, K.H., Gordian 1973, 380.
12.   Adda, J., Roger, S. and Dumont, I., In "Flavor of Foods and
        Beverages;" Charalambous, G., and Inglett, G., Academic
        Press, New York, 1978, 65.
13.   Langsrud, T., Reinbold, G., and Hammond E., J. Dairy Sci.
        1977, 60, 16.

14. Soumalainen, H.; Nykanen, L., J. Inst. Brew. 1972, 72, 469.
15. Seitz, E., Proc. 183 American Chemical Soc. Mtg., 1978, p. 57.
16. Vedamuthu, E., Dev. Ind. Microbiol, 1979, 20, 187.
17. Wood, B.I.; Cardenas, O.S.; Yang, F.M.; McNulty, D., In "Lactic Acid Bacteria in Beverages and Foods;" Carr, I.; Cutling, C.; Whiting, G. Eds.; Academic Press, New York, 1975, 325.
18. Patel, N.; Lackin, A.; Derelanko, P., Felix, A.; Eur. J. Biochem. 1979, 101, 401.
19. Oura, E.; Vickari, R. In "Fermented Foods;" Rose, A.H.; Ed.; Economic Microbiology Series Vol. 7, Academic Press, London 1982. pp 88-140.
20. Home, S.; Monograph Eur. Brew. Conv. 1982, 7, 79.
21. Schrier, P., Proc. Weurman Flav. Res. Symp. 1979, p. 175.
22. Tressl, R.; Apetz, M.; Arrieta, R.; Grunewald, K.G. In "Flavor of Foods and Beverages" Charalamabous, G.; Inglett, G., Eds. Academic Press, New York, 1978, 145-168.
23. Ribéreau-Gayon, P., In "Flavor of Foods and Beverages;" Charalambous, G.; Inglett, G.; Eds. Academic Press, New York, 1978, pp 355-380.
24. Van Straten, S.; Jonk, G.; van Gemert, L., Ibid.
25. Swan, J.S.; Howie, D.; Burtles, S., In "The Quality of Foods and Beverages - Chemistry and Technology," Charalambous, G.; Inglett, G., Eds. Academic Press, New York, 1981, p. 167.
26. Der Heide, R.; Schaap, H.; Wobben, H.; de Valois, P.; and Timmer, R., Ibid. p. 183.
27. Steinke, R.; Paulson, M., J. Agr. Food Chem. 1964, 12, 381.
28. Jounela-Eriksson, P.; Lehtonen, M., In "The Quality of Foods and Beverages - Chemistry and Technology," Charalambous, G.; Inglett, G.; Eds. Academic Press, New York, 1981, p. 167.
29. Jounela-Eriksson, P., In "Flavor of Foods and Beverages," Charalambous, G.; Inglett, G., Academic Press, New York, 1978, p. 339.
30. Omelianski, V.L., J. Bacteriol, 1923, 8, 393-419.
31. Sastry, K.S.M.; Singh, B.F.; Manavalan, R.; Singh, P.; Atal., Ind. J. Exp. Biol. 1980 18, 836-839.
32. Ibid. pp. 1471-1473.
33. Kaminski, E.; Stawicki, S.; Wasowicz, E.; Kasparek, M., Die Nahrung 1980, 24, 203.
34. Kempler, G.M. Adv. App. Microbiol. 1983, 29, 29.
35. Murray, K.E., Shipton, I., Whitefield, F.B., 1970 Chem. Ind. 897.
36. Demain, A.L., Jackson, M., Trenner, N.R., 1967 J. Bacteriol 94, 323.
37. Gerber, N., Dev. Ind. Microbiol 1979, 20, 225.
38. Soumulainen, H., J. Inst. Brew. 1981, 87, 50.
39. Nykanen, L. and Nykanen, I., J. Inst. Brew. 1977, 83, 30.
40. Hanssen, H.P.; Sprecher, E.; Klingenberg, A., Z. Naturforsch. 1984, 39, 1030-1033.
41. Drawert, F.; Barton, H., J. Agric. Food Chem. 1978 26(3) 765-766.
42. Hattori, S.; Yamaguchi, Y.; Kanisawa, T., Proc. IV Int. Cong. Food Sci. Tech. 1974, 1, p. 143.

43.   Koizumi, T., Kabuta, K., Ohsawa, R., Kodama, K. Nippon
      Noglikagaku Kaishi, 1982, 56, 757-763.

44.   Fischer, K.H., Senser, F., Crosch, W., Z. Lebensm. Unters.
      Forsch. 1983, 177, 336-338.

45.   Tahara, S.; Fujiwara, K.; Mizutani, J., Agric. Biol. Chem.
      1973, 37, 2855-2851.

46.   Tahara, S.; Mizutani, J., Agric. Biol. Chem. 1975, 39, 281-
      282.

47.   Collins, R.P., Lloydia. 1976, 39, 20.

48.   Hubball, J.A.; Collins, R.P., Mycologia 1978, 70.

49.   Lanza, E.; Ko, K.H.; Palmer, J.K., J. Agric. Food Chem. 1976,
      24, 1247-1249.

50.   Lanza, E.; Palmer, J.K., Phytochem. 1977, 16, 1555.

51.   Schindler, I., Bruns, K., German Patent No. 2840143 1980.

52.   Collins, R.P.; Halim, A.F., J. Agric. Food Chem 1972, 20,
      437-438.

53.   Hanssen, H.P.; Sprecher, E., In "Flavour '81" Schrier, P.,
      Ed. Walter de Gruyter, Berlin 1981 pp. 547-556.

54.   Collins, R.P.; Halim, A.F., Can. J. Microbiol. 1972 18, 63-
      66.

55.   Devys, M.; Bousquet, J.F.; Kollman, A.; Barbier, M., C.R.
      Acad. Sci. Paris 1978, 286, 457-458.

56.   Kieslich, K., "Microbial Transformations of Non-Steroid
      Cyclic Compounds" 1976, John Wiley & Sons.

57.   Geigert, J.; Neidleman, S.L., U.S. Patent 4,503,153, 1985.

58.   Ciegler, A.; Proc. 3rd Int. Ferm. Symp. 1969, p. 689.

59.   Wood, J.B., Proc. Biochem. 1969, 50.

60.   Schindler, J.; Schmid, R.D., Proc. Biochem. 1982, 17, 2-8.

61.   Bartholomew, W.H.; Reisman, H.B., In "Microbial Technology
      Fermentation Technology", Vol. II, Peppler, H., Perlman, D.,
      Eds. 1979, 466.

62.   Sharpell, F.; Stegmann, C., Proc 6th Int. Ferment. Symp. 1981,
      2, 71-77.

63.   Nelson, J.H., J. Agr. Food Chem. 1970, 49, 57.

64.   Kosikowski, F.; Jolly, R., U.S. Patent 4,122,895, 1979.

65.   Dwivedi, B.K., Kinsella, J.E., J. Fd. Sci. 1974, 39, 620.

66.   Izumi, Y., Chibata, I.; Itoh, T., Angew. Chem. Int. 1978, 17,
      176.

67.   Joensson, H.; Pettersson, H.E.; Andersson, K., Belgian Patent
      883,752, 1980.

68.   Farbood, M.; Willis, B.I., International Patent Application
      W083/01072, 1983.

RECEIVED May 5, 1986

Production of Secondary Metabolites in Plant Cell Cultures

Robert J. Whitaker, George C. Hobbib, and Leslie A. Steward

DNA Plant Technology Corporation, 2611 Branch Pike, Cinnaminson, NJ 08077

The initial enthusiasm for tapping the vast synthetic potential of cultured plant cells has largely given way to the realization that much needs to be learned about the biochemical and genetic regulation of plant secondary metabolism before cost-effective, industrial-scale production becomes feasible. However, the rapidly emerging technologies of plant tissue culture biotechnology are poised to have significant impact on advancing the commercialization of valuable plant secondary metabolites. A key to economically sound production is clearly the induction and selection of high-yielding cell cultures. Somaclonal variation has proven to be a powerful tool for uncovering useful genetic variation for the improvement of agriculturally important crop plants. In a similar fashion, somaclonal variation technology can be used to induce high-yielding cell cultures. Once producing variants have been selected, precise definition of growth and production media will further enhance production and maintain the genetic stability of producing cultures.

Plants are a valuable source of a vast array of chemical compounds including flavors, fragrances, pigments, natural sweeteners, industrial feedstocks, antimicrobials, and pharmaceuticals. These compounds belong to a rather broad group of metabolites collectively referred to as secondary products. The precise physiological function of secondary products has been a topic of debate among researchers. However, it seems clear that secondary products have not evolved to perform vital physiological functions, in the same manner as primary compounds like amino acids or nucleic acids, but rather seem to serve as a chemical interface between the producing plants and their surrounding environment. For example, plants may produce secondary products to ward off potential predators, attract pollinators, or combat infectious diseases (1).

A number of plant species commonly sought for their secondary products are native to very remote and sometimes politically unstable

0097–6156/86/0317–0347$06.00/0
© 1986 American Chemical Society

geographic areas of the world. Additionally as with any plant grown in the environment, the plants that produce valuable secondary products are subjected to a variety of climatic stresses that can ultimately determine the level and the quality of production. These factors greatly impinge upon the reliability of the particular product which, of course, is a major concern for industrial processors. Indeed, wild fluctuations in availability, quality, and price are very common (2). Clearly there is a need for dramatic improvements in the sourcing of many important plant-based chemicals.

The potential for the use of the emerging technologies of plant cell tissue culture for the production of valuable secondary compounds has been viewed with a sense of optimism and enthusiasm by biotechnologists. To understand the economic implications of plant tissue culture production of secondary compounds, one only has to note that, despite substantial advances in synthetic organic chemistry, plants are still the major source of twenty-five percent of all prescription medicines, provide the raw materials used extensively by the flavor and fragrance industries, and are the source of a number of natural sweeteners and insecticides (3). However, the initial optimism for plant tissue culture production of secondary metabolites has been somewhat tempered by the observation that cultured plant cells routinely yield very low concentrations of the commercially most important secondary products. It is evident that the future of this area of biotechnology depends upon the development of technologies that permit the induction and selection of genetic variants that over-produce particular secondary products and the design of culture systems tailored to the unique growth requirements of plant cells.

Expression of Secondary Metabolite Synthesis in Cell Cultures

Much of the difficulty in obtaining cell cultures that produce secondary products has been blamed on the lack of morphological differentiation in rapidly growing cell cultures (4). It has been postulated that if the synthesis and subsequent accumulation of a particular secondary product is in any way dependent on specialized cellular structure, then there is no chance to exploit plant cell cultures for chemical production unless those structural modifications can be induced (5). Many of the most desirable secondary metabolites are formed only in highly specialized tissues, i.e. roots, leaves, flowers. For example, the cardiac glycosides of Digitalis are principally found in leaf cells; quinine and quinidine are found in the bark of Cinchona trees; and tropane alkaloids are largely synthesized in the roots and translocated to the leaves in many Solanaceae species. It has been suggested that in plant cell cultures this level of morphological differentiation and maturation is largely absent and, therefore, secondary product synthesis is suspended.

An interesting investigation into the question of differentiation versus secondary product synthesis was performed by monitoring celery flavor synthesis in celery tissue cultures (6). Celery cultures at various stages of differentiation, including undifferentiated cells, globular, heart, and torpedo embryos, and differentiated plants were examined for flavor compounds. The less differentiated globular and heart-shaped embryos demonstrated no flavor compounds while the more differentiated torpedo-shaped embryos, which posses chlorophyll-containing plastids, did possess the characteristic celery phlthalide flavor compounds. There were no oil ducts in the torpedo-shaped embryos, so that these highly specialized cell types were not required for flavor compound production (7). It was thought that the appearance of chlorophyll and, hence, the maturation of the plastids, may in some way be associated with the formation of celery flavor compounds. A positive correlation was observed for the greening (development of chlorophyll) of celery petioles and the production of phlthalide compounds. By replacing 2,4-D in the growth medium with a structural analog, 3,5-dichlorophenoxyacetic acid, cell cultures of celery were isolated which began to show plastid development and chlorophyll synthesis. These green cultures also possessed phlthalide flavor compounds. Light microscopy of these cell cultures showed no signs of differentiation into organized meristems or embryo formation. A review on the role of morphological and cellular differentiation in the synthesis of flavor compounds has been presented by Collin and Watts (8).

While phlthalide biosynthesis in celery cannot be used as a general model for all secondary product synthesis, this example does point out the dangers in assuming that advanced morphological differentiation is a prerequisite to secondary product synthesis. While greening cell cultures imply a more advanced plastid development than is normally found in non-green cell suspensions, this certainly does not represent nearly the same level of morphological differentiation previously thought necessary to initiate flavor synthesis. Therefore, it is likely that at least some cell cultures characterized as "non- producers" can be induced or "turned-on" for secondary product synthesis with relatively minor modifications in media composition.

It has recently been reported that cell cultures of 18 of 19 species belonging to the genera Asperula, Galium, Rubia, and Sherardia produced anthraquinones at higher levels than those found in the intact plants (9). The concentration of sucrose and the types of substituted phenoxyacetic acid growth regulators were varied in an attempt to identify optimal conditions for anthraquinone production. In general, no consistent pattern of nutritional components were determined. Cell cultures of plants from the same family, genus, or species demonstrated quite different media requirements for anthraquinone production. The dramatic effects on secondary metabolite production precipitated by alterations in the media composition provides yet another vivid example of the importance of systematically defining optimal media requirements before deeming a cell culture non-productive for a specific secondary compound. The use of media manipulation and potential gene expression regulators to increase overall secondary product synthesis will be discussed in a later section of this chapter.

Selection of High-Producing Genetic Variants

Once production of the desired secondary compound is demonstrated in cell cultures, the emphasis can be shifted toward inducing and selecting genetic variants that synthesize increased levels of the compound. However, before one can embark on a program of genetic modification for enhanced production capacity, rapid, but sensitive assay procedures must be developed for the detection of the desired compounds (10). These methods should be geared to handle a large number of samples quickly, but should minimize the amount of tissue required for analysis.

The selection of source material for explants and callus initiation can be vital to obtaining productive cultures. Deus and Zenk (10) stress the importance of using high-yielding differentiated plants as a source of explant material for establishing cell cultures. These authors demonstrated statistically that high alkaloid producing plants were more likely to give rise to cell cultures with high alkaloid contents. Kinnersley and Dougall (11,12) have found the same relationship investigating nicotine production in cell cultures of Nicotiana tabacum.

Plant Tissue Culture Biotechnology. Plant tissue culture biotechnology is comprised of a number of rapidly emerging and powerful technologies that can be used to: (1) reproduce identical genetic copies of elite plants or plant cell lines, (2) generate genetic variation from explants of cultured somatic tissue, (3) create genetically homogeneous breeding lines via anther culture and the regeneration of haploids, and (4) combine desirable characteristics from two individual plants by protoplast fusion. These technologies are currently being performed with a wide range of agriculturally important crop plants and the regenerated plants are carefully being integrated into ongoing breeding programs.

The genetic variation that is routinely observed in plants regenerated from somatic tissue has been termed somaclonal variation. Somaclonal variation can be attributed to both pre-existing genetic variation in the somatic explants or variation induced by the cell culture and regeneration procedure (13). These genetic changes can be inherited by either Mendelian or non-Mendelian mechanisms and the nature of these genetic changes has been attributed to single gene mutations, chromosomal rearrangements, mitotic crossing over, and organelle mutation and segregation (13). A review of the genetic variability in plants regenerated from somatic tissue has been presented by Reisch (14).

Somaclonal variation has already proven to be an invaluable tool of the biotechnologist for introducing genetic variation into elite breeding lines. New, improved breeding lines of tomato, tobacco, oil palm, rice, and wheat have been obtained as the direct result of somaclonal variation programs (13,15). The most detailed characterization of somaclonal variation has been carried out in tomato. Evans and Sharp (16) reported: (1) the recovery of 13 discrete nuclear gene mutations in different tomato breeding lines including recessive mutations for male sterility, jointless pedicel, tangerine-virescent fruit and flower color, chlorophyll deficiency, virescence, and mottled leaf appearance and dominant mutations for fruit ripening and growth habit; (2) that single gene mutations, derived from somaclonal variation, occur at a frequency of about one mutant in every 20-25 regenerated plants; and (3) evidence suggesting the recovery of new mutants not previously reported using conventional mutagenesis procedures.

Somaclonal Variation and Secondary Product Synthesis. The induction and recovery of genetic variants by somaclonal variation technology can have a profound impact on the economic feasibility of secondary metabolite production. While most of the discussion up to this point has focused on cell culture production of secondary compounds, there are clearly a number of instances where whole plant production is both more efficient and economically prudent. This is especially true for those compounds that are of moderate to low cost but have large and potentially expandable markets. Compounds that fit this scenario might be the natural sweeteners of Stevia rebaudiana, stevioside and rebaudioside, the natural pyrethrins of Chrysanthemum, special composition oils from any of the typical specialty oil plants like soybean or rapeseed, and essential oil production by plants used in the flavor and fragrance industry. In these examples, successful commercialization would require very large quantities of the secondary product. While bioreactor technology for large-scale production using cell cultures is being developed, it is unlikely that such high levels of production in vitro would be economically feasible in the near future.

Somaclonal variation can be used to induce and select new plant varieties with increased levels of accumulation for specific secondary metabolites. Interestingly, many of the genetic changes that have been attributed to somaclonal variation are manifested as alterations in the chemical composition of the regenerated somaclone or the selfed progeny of that plant. Somaclonal variants with altered levels of carrotenoids, chlorophylls, anthocyanins, terpenes, alkaloids, and sugars are routinely observed. In practice, a selected somaclonal variant, with an increased level of a specific secondary metabolites, could be grown as a field crop, harvested and processed to obtain the chemical of interest. This approach represents a technically feasible and a more immediate solution to obtaining required amounts of some low-cost, large market plant secondary compounds.

However, in many cases, the more long-term approach of bioreactor production of secondary metabolites using plant cell cultures is more desirable for practical, economic, and proprietary reasons. Plant cell cultures can be established from an impressive array of plant species, including most of those that produce secondary products of commercial interest (4). To date, a number of cell cultures have been established that produce secondary products at levels in excess of those found in the intact plant (Table I). However, in most instances, high-yielding lines have been described as arising spontaneously and not as the result of a tissue culture program designed to optimize the induction of genetic variation. Two striking exceptions to this scenario are the selection of variants for increased nicotine synthesis in cell lines of Nicotiana tabacum (18) and high anthocyanin producing cell cultures of Euphorbia millii (19).

Once the appropriate assay technique has been chosen, the production and growth media defined, and the source of explant material determined, the process of inducing and selecting genetic variants for increased secondary product synthesis can begin (Figure 1). Callus cultures, arising from somatic explants, can be screened for chemical production and suspension cultures established from those identified as producing the desired metabolite. The advantage of using cell suspension cultures is that it promotes the generation of a large number of cell aggregates that can be replated and screened for production. The process of selecting cell aggregates that overproduce specific metabolites has been termed cell aggregate cloning (18). Cell aggregate cloning has been used successfully to select for photoautotrophic cells (19), high vitamin-producing cells (20), high pigment cells (21), and high alkaloid containing cells (22,23) in various plant species. An illustrative example of cell aggregate cloning is the selection of high anthocyanin-producing

Table I. Secondary metabolites accumulated in high levels by plant tissue

Secondary Product	Plant Species	PRODUCT INFORMATION Cell Culture (% of Dry Weight)	Plant
Shikonin	Lithospermum erythrorhizon	12	1.5
Ginsengoside	Panax ginseng	27	4.5
Anthraquinones	Morinda citrifolia	18	2.2
Ajmalicine, Serpentine	Catharanthus roseus	1.8	0.8
Rosmarinic acid	Coleus blumei	15	3
Ubiquinone-10	Nicotiana tabacum	0.036	0.003
Diosgenin	Dioscorea deltoides	2	2

Note: Data from references 17 and 3.

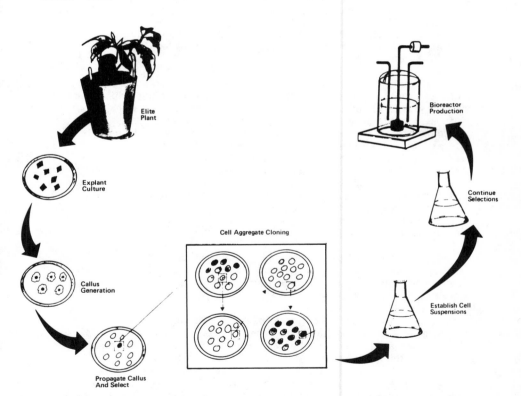

Figure 1. Somaclonal variation for development of high-producing cell lines.

cell lines from cultured cells of Euphorbia millii by Yamamoto et al (21). In these experiments, cell aggregates were plated on solid media and cultured. The resulting calli were then divided; one half for continued growth and the other for analysis for anthocyanin content. The highest pigmented calli were continually subcultured in this manner for 24 selections. The amount of pigment found in the highest producing lines after this period of time was seven times greater than the original cells.

The basis for the observed chemical accumulation in some cell aggregates is the result of spontaneous genetic variation that was previously masked in the original explant material or variation that was induced during the process of tissue culture initiation and subculture, i.e. somaclonal variation. Therefore, the same technology that has been successfully applied to some agricultural crop plants to select agronomically improved varieties is also widely applicable to the induction of cell cultures that produce high levels of secondary metabolites. The only difference between the two applications is, of course, the material being evaluated for genetic alteration. Somaclonal variation technology, as it is applied to crop plants, relies on the regeneration of whole plants, their self-fertilization, and the analysis of the resulting F_1 progeny. The application of somaclonal variation for secondary product synthesis in cell cultures is reliant on the generation of cell aggregates that can be individually selected and evaluated for specific chemical production.

Genetic Stability of High-Producing Cell Lines

The genetic stability of high-producing cell cultures greatly affects the economic potential of secondary metabolite production by plant tissue cultures. While stable, high-producing cell lines have been reported when repeated screening has been employed (21,24), this approach has been ineffective in stabilizing alkaloid production in Catharanthus (17) or anthocyanin production in Daucus cultures (25). It has been suggested that instability is a function of the genetic heterogeneity of the cell population in a given suspension culture (26). Cell suspension cultures typically exist as mixtures of various cell types possessing a number of shapes and sizes in various degrees of aggregation. It is expected that this morphological variation represents cell types with different genetic and biochemical capacities for secondary metabolite production. As secondary metabolite synthesis is often associated with senescent cells, the observed instability may reflect the washing out of high-producing cells due to their inherently slower growth rates relative to non-producing cells. Therefore, media compositions and cultural practices need to be tailored to enrich for producing cell types.

Sato and Yamada (27) have recently reported the establishment of high berberine-producing cell cultures of Coptis japonica. It was stressed that the stability of the producing cell lines was highly dependent on the repeated cloning of cell lines that demonstrated berberine synthesis. The authors also point out the importance of culture conditions to the maintenance of stable production levels. Fluctuations in berberine production were correlated to changes in physiological conditions and the nutritional make-up of the culture medium.   It, therefore, seems likely that stable secondary metabolite synthesis is as much a function of cultural practice and metabolic regulation as repeated clonal selection once high-producing lines have been established. The use of selective agents that favor specific cell types, with particular biochemical capabilities, could prove useful in maintaining stable, highly productive cell cultures.

Scale-Up for Secondary Metabolite Production

Bioreactor Production of Secondary Metabolites: The scale-up of plant cell cultures for the production of secondary products presents a number of technical challenges. While there are some similarities between the large-scale culture of microbes and plant cells, the striking number of physiological differences between these cell types precludes the use of microbial fermentation systems for plant cell culture (28). Plant cells are generally much larger than microbial cells, and they possess rigid cell walls. These features render plants cells much more susceptible to the shear forces that develop in conventionl microbial blade fermentors. Plant cells also grow much slower than microbial cells, and therefore, stringent aseptic conditions need to be observed. Due to the tendency for plant cells to aggregate, settling becomes a problem in large-scale cultures in terms of oxygen transfer. Most importantly, plant cells generally do not excrete their secondary products, but retain them inside the cell vacuole. Therefore, a destructive harvest would be necessary to release intercellular secondary compounds. This would preclude long-term recycable cell cultures and increase the overall cost of production.

There are a number of potential designs for plant cell bioreactors including: immobilized cell bioreactors, hollow fiber systems, air-lift vessels, and spin-filter bioreactors. Undoubtedly, no single design will be sufficient for all applications using plant cells for secondary product synthesis. However, two systems have received a great deal of experimental attention in the area of plant secondary product synthesis: immobilized cell bioreactors and spin-filter bioreactors. Imobilized cell bioreactors exploit the physical advantages of cell entrappment. Cells have been immobilized in gels of calcium alginate (29,30), carrageenen (31), polyacrylamide, and agarose (31). The mild conditions associated with the immobilization procedure generally yields normal cell viability (28). Indeed, plant protoplasts have been successfully immobilized in alginate-based gels (32,33). Immobilization is often reversible which allows for cell harvest and cell mass measurements. A review of plant cell immobilization and its uses for plant secondary product synthesis is presented by Brodelius (34).

Immobilized plant cells have respiration and bioconversion rates that are very similar to plant cells in suspension cultures but also have the advantage of the physical protection of the immobilizing matrix (28). The effects of the shearing forces created by the movements of the media is greatly reduced in immobilized systems. Cell immobization also permits the operation of a continuous culture bioreactor at dilution rates in excess of the maximum growth rate of the culture as entrapped cells are not as susceptible to washout as those in batch cultures. Immobilized cells are often set up in a fluidized-bed configuration or a packed-bed system. Both systems have been described in a review by Prenosil and Pedersen (28). The ease in which the media can be circulated through the immobilized cell bed and collected has led several investigators to this system for biotransformation studies. Indeed, the conversion of inexpensive precursors to the more valuable end-product by immobilized cells has been studied for a number of secondary products listed in Table II.

A detraction of the immobilized cell reactors is the requirement that the plant cell must excrete its secondary compounds. Theoretically, the production media would flow across the cell bed providing nutrition to the cells and removing the chemical products. At some point, the medium would be exchanged and the desired product extracted from the spent medium. However, plant cells do not normally excrete secondary products into the medium, but rather, they sequestor these compounds in the vacuole. Studies on plant cell excretion have suggested that accumulation and excretion of secondary products may be conveniently modulated by strict control of external pH (40). Precise definition of specific excretion and accumulation parameters for individual cell lines would undoubtedly prove invaluable in designing cost-effective bioreactors for scale-up of producing cell cultures. Recently, non-destructive permeabilization of immobilized plant cells by treatment with organic chemicals has been reported (41).

Another potentially exciting system for plant secondary product synthesis is the spin-filter bioreactor. Originally designed for mammalian cell culture (42), the spin-filter bioreactor permits continuous culture of plant cells at very high densities without cell washout. This is accomplished by the use of a spinning filter which allows for the removal of spent media and the introduction of fresh media without washing out cell mass. The filter rotates so as to prevent clogging, but the rate of rotation is slow enough to avoid cell damage. Therefore, this design facilitates increased nutrient feed rates without reducing the cell growth rate, a significant difference over conventional batch cultures.

One drawback of the spin-filter bioreactor might be the formation of cell aggregates. The inner most cells of the cell aggregate are distinctly different from those on the outer edge (28,43) Metabolic release of chemicals also varies between single cells and aggregates as well as within the aggregate itself. It has been shown that single cells are a more reliable source of chemicals, therefore, the accumulation of cell aggregates must be controlled. It has been observed that the ratio of cell types within a culture can be controlled by the degree of physical mixing within the bioreactor. The use of air spargers and modified paddle-type impellers limit the accumulation of aggregates by providing aeration and agitation. Whatever the specific design of the bioreactor, it is likely that tissue culture biotechnology can be employed to further increase secondary product synthesis in high-yielding cell lines. These manipulations can have significant impact on the economic feasibility of plant tissue culture production of secondary metabolites, and in many cases, are specifically directed to the technical problems associated with the scale-up of plant cell cultures.

Production Medium. It is vitally important to define a culture medium that promotes the production of the secondary metabolite of interest. As a production medium is unlikely to support the level of growth required to obtain the appropriate biomass, it may be desirable to develop a two-step approach whereby one medium is utilized solely for growth, and a second is employed for secondary product synthesis and accumulation (44). The

Table II. Biotransformation for the production of plant secondary
metabolites.

Plant Species	Substrate	Product	Reference
Digitalis lanata	digitoxin	digoxin	29,30
Daucus carota	digitoxigenin	hydroxylated derivatives	35
Solanum tuberosum	solavetivone	sequiterpene	36
Papaver somniferum	thebaine	neopine	37
	codeinone	codeine	38
Citrus	valencene	nootkatone	39

effect of altering the basic components of plant tissue culture media on the production of secondary compounds has been extensively reported in the literature (45,46). When formulating a production medium, one must consider the dramatic effects on secondary product synthesis incurred by altering even minor components of the medium. Carbon source (sucrose, glucose, fructose, etc.), nitrogen source (organic or inorganic), vitamins, and ions have all been shown to play significant roles in altering the expression of secondary metabolic pathways.

The presence or absence of phosphate ions plays an important role in the expression and accumulation of some secondary products. Zenk et al. (47) have demonstrated a 50% increase in anthraquinone accumulation in cell cultures of Morinda citrofolia when phosphate was increased to a concentration of 5g/l. In suspension cultures of Catharanthus roseus, the overall accumulation of secondary metabolites like tryptamine and indole alkaloids has been shown to occur rapidly when cells were shifted to a medium devoid of phosphate (48,49). A study on the uptake of phosphate and its effect on phenylalanine ammonia lyase and the subsequent accumulation of cinnamoyl putrescine in cell suspension cultures of Nicotiana tabacum demonstrated marked sensitivity to phosphate concentration (50). Enhanced phenylalanine ammonia lyase activity and increased production of cinnamoyl putrescine was induced by subculture onto phosphate-free medium while suppression of these effects and stimulation of growth was observed with phosphate concentrations of 0.02-0.5uM. Interestingly, phenylalanine ammonia lyase activity is stimulated by increasing phosphate concentrations in cell suspension of Catharanthus roseus (51).

An additional insight into the importance of specific media components on production of secondary products can be gained by examining the case history of shikonin production. It had been shown that callus cultures of Lithospermum erythrorhizon could be induced to produce shikonin on Linsmaier-Skoog medium supplemented with 1μM indole acetic acid (IAA) and 10μM kinetin (KIN) (52). The effects of specific nutritional components of the tissue culture medium on growth of the cell cultures and production of shikonin were also investigated (53). Fujita et al. (54,55) found that the levels of NO_3^-, Cu^{++}, and SO_4^{--} had profound effects of shikonin biosynthesis. Optimal concentrations were identified for each ion (18) as well as optimal levels of key organic components. The resultant medium supported production of shikonin at a rate approximately 13 times that obtained on previous media formulations.

Analogs. Synthesis of chemical compounds has been induced in cell suspensions by the addition of structural analogs. For instance by substituting an analog for the natural amino acid in a biosynthetic pathway, metabolic activity in that particular biosynthetic pathway can be increased or decreased. The amino acid analog 4-methyltryptophan, has been used to select Cartharanthus roseus cell lines that produce increased concentrations of tryptamine (56). This is accomplished by relieving feedback inhibition control over the tryptophan biosynthetic pathway thus increasing carbon flow toward tryptamine synthesis. These cells also overproduce the alkaloid, ajmalicine. Though successful chemical production increases have been attained using a number of amino acid analogs, many high producing cell lines have proven genetically unstable, and continuous selection is required to maintain increased production rates (56).

Growth Regulators. Phytohormones and other growth regulating compounds have long been known to influence morphological and, therefore, physiological development of cultured plant cells or whole plants. The commonly used phytohormone analog, 2,4-D has been implicated in a number of systems as interfering with secondary metabolism (57). In carrot cell cultures the presence of exogenous giberellic acid blocks anthocyanin synthesis by preventing chalcone-biosynthesis (58). The chemical agent 2-diethylaminoethyl-2,4-dichlorophenylether and its derivitives have proven to have a profound effect on alkaloid production in C. roseus. A 20% increase in total alkaloids with a similar increase in ajmalicine, and catharanthine was noted (5,52). Understanding hormonal influences at the molecular level will be necessary to unlock the secret of these powerful regulators and to use them to specifically modulate gene activity.

Microbial Insult/Fungal Elicitors. It is common in nature to find that microbial insult of whole plants leads to the production of specific secondary metabolites. The molecules responsible for stimulating secondary product synthesis are referred to as elicitors. Fungal elicitors are the best studied elicitors, and their active regulatory molecules have been identified as being glucan polymers, glycoproteins, and low molecular weight organic acids (59). Albershiem and his colleagues refer to these regulatory molecules as oligosaccharins (60-61).

An example of an elicitor inducing a latent biosynthetic capability is found in parsley, where synthesis of the coumarin compound psoralen has been induced by fungal elicitors (62). Derivatives of psoralen are used in the treatment of psoriasis and as ingredients of photosensitizing suntan lotions. Treatment of parsley cells by the addition of a cell wall fraction of the fungus Phytopthora megasperma f. sp. glycinea resulted in the production of the mRNA's encoding two enzymes of phenylpropanoid metabolism, namely phenylalanine ammonia-lyase (PAL) and 4-coumarate (CoA-ligase) (63).

Another instance of the utility of fungal elicitors is the use of autoclaved fungal mycelia to increase yields of diosgenin in cell suspensions of Mexican Yam (64). Diosgenin is a steriodal saponin, an important precursor in the preparation of oral contraceptives and other medical steroids. Clearly, fungal elicitors have become an attractive tool for regulating secondary product synthesis. Fungi can be easily grown, recovered, and harvested, and crude preparations can be conveniently screened for inducing chemical synthesis in cultured cells (56).

Physical Stress. Physical stress factors such as temperature variations, pH change, and light exposure have in some cases resulted in inhibition of chemical production and in other systems stimulated secondary synthesis. Cell suspensions of Papavar bracteatum grown at 36°C have been shown to accumulate protopine, sanguinarine, isothebaine, and orientalidine. Reducing this temperature to 17°C caused growth inhibition and a further reduction to 5°, caused the release of thebaine into the media. As chilling stress is known to severely inhibit photosynthetic reactions and disrupt organelles (65), the thebaine was most likely produced during growth inhibition and released due to the stress caused by chilling (66).

Light induced synthesis of secondary metabolites has been exploited in several cell cultures. The previously described effects of fungal elicitors that induce mRNA production encoding for PAL and CoA-ligase in parsley, is duplicated when these cells are irradiated with UV light (63). Anthocyanin synthesis in many plant tissues has long been known to be promoted by light. More recent work has revealed that red light and far-red light accentuates and reduces, respectively, anthocyanin synthesis in apple fruit skin and poinsetta. It is evident that a phytochrome is involved in a reversible reaction in regulating anthocyanin (67) synthesis.

Protoplast Fusion for Fine-Tuning Producing Cultures: Although much more speculative than the application of somaclonal variation for the induction and selection of high-yielding cell lines or whole plants, the technology of protoplast fusion may play a vital role in genetically fine-tuning cell cultures for increased production. Where the addition of exogenous phytohormones have been shown to impair secondary metabolism, fusion of protoplasts from the producing cell line with protoplasts of a hormone autotrophic tumor cell might result in fusion products that demonstrate growth and increased production in a hormone-free medium.

An intuitively appealing idea is the prospect of culturing photoautotrophic cells for secondary metabolite production. These cultures would have the economic advantage of using solar energy directly (i.e. photosynthesis) and would, therefore, be capable of growth and production on a medium with little or no exogenous sugar. The scale-up of cell cultures for the production of secondary compounds will require aseptic conditions for prolonged periods of time. As plant tissue culture media is relatively expensive and inviting to microbial contamination, any modification that would reduce both cost and contamination problems would have significant impact on the feasibility of tissue culture production of secondary metabolites.

Selection procedures for isolation of photoautotrophic cell lines have been reviewed by Yamada and Sato (68). It should be noted that the culture of photoautotrophic cells has the potential for increasing secondary product synthesis. In instances where the synthesis of a particular secondary metabolite is regulated by the level of cellular differentiation, it is anticipated that the culture of photoautotrophic cells will have significant effects on production. Indeed, some alkaloids (69) vitamins (20), and components of volatile essential oils (70) have been produced by cultured green cells.

Photoautotrophy may be introduced into a producing cell line by either of two methods: (a) producing cell lines can be selected for photoautotrophy, or (b) protoplasts from a photoautotrophic cell line can be fused with protoplasts of the producing cell line and selection for both characteristics performed on subsequent fusion products.

Concluding Remarks

The rapidly emerging technologies of plant tissue culture biotechnology are poised to have a significant impact on the production of valuable secondary metabolites by plant cell culture or whole plants. Somaclonal variation has proven to be a powerful tool for inducing useful genetic variants for the improvement of agriculturally important crop plants. In a similar fashion, somaclonal variation technology can be used to induce high-yielding cell cultures. Once high producing variants have been selected, precise definition of growth and production media will further increase production and facilitate the maintenance of stable producing cultures. These advancements will eventually permit the cost-effective, industrial scale production of plant secondary metabolites.

Literature Cited

1. Bell, E A., In: "Secondary Plant Products"; Bell, E. A. and Charlwood, B. V. eds.; Springer Verlag: New York, 1980; pp. 11-21.
2. Curtin, M. E. Biotechnol 1983, 1, 649-657.
3. Rhodes, M. J. C., Kirsop, B. H. Biologist 1982, 2, 134-140.
4. Berlin, J. In: "Endeavour"; 1984. p. 8.
5. Constabel, F., Gamborg, O. L., Kurtz, W. G. W., Steck, W. Planta Med. 1974, 25, 158-165.
6. Al-Abta, S., Galpin, I. J., Collin, H. A. Plant Sci. Lett. 1979, 16, 129-134.
7. Al-Abta, S., Collin, H. A. New Phytol. 1978, 80, 517-521.
8. Collin, H. A. and Watts, M. In: Handbook of Plant Cell Culture"; Evans, D. A., Sharp, W. R., Ammirato, P. V., Yamada, Y. eds.; Macmillan: New York, 1983; pp. 729-747.
9. Schulte, U., El-Shagi, H., Zenk, M. H. Cell Reports 1984, 3, 51-54.
10. Deus, B., Zenk, M. H. Biotech. and Bioeng. 1982, 24, 1965-1974.
11. Kinnersley, A. M., Dougall, D. K. Planta. 1982, 154, 447-453.
12. Kinnersley, A. M., Dougall, D. K. Planta. 1980, 149, 205.
13. Evans, D. A., Sharp, W. R., Medina-Filho, H. P. Amer. J. Bot. 1984, 71, 759-774.
14. Reisch, B. In: "Handbook of Plant Cell Culture"; Evans, D. A., Sharp, W. R., Ammirato, P. V., and Yamada, Y. eds.; Macmillan: New York, 1983; pp. 748-781.
15. Sharp, W. R., Evans, D. A., Ammirato , P. V. Eur. Chem. News. May 1984.
16. Evans, D. A., Sharp, W. R. Science. 1983, 221, 949-951.
17. Zenk, M. H. In: "Frontiers of Plant Tissue Culture"; Thorpe, T. A. ed.; International Association of Plant Tissue Culture: Calgary, 1978; pp. 1-14.
18. Ogino, T., Hiaoka, N., and Tabata, M. Phytochem. 1978; 17, 1907-1910.
19. Yasuda, T., Hashimoto, T., Sato, F., and Yamada, Y. Plant Cell Physiol. 1980.
20. Watanabe, K. and Yamada, Y. Phytochem. 1982, 21, 513-516.
21. Yamamoto, Y., Mizuguchi, R., and Yamada, Y. Appl. Genet. 1982, 61, 113-116.
22. Zenk, M. N., El-Shagi, H., and Ulbrich, B. Nuturwiss. 1977, 64, 585-586.
23. Yamada, Y. and Hashimoto, T. Plant Cell Rep. 1982, 1, 101-103.

24. Ohta, S. and Yatazawa, M. In: "Plant Tissue Culture"; Fujiwara, A. ed.; Japanese Association for Plant Tissue Culture: Tokyo, 1982; p. 321-322.
25. Dougall, D. K., Johnson, J. M., and Whitten, G. H. Planta. 1980, 149, 292-297.
26. Ellis, B. Can. J. Bot. 1984, 62, 2912-2917.
27. Sato, F. and Yamada, Y. Phytochem. 1984, 23, 281-285.
28. Prenosil, J. E. and Pedersen, H. Enzyme Microb. Technol. 1983, 5, 323-331.
29. Alfermann, A. W., Schuller, I., Reinhard, E. Planta Med. 1980, 40, 218-223.
30. Brodelius, P., Deus, B., Mosbach, K., and Zenk, M. H. FEBS Lett. 1979, 103, 93-97.
31. Brodelius, P. and Nilsson, K. FEBS Lett. 1980, 122, 312-316.
32. Scheurich, P., Schnabel, H., Zimmerman, Y., and Klein, J. Biochem. Biophys. Acta. 1980, 598, 645-651.
33. Brodelius, P. and Mosbach, K. In: "Advances in Applied Microbiology"; Laskin, I. ed.; Academic Press: New York, 1982; Vol. 28, pp. 1-26.
34. Brodelius, P. In: "Handbook of Plant Cell Culture"; Evans, D. A., Sharp, W. R., Ammirato, P. V., and Yamada, Y. eds.; Macmillan Press: New York, (in press).
35. Jones, A. and Veliky, I. A. Appl. Microbiol. Biotech. 1981, 13, 84-89.
36. Zacharius, R. M. and Kalan, E. B. Plant Cell Rep. 1984, 3, 189-192.
37. Tam, W. H. J., Kurz, W. G. W., Constabel, F., and Chatson, K. B. Phytochem. 1982, 21, 253-255.
38. Furuya, T., Yoshikawa, T., and Taira, M. Phytochem. 1984, 23, 999-1001.
39. Drawert, F., Berger, R. G., and Godelmann, R. Plant Cell Rep. 5, 37-40.
40. Renaudin, J. P. Plant Sci. Lett. 1981, 22, 59-69.
41. Brodelius, P. and Nilsson, K. Eur. J. Appl. Microbiol. Biotechnol. 1983, 17, 275-280.
42. Himmelfarb, P., Thayer, P. S., and Martin, H. E. Science. 1969, 164, 555-557.
43. Shuler, M. L., Sahai, D. P., Hallsbey, G. A. Annals NY Acad. Sci. 1983, 413, 373-382.
44. Sahai, O. P. and Shuler, M. L. Biotechnol. Bioeng. 1984, 26, 27-36.
45. Dougall, D. K. In: "Plant Tissue Culture as a Source of Biochemicals"; Staba, E. J. ed.; CRC Press: Boca Ration, Florida, 1980; pp. 21-58.
46. Delfel, N. E. and Smith, L. J. Planta Medica. 1980, 40, 237-244.
47. Zenk, M. N., El-Shagi, N., and Schulte, Y. Planta Medica. Suppl. 1975.
48. Knobloch, K. H., and Berlin, J. Z. Naturforsch. 1980, 35, 551-556.
49. Knobloch, K. H., Beutnagel, G., and Berlin, J. Z. Naturforsch. 1981, 36, 40-43.
50. Knobloch, K. H. and Berlin, J. Plant Cell Reports. 1982, 1, 128-130.
51. Knobloch, K. H. and Berlin, J. Plant Cell Tissue Organ Culture. 1983, 2, 333-340.
52. Tabata, M., Migukami, H., Kirasoka, N., and Konoshima, M. Phytochem. 1974, 13, 927-932.
53. Mizukami, H., Konoshima, M., and Tabata, M. Phytochem. 1977, 16, 1183-1186.
54. Fujita, Y., Hera, Y., Ogino, T., and Suga, C. Plant Cell Reports. 1981, 1, 59-60.
55. Fujita, Y., Hara, Y., Suga, C., and Morimoto, M. Plant Cell Reports. 1981, 1, 61-63.

56. Zenk, M. H., El-Shagi, H., Arens, H., Stockigt, J., Weiler, E. W., and Deus, B. In: "Plant Tissue Culture and its Biotechnological Application"; Barc, W., Reinhard, E., Zenk, M. H. eds.; Springer Verlag: Berlin, pp. 27-43.
57. Esau, K. In: "Anatomy of Seed Plants"; John Wiley and Sons: 1977, pp. 448-471.
58. Hinderer, W., Petersen, M., Seitz, H. V. Planta. 1984, 160, 544-549.
59. DiCosmo, F. and Talleri, S. G. Trends Biotechnol. 1985, 3, 110-111.
60. Albersheim, P., Darvill, A. G. Scientif. American. 1985, 253, 58-64.
61. Van, K. T. T., Toubart, P., Cousson, A., Darvill, A., Gollin, D., Chelf, P., Albersheim, A. Nature. 1985, 314, 615-617.
62. Tietjen, K. J., Hunkler, D., and Matern, U. Eur. J. Biochem. 1983, 131, 401-407.
63. Kuhn, D. N., Chappell, J., Boudet, A., Hahlbroch, K. Proc. Natl. Acad. 81 USA. 1984, 81, 1102-1106.
64. Rochem, J. S., Schwarzberg, J., Goldberg, I., Plant Cell Reports. 1984, 3, 159-160.
65. Oquist, G. Cell Environment. 1983, 6, 281-300.
66. Lockwood, G. B. Z. Pflanzenphysiol. 1984, 114, 361-363.
67. Kadkade, P. G. In: "Tenth Annual Meeting Plant Growth Regulator Society of America"; 1983; pp. 132-138.
68. Yamada, Y. and Sato, F. In: "Handbook of Plant Cell Culture"; Vol. 1, Evans, D. A., Sharp, W. R., Ammirato, P. V., and Yamada, Y. eds.; Macmillan: New York; 1983: pp. 489-500.
69. Hartman, T., Wink, M., Schoofs, G. and Teichmann, S. Plant Medica. 1980, 39, 282
70. Corduan, G. and Reinhard, E. Phytochem. 1972, pp. 917-922.

RECEIVED April 4, 1986

Essential Oil Production
A Discussion of Influencing Factors

Brian M. Lawrence

R. J. Reynolds Tobacco Company, Bowman Gray Technical Center, Reynolds Boulevard, Winston-Salem, NC 27102

The traditional view of essential oil production
is that of simple farming followed by oil removal
from those parts which are harvested. This is
essentially true, however, there are certain intrinsic factors (genotype and ontogeny) and extrinsic factors (light, temperature, water and
nutrients) that strongly influence oil production
optimization. In this presentation, these influencing factors will be discussed. Particular
attention will be given to the diurnal fluctuation of oil production as influenced by light and
water.

The amount of secondary metabolite or active principle produced
by a plant during its life cycle is a balance between formation
(biosynthesis) and elimination (catabolism and chemical and/or
physical loss). These two opposing functions are directly
controlled by two main groups of factors. The first group is
comprised of all internal, hereditary or intrinsic factors or
properties (e.g. genotype and ontogeny), while the second group
is comprised of all other external, environmental or extrinsic
factors or properties (e.g. pressure, wind, light, temperature,
soil, water, nutrients). Within a single clone the intrinsic
factors are fixed while the extrinsic factors are not.

Intrinsic Factors

The lack of similarity in oil composition between phenotypes grown
in the same environment is a manifestation of genotypic differences.
Thus, in plants at the same stage of development (ontogeny) and
of the same genotype, extrinsic factors can assume a quantitative
modifying effect which can cause both qualitative and quantitative
variation in oil content.

0097–6156/86/0317–0363$06.00/0
© 1986 American Chemical Society

A qualitative difference in essential oil composition between two or more phenotypes, which are raised in the same environment, indicates that one or more factors in one of the original specific environments was either above or below the influence threshold for the production of a specific essential oil (4).

By contrast, if a qualitative difference in oil composition is found between two morphologically identical species grown in a single environment but originating from two dissimilar environments, then the difference in chemical composition is genotypic rather than phenotypic. All strains of a single species are not formed quantitatively alike. Such is the situation with chemotypes or chemical races which should be viewed from the biosynthetic perspective in order that true chemotypes and sub-chemotypes be differentiated (19).

It has been reported (2,5) that the formation of active principles occurs predominantly during the periods of vigorous growth or during a time of intensive metabolic processes such as when a plant is flowering or fruiting. Before examining some examples of ontogenetic variation, it is worth pointing out that essential oils vary drastically within a single species depending upon the organ from which it is obtained (13). An example of the change in chemical composition of Coriandrum sativum during its growth cycle demonstrates an extremely drastic change in composition (see Table I). Similar compositional changes have been observed with Matricaria chamomilla (18), Anethum graveolens (20), Tagetes minuta (23) etc. It must be noted that many different oil yields of the same genotype can be found. Thus, for example the oil content of Foeniculum vulgare and Daucus carota has been observed to vary from 1.3-9.8% and 0.05-7.15% respectively. Hence for economic exploitation, it is advisable to screen each genotype to maximize oil content.

Extrinsic Factors

In considering the various extrinsic factors such as pressure, wind, light, temperature, soil, water and nutrients and their bearing on essential oil composition, it can be readily understood that the yield of oil obtained from a specific clone can often be influenced by the differing conditions experienced from one season to another. According to Fluck (5), the most important extrinsic factors affecting essential oil production are climate (temperature and light) and soil (chemical and water). It is the view of this author that those taxa with glandular hair reservoirs (e.g. Labiatae, Verbenaceae, Geraniaceae, Myrtaceae and Rutaceae), in which the essential oil is accumulated, are affected most by extrinsic factors such as air temperature, relative humidity, cloudiness (availability of UV light), rainfall, barometic pressure and wind speed. Taxa in which the oil is found in schizogenous ducts occurring in the leaves, calyces and stems are less affected by changes in meteorological conditions (e.g. Lauraceae and Compositae). Oils found in secretory ducts called vittae, which are schizogenously formed on the fruits and roots (e.g.

Table I.   Comparative Chemical Composition of Coriandrum Sativum
At Various Stages of Maturity

	Compound	Stages of Maturity					
		1	2	3	4	5	6
1.	Octanal	1.20	11.20	0.85	10.66	0.44	0.35
2.	Nonanal	0.51	0.20	0.11	0.05	0.05	0.08
3.	Decanal	30.0	18.0	11.9	6.3	6.2	1.6
4.	Camphor	0.08	TRACE	0.52	1.26	2.18	2.44
5.	Trans-2-Decenal	20.6	46.5	46.5	40.6	30.2	3.9
6.	Dodecanal	3.30	1.67	0.96	0.64	0.52	0.41
7.	Trans-2-Undecenal	2.56	2.17	1.39	-	-	-
8.	Tridecanal	3.07	1.87	2.02	0.92	1.08	0.46
9.	Trans-2-Dodecenal	7.63	8.14	5.95	4.59	4.78	2.49
10.	Tetradecanal	0.68	0.30	0.12	0.15	0.11	0.15
11.	Trans-2-Tridecenal	0.49	0.21	0.14	0.09	0.09	0.13
12.	Trans-2-Tetradecenal	4.45	2.57	1.73	1.53	1.59	1.73
13.	Linalool	0.34	4.27	17.47	30.05	40.88	60.37
14.	Geraniol	0.19	0.11	0.35	0.71	0.93	1.42
15.	Geranyl Acetate	4.17	0.78	0.76	0.69	0.69	0.66

Stage 1:   Flower beginning to open
      2:   Nearly full flowering
      3:   Full flowering, primary umbel young green fruit
      4:   Past full flowering 50% flower, 50% fruit
      5:   Full green fruit
      6:   Brown fruit on lower umbels, green fruit on upper
           umbels

Umbelliferae and Zingiberaceae), are the least affected by changes
in meteorological conditions, and oils produced by plants bearing
these latter type of oil glands vary little from one season to
another.

It is difficult to separate the effects of temperature from
light; however, of the various factors which affect the environment, light has probably the greatest effect because (with few
exceptions) it varies greatly over a single twenty-four hour
period. In contrast, other climatic variants such as temperature
and available moisture do not vary quite so drastically on a
regular basis. Hence, changes in light are often accompanied by
a diurnal fluctuation in active principles.

In 1941, Allard (1) reported that for optimum peppermint oil
production and satisfactory growth cycle, the plant should only be
grown in long day length environments. Thirteen years later,
Langston and Leopold (3) showed that a true photoperiodic effect
was responsible for determining the growth habit and intensive
metabolic changes in peppermint. Thus, short-day growth resulted
in a plant with many stolons and small leaves.

In 1967, Burbott and Loomis (8) performed photoperiod growth
chamber experiments on peppermint; and in this study they found
that the long day conditions enhanced plant growth and caused an
overall increase in monoterpenes or essential oil. They further
found that either short nights or cool nights combined with full
light intensity during the day enhanced the formation of menthone
and depressed the accumulation of menthofuran. From these results,
Burbott and Loomis concluded that the level of oxidized or reduced monoterpenes was a direct reflection of the levels of
oxidized and reduced cofactor levels of the terpene producing cells
which in turn depends upon the photosynthate levels in the cells.

Furthermore, they found that warm nights caused respiratory
substrate (photosynthate) depletion resulting in oxidizing
conditions and the accumulation of menthofuran. In contrast, cool
nights allowed the preservation of high levels of respiratory
substrates and thus maintained reducing conditions which favored
the production of menthone and menthol. More recently, Clark and
Menary (17) confirmed the findings of Burbott and Loomis and also
showed that long days and high night temperatures favored high oil
yields with an unusually high menthofuran content.

In 1973, Tatro et al (11) examined the qualitative and
quantitative changes in essential oil composition of leaf oils
obtained from Juniperus occidentalis, J. osteosperma and J.
californica, and found that neither growth medium or seasonal
variation had a significant effect on oil composition. By contrast,
they observed a definite diurnal cycling which they attributed to
the fluctuation in air temperature over a twenty-four hour period.
Six years later, Hopfinger et al (16) examined the diurnal
variation in the leaf oil of Citrus sinensis (L.) Osbeck. The
authors found that there was a two-fold diurnal change in oil

content which was dependent on the season of harvest. The highest yield of oil was obtained when the day temperature was hot (22-36° C) and the night temperature was cold (8-15°C). During the seasons examined there was a variety of humidities experienced; however, no correlation between yield of oil and humidity even in association with day and night temperatures could be found. The authors speculated that since natural volatilization would cause a differential decrease in the more volatile components of the oil, a phenomenon observed by Dement et al (12), then the changes in pool size to compensate for the losses must reflect a more bio-chemically important function than was once thought.

It was stated earlier that the formation of essential oils occurs predominantly during periods of active growth, thus, to enhance active growth the use of fertilizers might be recommended. This is not necessarily true, however, for all plants in all situations. To examine this question a little more closely a literature survey was undertaken. The results of this survey re-vealed that the effect of nitrogen, potassium or phosphorus, singly or in combination, was found to cause an increase, decrease or to have no effect on the oil yield of a specific plant. For example, Skrubis (7) showed that all three nutrients had a positive effect on peppermint oil production, whereas, Hornok (20) showed that only nitrogen and phosphorus had a positive effect while potassium had a negative effect. Such discrepancies are expected because if the nutrient needs for optimum growth of a plant in a specific environment are less than the available nutrients, the plant will either not respond to additional nutrients or it will respond adversely.

The magnitude and overall effect of a macronutrient on oil yield is dependent upon environment, available water, specific plant type and its stage of development. For example, nitrogen will generally increase the mass of plant material produced per unit area independent of soil type. Also, it has been determined that for herbaceous plants the oil content of particular selection remains constant irrespective of plant size. Therefore, as nitrogen can cause an increase in overall dry matter, it can have a positive effect on oil production. Phosphorus and potassium can also cause an increase in dry matter, however, their presences in soil does not necessarily mean that they are available to the plant. In acid soils, for example, phosphorus is not readily accessible while potassium is easily leached away. Generally, in most neutral or alkaline soils the effect of phosphorus and potassium on oil content is less dramatic than nitrogen unless a true deficiency is found.

The type of plant can influence the effect of a macronutrient. For example, although nitrogen will cause an increase in vegetative growth, this increase is only of value for herbaceous plants such as those found in the Labiatae and Compositae families. Umbelliferous plants are of value for their seed essential oils. The addition of nitrogen to them will cause an increase in leaf production thereby resulting in a decrease in seed production and an eventual decrease in oil content.

Water is both the medium in which cellular transformations take place and the medium for transporting all soluble substance. Water utilization of a plant is dependent upon the annual climatic conditions of the microenvironment, the water holding capacity of the soil, and the water requirements of the plant during its many stages of development. A review of the literature on the effect of water on the yield of essential oil has indicated that the optimum soil water content was found to be 80-90% (6). It was found that both a lack and excess of water caused a decreased oil yield. Krupper et al (9) found that irrigation of peppermint did not effect the yield of oil; however, it was recommended that irrigation be so arranged that the soil water content is held between 65-80% of total field capacity.

Nelson et al (10) reported that sprinkler irrigation of peppermint when ambient temperature exceeded 30°C cooled the plants by the latent heat of evaporation which in turn lowered the concentration of menthofuran in the oil. The authors suggested that the evaporative cooling had the same effect as cool nights. In 1980, Clark and Menary (17) reviewed the work of Nelson and compared it to their own experiments on the effect of temperature, photorespiration and dark respiration on oil composition. They concluded that evaporative cooling would increase the net CO_2 fixation by decreasing both photorespiration and dark respiration whereas cool nights would only decrease dark respiration. These same authors also found that the timing of irrigation was very important. For example, an increased rate of irrigation on peppermint during the last half of its growing season proved to be most effective for increasing the overall oil yield per hectare.

Loomis (14) proposed that the yield of peppermint oil could be maximized by carefully controlling the water stress of the plants. He stated that water stress and other factors interacting with it control the balance between growth and photosynthesis and determine whether photosynthate is used directly for growth, flowering or biosynthesis and essential oil maturation. The following year, Gershenzon et al (15) concurred with the earlier findings and found that Salvia douglasii grown under stressed intake produced larger positive monoterpene yield differences than non-stressed plants.

Franz (22) reported that the yield and composition of the oil of Artemisia dracunculus increased with less frequent water regime. He found that the plants really only needed to be irrigated during critical stages of growth such as shoot formation, secondary shoot formation, flower bud formation and after harvesting.

It has been observed by this author that the adaptability of an essential oil bearing plant to a dry habitat (xerophyte) or an environment that is neither too wet or too dry (mesophyte) has a profound effect upon the influence of water stress and intake on the oil yield. For example, xerophytic plants such as Coriandrum sativum, Salvia sclarea, Lavandula vera, Matricaria chamomilla, etc. produce an increased oil yield under moisture stress. In contrast, mesophytic plants such as Carum carvi, Levisticum officinale, Anethum graveolens, Ocimum basilicum etc. produce a decreased oil yield under moisture stress. They require a regulated water supply throughout their growth cycle to maximize oil yield. If such a regimen were applied to xerophytic plants their oil yield would be lower than if they were under moisture stress.

In summary, it can be stated that all seed planted (non-clonally reproduced) essential oil bearing plants possess a high degree of variability in essential oil content. Some of these plants can also exhibit a genetic variability which is expressed in the occurrence of infraspecific chemical differences. Finally, depending upon the type of oil glands present and the type of environment from which the plant originated, the influence of extrinsic factors such as time of day, water intake, light and nutrient intake can have a profound effect on the oil yield and composition.

Literature Cited

1. Allard, H. A. J. Ag. Res. 1944, 63, 55-64.
2. Boshart, B. Forschungdienst Somderh. 1942, 16, 543-549.
3. Langston, R. G.; Leopold, A. C. Proc. Amer. Soc. Hort. Sci. 1954, 63, 347-352.
4. Fluck, H. J. Pharm. Pharmacol. 1955, 7, 361-383.
5. Fluck, H. In "Chemical Plant Taxonomy"; Swain, T., Ed.; Academic: London, 1963; Chap. 7.
6. Schrodar, H. Pharmazie. 1963, 18, 47-58.
7. Skrubis, B. Perf. Essent. Oil. Rec. 1964, 55, 655-657.
8. Burbott, A. J.; Loomis, W. D. Plant Physiol. 1976, 42, 20-28.
9. Krupper, H., Lossner, G.; Schrodar, H. Pharmazie. 1986, 23, 192-198.
10. Nelson, C. E.; Early, R. E.; Mortensen, M. A. Washington Ag. Expermt. Stn. Circular No. 541, 1971, 1-19.
11. Tatro, V. E.; Scora, R. W.; Vasek, F. C.; Kumamoto, J. J. Amer. Bot. 1973, 60, 236-241.
12. Dement, W. A.; Tyson, B. J.; Mooney, H. A. Phytochem. 1975, 14, 2555-2557.
13. Stahl, E. Ph.D. Thesis University Hamburg, Hamburg 1977.
14. Loomis, W. D. Proc. 28th Ann. Meetg. Oregon Essent. Oil Growers League. 1977, 13-14.
15. Gershenzon, J.; Lincoln, D. E.; Langenheim, J. J. Biochem. System. Ecol. 1978, 6, 33-43.
16. Hopfinger, J. A.; Kumamoto, J.; Scora, R. W. Amer. J. Bot. 1979, 66, 111-115.
17. Clark, R. J.; Menary, R. C. Aust. J. Plant Physiol. 1980, 7, 685-692.
18. Repcak, M.; Halasova, J.; Honcariv, R.; Podhravsky, D. Biologia Plantarum (Praha). 1980, 22, 183-191.
19. Lawrence, B. M. In "Essential Oils"; Mookherjee, B. D.; Mussinan, C. J.; Eds.; Allured Press: Wheaton, Ill. 1981, 1-81.
20. Hornok, L. Acta. Hort. 1983, 132, 239-237.
21. Porter, N. G.; Shaw, M. L.; Shaw, G. J.; Ellingham, P. J. N. Z. J. Ag. Res. 1983, 26, 119-127.
22. Franz, Ch. Acta Hort. 1983, 132, 203-215.
23. Lawrence, B. M. 1984, unpubl. information.

RECEIVED March 25, 1986

28

Production of a Romano Cheese Flavor by Enzymic Modification of Butterfat

Kuo-Chung M. Lee, Huang Shi[1], An-Shun Huang[2], James T. Carlin[3], Chi-Tang Ho, and Stephen S. Chang

Department of Food Science, Cook College, New Jersey Agricultural Experiment Station, Rutgers University–The State University of New Jersey, New Brunswick, NJ 08903

A Romano cheese-like aroma was produced from a butter-fat emulsion by treating it with a crude enzyme mixture isolated from Candida rugosa. The emulsion consisted of 20% butterfat and 1.5% Tween 80 in a buffer solution. The treated emulsion was held at 37°C for three hours and then aged at room temperature for three days to develop the cheese-like flavor. The volatile flavor components were isolated from both the enzyme modified butterfat (EMB) and a commercial sample of Romano cheese. The flavor isolates were separated into acidic and nonacidic fractions and analyzed by gas chromatography-mass spectrometry. The results showed good correlation between the acidic fractions of the two samples. The acidic fractions contained similar relative concentrations of eight short-chain fatty acids (C_2 - C_{10}). Methyl ketones and esters were major components in the nonacidic fraction of the EMB.

The characteristic flavor of various cheeses is primarily due to the enzymatic action of microbial flora contained in the curd. Enzymes extracted from these microorganisms and reacted with corresponding substrates may also produce a specific cheese flavor. The flavor produced may be economical and could be classified as "natural".

Several varieties of the popular Italian cheeses owe their characteristic flavor to the action of lipolytic enzymes. Romano is a very hard, ripened cheese. Originally, it was made from ewe's milk; it is now also made from cow's and goat's milk. The sharp, peppery-like flavor, traditionally termed "piquant", results from extensive lipolysis (1). Long and Harper (2) and Arnold et al. (3)

[1]Current address: Scientific Research Institute of Fermentation Industry, Ministry of Light Industry, Beijing, People's Republic of China
[2]Current address: Nabisco Brands, Inc., Morristown, NJ 07960
[3]Current address: Thomas J. Lipton, Inc., Englewood Cliffs, NJ 07632

reported that the development of a desirable, characteristic Romano
cheese flavor is related to the short-chain fatty acids, especially
butanoic acid. The production of butanoic acid closely paralleled
the development of flavor. The lipase system which failed to produce a high concentration of butanoic acid also failed to produce
the desirable Romano cheese flavor.

Crude rennet pastes or dried glandular enzymes of suckling
young mammals are used for coagulation of milk in the production of
Italian cheeses. The lipolytic enzyme responsible for the production of the characteristic "piquant" flavor is a pregastric esterase
in the rennet. Rennets from microorganisms, such as Aspergillus
niger and A. oryzae, also have been used in making Italian cheeses.
A microbial esterase, Mucor miehei esterase, has been extensively
studied for flavor development in cheese and was found to produce
flavor notes resembling those of Fontina and Romano cheeses (4).

The technology has been developed for production of flavor systems via controlled enzyme modification of butterfat (EMB). Lipases
and esterases from various sources are used (5). Nelson (6) described the essential steps for producing lipolyzed butterfat products. Arnold et al. (3) published a comprehensive review on the
application of lipolytic enzymes.

The purpose of this study is to investigate the production of
a Romano cheese-like flavor by enzyme modification of butterfat.
Candida rugosa was selected for enzyme modification of butterfat
since it possesses a high lipase activity.

Materials and Methods

Enzyme Preparation. A crude enzyme mixture was prepared from the
fermentate of Candida rugosa (American Type Culture Collection, ATCC
No. 14830. The C. rugosa was revived and cultivated in YM Agar
slants at 24°C for one week. The growth on one YM Agar slant was
transferred aseptically to a two-liter flask containing 400 ml sterilized fermentation medium consisting of 2.0% defatted soyflour,
1.0% potato starch, 1.0% maltose, 0.5% K_2HPO_4, 0.1% $MgSO_4 \cdot 7H_2O$ and
0.1% $(NH_4)_2SO_4$. Fermentation was carried out in an incubator shaken
at 200 rpm and 25°C for 20 hours.

The fermented broth was filtered through several layers of
cheesecloth and was centrifuged at 600xG for 10 minutes to obtain
the cell-free supernatant. The supernatant was precipitated with
cold acetone and separated by ultracentrifugation at 25,000xG for
2 hours. The precipitate was then freeze-dried.

The enzyme powder obtained was assayed to determine the optimum
pH and temperature for fatty acids production (7). A 20% butterfat
emulsion was used as the enzyme substrate instead of an olive oil
emulsion. The optimum pH for fatty acid production was 7.7. The
optimum temperature was 37°C.

Flavor Development. The crude enzyme mixture isolated from C. rugosa was allowed to react with a butterfat emulsion consisting of
20% butterfat and 1.5% Tween 80 (Polyoxyethylenesorbitan monooleate)
in a 0.1 M sodium phosphate buffer solution. Butterfat was prepared from Land O'Lakes unsalted butter (Arden Hills, MN). Butter
was melted and washed with hot water (50°C) in a separatory funnel

several times until the aqueous phase became clear. The washed butterfat was centrifuged at 600xG for 10 minutes to separate and remove water. The emulsion (pH 7.7) was prepared by passing the ingredients through a hand homogenizer at least twice. Enzyme powder
consisting of 1.5% of the emulsion was dissolved into the buffer
solution and homogenized with the butterfat emulsion. It was incubated at 37°C for three hours. This resulted in the development of
a cheese-like flavor. The emulsion was then aged for three days at
room temperature to complete the flavor development.

The aroma and flavor-by-mouth of the EMB were evaluated by a
panel of five trained flavorists. A neat sample of the EMB was
evaluated for aroma only. The aroma and flavor-by-mouth of the sample were evaluated in a 0.5% NaCl solution at a level of 0.6%.
Panelists independently provided a list of product descriptions.
The panel summarized its results in a concensus discussion following
the evaluation.

Isolation of Volatile Flavor Compounds from the EMB Sample,
Romano Cheese and Butterfat. The volatile flavor compounds were
isolated from the EMB sample, a commercial sample of Romano cheese
and a butterfat control sample by vacuum steam distillation. Volatiles were isolated from 2.5L EMB in five batch isolations. The EMB
was mixed in a Waring blender prior to each isolation. Romano
cheese was obtained from a commercial source (Stella Romano cheese,
Universal Foods Corp., Milwaukee, WI). Volatiles were isolated from
700 gm Romano cheese in five batch isolations. One hundred and forty grams of cheese were cut into pieces for each isolation and slur-
ried with 360 ml 0.1% sodium phosphate buffer solution in a blender.
Volatiles were also isolated from 500 ml of butterfat emulsion control sample (20% butterfat).

The slurry samples of EMB, Romano cheese and butterfat were
vacuum distilled at 50°C for 8 hours. The volatiles were condensed
in a series of cold traps cooled with a dry ice-acetone slurry. The
condensates collected in the traps were combined, saturated with
sodium chloride and extracted with ethyl ether.

The ether extracts were separated into acidic and nonacidic
fractions by extraction with a 10% aqueous sodium carbonate solution. The ether solutions of acidic and nonacidic fractions were
dried over anhydrous sodium sulfate and subjected to a preliminary
concentration using a 30-plate Oldershaw distillation column. The
ether extracts from both fractions were concentrated to final volumes of 5 ml using a spinning band distillation apparatus.

GC-MS Analysis and Identification. A portion of the acidic fractions were quantitatively converted to methyl esters using a BF_3-
methanol reagent. The procedure used was that of Metcalfe and
Schmitz (8). The esterified acidic fractions were analyzed using a
coupled gas chromatograph-mass spectrometer (GC-MS) system consisting of a Varian Moduline 2700 gas chromatograph equipped with a
flame ionization detector interfaced by a jet separator to a Du Pont
Instruments Model 21-490 mass spectrometer. A 10 ft x 1/8 in
o d stainless steel column packed with 10% stabilized DEGS on 80/
100 mesh Chromosorb W AW DWCS was used. The He flow rate was
30 ml/min. The column temperature was held 5 minutes at 30°C, then

programmed at 4°C/min to a final temperature of 200°C. Mass spectra
were obtained at 70eV and a source temperature of 250°C. The gas
chromatograms generated from the GC-MS analyses were recorded and
integrated using a Hewlett-Packard 5840A GC terminal.

The nonacidic fractions were analyzed using a coupled GC-MS
system. The system consisted of a Hewlett-Packard 5840A gas chromatograph and a Hewlett-Packard 5985 mass spectrometer. A J&W
fused silica capillary column (30m x 0.25mm)coated with DB-1 (methyl
silicon) was used for the analysis. The column was held for 2 min
at 40°C and then temperature programmed to 250°C at 4°C/min. Mass
spectra were obtained at 70eV and a source temperature of 200°C.

Compounds were identified by comparing the mass spectra obtained with those of published reference spectra. Approximate relative percentages of the compounds identified in the acidic fractions
were calculated based on the concentration of the most abundant compound identified.

Results and Discussion

A crude enzyme mixture was isolated from the fermentate of Candida
rugosa (ATCC No. 14,830),which is reported to produce high activity
lipases (9). The enzyme mixture was added to a 20% butterfat emulsion. A cheese-like flavor developed after 3 hours of incubation at
37°C. A desirable Romano cheese note developed after continued incubation at room temperature for three days. Nelson (6) studied
lipolyzed butterfat flavor and concluded that the surface active
characteristics of both fatty acids and mono- and diglycerides were
important in the lipolyzed system. A tempering period of hours or
even days was usually required to establish equilibrium at the interface of aqueous and fat phases. He pointed out that the lipolyzed flavor appeared to intensify as the equilibration proceeded
and that this intensification was sometimes mistaken for residual
lipolytic activity.

Sensory Evaluations. Summary descriptions of the aroma and flavor-
by-mouth of the EMB were established by a panel of five trained
flavorists. The aroma of the neat sample was described as strong
cheesy, strong Romano cheese-like with slight milky, creamy, ketonic
and soapy notes. The aroma of the sample in a 0.5% NaCl solution at
a level of 0.6% was described as soapy and milky with slight Romano
cheese-like, acid and ketonic notes. The flavor-by-mouth of the
sample in the NaCl solution was described as containing slight
sharp cheese, bitter and waxy notes.

Tween 80 has an astringent and bitter taste (10). However a
related study showed that the bitter note in the EMB was not due to
the Tween 80. Gum Arabic, which does not have a bitter taste (10)
was used to replace Tween 80 as the emulsifier. The sample produced still had the bitter and soapy note. A bitter, soapy character is generally found in lipolyzed products (11) and may be related
to the mono- and diglycerides formed during lipolysis.

Acidic Components. The Romano cheese-like flavor of the enzyme-
modified butterfat led to a study of its volatile flavor compounds.
Table I lists the compounds identified, their absolute concentra-

tions and the approximate relative percentages at which they were
present based on the concentration of butanoic acid in each sample
of EMB and commercial Romano cheese.

The composition of the samples is very similar. Both contain eight
n-fatty acids (C_2 - C_{10}). In addition, sorbic acid, a preservative,
was present in the commercial product. The quantity of the acidic
components isolated from the volatiles of the EMB sample was more
than three times greater than that of the commercial Romano cheese.
Harper (12) reported that butanoic acid and other higher fatty acids
may be related to the intensity and character of Romano cheese flavor.

Nonacidic Components. Figure 1 shows the total ion chromatograms
obtained from the nonacidic fractions of the flavor isolates. A
total of 22 compounds were identified in the EMB and 12 in the
Romano cheese. Table II lists the volatile flavor compounds identified in the nonacidic fractions.
     The odd-numbered ketones identified in both samples are common-
ly found in mold-ripened cheeses, such as Blue cheese, and are responsible for their characteristic aroma. Large quantities of these
compounds also have been found in nonmold-ripened cheeses, such as
Cheddar, Swiss and Romano. These compounds may arise by β-oxidation
of the appropriate fatty acids (14), or by decomposition of β-keto
acids.
     A series of ethyl esters of fatty acids, from butanoate to
tetradecanoate, were identified in the EMB. Two esters, ethyl oc-
tanoate and ethyl nonanoate, were found in Romano cheese. Esters
are important flavor compounds in cheeses; however, a high concentration of esters may cause a "fruity" defect in cheese flavor.
γ- and δ-dodecalactone were identified in the EMB sample as well as
in Romano cheese. Lactones are well distributed in food flavors.
In cheese, a series of γ- and δ-lactones have been identified (11).
     p-Nonylphenol was identified in the EMB sample. Phenolic com-
pounds usually cause a "phenolic" defect in Gouda cheese, and could
be generated from certain variants of lactobacilli. However, it is
possible that phenolic compounds also play a role in cheese flavor.
Phenol and p-cresol were found in Cheddar cheese and were included
with guaiacol in a synthetic Cheddar cheese flavor formulation (15).
     2-Phenylethanol was identified in the EMB sample and has a rose-
like odor and taste. It can be produced through fermentation and is
present in many foods including wine (16) and bread (17).
     n-Hexylfuran was identified in the EMB sample and may arise
from a lipid source. 2-Ethyl-4,5-dimethyloxazole was also identified in this sample. This compound has been previously identified
only in food systems subjected to heat treatment.
     Benzaldehyde and phenylacetaldehyde were identified in Romano
cheese. These two compounds also have been found in Cheddar, Swiss
and Blue cheeses. Benzaldehyde has a powerful, sweet, almond-like
odor and exists in many foods. Phenylacetaldehyde has a strong,
sweet floral, penetrating aroma and is found in many cooked foods.
In dairy products, phenylacetaldehyde probably is formed from
phenylalanine in milk protein through enzymatic transamination, followed by decarboxylation. This compound also contributes to the
malty defect of cultured milk products (18).

Table I. Volatile Flavor Compounds Identified in the Acidic
Fractions of Enzyme Modified Butterfat (EMB),
Romano Cheese and Untreated Butterfat*

Fatty Acid	Absolute Concentration (mg/kg of fat)			Relative Concentration (%)**		
	EMB	Romano	Untreated Butterfat	EMB	Romano	Untreated Butterfat
Acetic	1.3	3.1	-	0.05	0.35	-
Propanoic	3.7	0.8	-	0.14	0.09	-
Butanoic	2620.0	896.0	56.5	100.00	100.00	100.00
Pentanoic	13.4	4.5	-	0.51	0.50	-
Hexanoic	743.6	285.2	18.8	28.38	31.83	33.19
Heptanoic	2.6	2.0	-	0.10	0.22	-
Octanoic	41.9	26.6	10.8	1.60	2.97	19.12
Decanoic	0.3	0.1	0.2	0.01	<0.01	0.35

*Identified as their methyl esters.
**Calculated based on the concentration of butanoic acid in each sample.

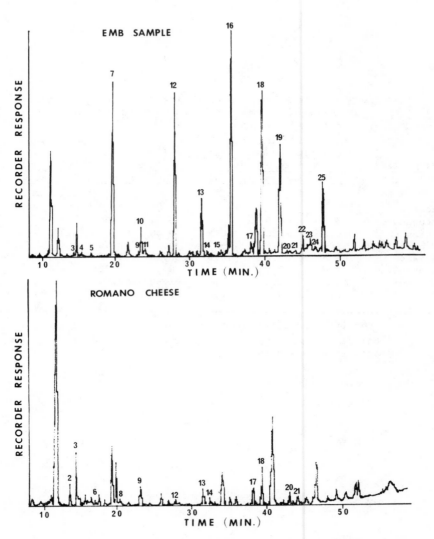

Figure 1.   Total Ion Chromatograms of the Nonacidic Fractions from the Enzyme Modified Butterfat (EMB) and Romano Cheese.

Table II. Volatile Flavor Compounds Identified in the
Nonacidic Fractions from Enzyme-Modified Butterfat (EMB)
and Romano Cheese

Peak No.	Compound	EMB	Romano Cheese	Previously Reported in Butterfat (13)
	Hydrocarbon			
2	Ethyl benzene	-	+	+
	Alcohols			
11	2-Phenylethanol	+	-	-
24	p-Nonylphenol	+	-	-
	Carbonyls			
6	Benzaldehyde	-	+	+
8	Phenylacetaldehyde	-	+	-
3	2-Heptanone	+	+	+
9	2-Nonanone	+	+	+
13	2-Undecanone	+	+	+
17	2-Tridecanone	+	+	+
23	2-Pentadecanone	+	-	-
	Esters and Lactones			
1	Ethyl butanoate	+	-	+
4	Ethyl 3-methylbutanoate	+	-	-
7	Ethyl hexanoate	+	-	+
10	Ethyl heptanoate	+	-	-
12	Ethyl octanoate	+	+	+
14	Ethyl nonanoate	+	+	-
16	Ethyl decanoate	+	-	-
19	Ethyl dodecanoate	+	-	-
22	Ethyl tridecanoate	+	-	-
25	Ethyl tetradecanoate	+	-	-
20	γ-Dodecalactone	+	+	+
21	δ-Dodecalactone	+	+	+
	Miscellaneous			
5	2-Ethyl-4,5-dimethyl-oxazole	+	-	-
15	n-Hexyl furan	+	-	-
18	BHT[*]	+	+	-

+ : identified in sample.
- : not identified in sample.
* : BHT was a preservative used in carrier solvent.

Comparison of the acidic volatile profiles indicates that the EMB is similar to Romano cheese. This correlates well with the flavorists' impressions of the aroma and flavor-by-mouth of this sample. Nonacidic volatile flavor profiles of EMB and Romano cheese are dissimilar. This indicates different formation pathways for these two samples. Current studies are investigating applications of the EMB as a flavoring agent and the feasibility for commercial application.

Acknowledgments

New Jersey Agricultural Experiment Station, Publication No. D-10412-1-85 supported by State funds and a grant-in-aid from Firmenich, Inc., Princeton, NJ. We thank Mrs. Joan B. Shumsky for her secretarial aid.

Literature Cited

1.   Peppler, H. J.; Dooley, J. G.; Huang, H. T.   J. Dairy Sci. 1976, 59, 859.
2.   Long, J. E.; Harper, W. J.   J. Dairy Sci.   1956, 39, 245.
3.   Arnold, R. G.; Shahani, K. M.; Dwivedi, B. K.   J. Dairy Sci. 1975, 58, 1127.
4.   Huang, H. T.; Dooley, J. G.   Biotechnology and Bioengineering 1976, 18, 909.
5.   Nelson, J. H.; Jensen, R. G.; Pitas, R. E.   J. Dairy Sci. 1977, 60, 327.
6.   Nelson, J. H.   J. Amer. Oil Chem. Soc.   1972, 49, 559.
7.   Decker, L. A.   "Worthington Enzyme Manual, Enzymes, Enzyme Reagents Related to Biochemicals"; Worthington Biological Company: Freehold, NJ, 1977; p. 125.
8.   Metcalfe, L. D.; Schmitz, A. A.   Anal. Chem.   1961, 33, 363.
9.   Yamada, K.; Machide, H.   U. S. Patent 3 189 529, 1965.
10. Lee, K.-C. M., unpublished data.
11. Ney, K. H.   In "The Quality of Foods and Beverages"; Charalambous, G.; Inglett, G., Ed.; Academic Press: New York, 1981, Vol. I, p. 389.
12. Harper, W. J.   J. Dairy Sci.   1947, 40, 556.
13. Siek, T. J.; Lindsay, R. C.   J. Dairy Sci.   1970, 53, 700.
14. Hawke, J. C.   J. Dairy Res.   1966, 33, 225.
15. Pintauro, N. D.   "Flavor Technology"; Noyes: Park Ridge, NJ, 1971.
16. Webb, A. D.; Muller, C. J.   Adv. Apply. Microbiol.   1972, 15, 75.
17. Maga, J. A.   Crit. Rev. Food Technol.   1974, 5, 55.
18. Sheldon, R. M.; Lindsay, R. C.; Libbey, L. M.; Morgan, M. E. Appl. Microbiol. 1971, 22, 262.

RECEIVED January 3, 1986

SUMMARY

29

Biogeneration of Aromas
In Summation

Ira Katz

International Flavors & Fragrances, Union Beach, NJ 07735

Many important food aromas arise via biochemical pathways. These pathways include microbial fermentation,
exogenous and endogenous enzymic action and plant metabolism. In the past, flavor research was concentrated on identifying the important aroma chemicals
responsible for the specific food aroma. Using this
knowledge, present day research is focusing on eluci-
dating the biochemical pathways responsible for important aroma chemical production. In addition, advances in biotechnology is providing opportunities to
produce important aroma chemicals. It is anticipated
that increasing our knowledge of bioaroma generation
will ultimately lead to foods with intensified flavor,
production of specific food aromas and production of
previously unavailable aroma chemicals.

This book on the biogeneration of aromas comes at an opportune
time. During the 1960's and 70's research in this area at the
industrial and academic level was neglected in favor of classical
analytical research aimed at unravelling the complexities of the
chemical mixtures responsible for the aroma of food. In retrospect
it appears that this was the proper way to approach this problem,
since we had to identify the important aroma contributors before
considering their biological origin.

Factors Stimulating Bioaroma Research

Ten years ago bioaroma generation was not a commonly discussed research topic and most scientists did not consider it to be a viable
means of aroma production. Today, there is obviously a resurgence
in this area of research and a number of factors appear to be responsible for the renewed interest in bioaroma research.
    The "new biology", being led by advances in genetic research
and genetic engineering permits forward thinking food scientists to
explore bioaroma production in practical terms and experimentally
pursue the concept.
    Bioaroma generation may provide the ability to produce impor-

0097–6156/86/0317–0380$06.00/0

tant secondary metabolites that are not available by other conventional production procedures. It is difficult to separate this concept from economic considerations since many available aroma chemicals are too expensive to use. Conceptually, bioaroma generation
holds the promise of significantly lowering the cost of many important aroma chemicals and thus allowing their broad, general use.
    There appears to be a realization that we must understand the
biological world we live in. The accumulation of basic knowledge
dealing with bioaroma generation is part of this and, in addition to
using it in its own right, can lead to information that is important
in other areas of food research.
    Another stimulation for research on the biogeneration of aromas
comes from the market place where consumers appear to be rejecting
the artificial flavor label and requesting the natural label.
    Regardless of the motivation, and whether it will be long last-
ing, the scientific aspects of this subject are important and worthy
of study. The aroma or odor of food is an important part of the
hedonic value of the food and thus becomes part of our food selection process. Since our nutritional state and well being are partially related to food selection, understanding the biological generation of aromas is not conducted only to satisfy the current need
of the marketplace, but takes on an important scientific and nutritional role.

Origins Of Aroma

Approximately 5,000 volatile chemicals have been identified in our
food supply and there is little doubt that others exist and will ultimately be identified. These complex chemical mixtures, responsible for food aroma, can number as high as 1,000 for an individual
food, and originate via a variety of processes that include enzymic
action, autoxidation, microbial action, food processing, cooking and
chemical interactions. It is no wonder there are over 5,000 chemicals contributing to food aroma.
    A convenient way to consider these chemicals is to divide them
into three broad groups.
    (1) One is the heat derived or, Maillard Browning aroma chemi-
cals that are formed when food is cooked or heat processed. These
are responsible for the aroma of cooked foods such as meat, coffee,
poultry and similar products. They also contribute to the aroma of
heat processed fruits and vegetables.
    (2) A second group of chemicals are those that form during
heat processing but are dependant upon chemical precursor formation
during a deliberate fermentation step. Examples of these are cocoa
and bread. The non-volatile chemical precursors formed during microbial fermentation react via the Maillard reaction to form the aroma
chemicals responsible for the typical aroma of the product.
    (3) The third group, and the subject of this book, are the
biologically derived aroma chemicals, often referred to as secondary
metabolites. This diverse and important class of aroma chemicals
generally arise via:
    (a) microbial fermentation;
    (b) enzymic action from endogenous enzymes;
    (c) enzymes added during processing;
    (d) end products of plant metabolism.

Examples of products whose aroma are primarily biologically derived
are vegetables, fruits, berries, essential oils, fermented dairy
products and alcoholic beverages.

Consumer Perception of Flavors

Several of the chapters in this book present arguments that a primary reason for understanding and ultimately manufacturing biologically derived aromas is the consumers demand for natural flavors.
There is little doubt that in addition to the scientific aspects, a
major impetus behind the interest in the biological origin of aromas
is the present day consumer demand for natural flavors. Also, the
present day health orientated, better educated consumer understands
that there is a relationship between health, diet and nutritional
well being. As a result, this has manifested itself as a demand for
natural flavors and foods or, perhaps more importantly, a rejection
of the word artificial. It is not the purpose of this book to argue the merits of consumer attitudes or to consider the complex marketing problems this has presented to the food industry. However,
we recognize that the demand for "natural flavor" is partially responsible for the present day intensified research on the biological origin of aromas. Therefore, it is important that we understand the legal definition of natural and artificial flavors as it
applies to the United States.

Definitions

In the Federal Register, the term natural flavor or natural flavoring means the essential oil, oleoresin, essence or extractive, protein hydrolysate, distillate or any product of roasting, heating or
enzymolysis, which contains the flavoring constituents derived from
a spice, fruit or fruit juice, edible yeast, herb, bark, bud, root,
leaf or similar plant material, meat seafood, poultry, eggs, dairy
products, or fermentation products thereof, whose significant function in food is flavoring rather than nutritional.
    The term artificial flavor or artificial flavoring is the oppo-
site and is defined as meaning any substance, the function of which
is to impart flavor, which is not derived from a spice, fruit or
fruit juice, vegetable or vegetable juice, edible yeast, herb, bark,
bud, root, leaf or similar plant material, meat, fish, poultry,
eggs, dairy products, or fermentation products thereof. It then
goes on to list the permitted substances that can be used in artificial flavors.
    It is apparent that the above definition of natural is broad
and the fine points are subject to legal interpretation. What is
important, from the standpoint of those interested in the biological
derivation of aroma, is that products derived by enzymolysis and
fermentation are considered to be natural substances.

Historical Perspective

The use of natural flavors by the flavor industry is not new. Historically, the industry began by the creative use of natural substances derived from botanical sources. During the 1960's, we experienced an intense effort in analytical research that signifi-

cantly increased our knowledge of the chemical constituents of food
aroma. As the flavor industry grew and cost effective synthetic
chemicals, structurally identical to the natural became available,
the synthetics were combined with the naturals to create cost effective and improved flavors. As our knowledge base increased, and
more synthetic aroma chemicals became available, there was an increase in the creation and use of totally artificial flavors and a
concommitant decrease in the use of naturals. Many food scientists
predicted that most natural flavors and botanicals would ultimately
be replaced by the less expensive and more cost effective artifi-
cials. Undoubtedly, this would have occurred to some extent, were
it not for the consumer demand for naturals that began in the late
70's and accelerated through the 80's. Ultimately, the replacement
may still occur since the underlying economic and technical factors
are still in place.

Biochemical Pathways

Our knowledge of the aroma chemical composition of foods is extensive compared to what we know about the biosynthetic pathways which
result in the formation of many of the important aroma chemicals in
these products. The excellent work presented by Tressl points out
the complexity of the subject and our lack of fundamental knowledge.
His elegant approach of incubating tissue slices with isotopically
labelled precursors and subsequently separating the aroma compounds,
including enantiomers, has greatly contributed to our understanding
of aroma chemical biosynthetic pathways.
      The biosynthetic formation of the mono- and sesquiterpenes is
reviewed by Croteau. In addition, he proposes pathways for the formation of oxygenated terpenes that are the dominating aroma chemicals
in many essential oils and citrus products. The carotenoids are a
class of compounds important not only for their vitamin A activity
but also as aroma precursors. Weeks describes the degradation of
these compounds to produce numerous volatile products including the
C13 ionones, important in fruit aromas. There are many ionones that
occur in trace quantities in fruits that greatly contribute to the
overall quality and perception of the aroma. Because of the unique
aroma effect of these ionones it is interesting to speculate on the
aroma of the end product if we could increase the concentration of
the important ionones in fruits and berries by breeding or processing. These fruits could have significantly enhanced aromas or,
could be perceived as new, unique foods.

Precursor Studies

Two papers delve into the topic of non-volatile bound forms of aroma
compounds. Schreier and Winterhalter show that terpenoids in papaya
are present as water soluble glycosides. Acree's group investigated
the non-volatile precursor of β-damascenone, a chemical with one of
the lowest known aroma thresholds. The latter were able to purify
the precursor by a factor greater than 22,000. It is interesting to
speculate on how many other important aroma compounds are bound
through oxygen linkages and how we might be able to take advantage
of such aroma chemical precursors.

Present Day Applications

A number of chapters deal with the emerging field of biotechnology
and its possible application to flavor and aroma chemical production. If economic barriers can be overcome, there is little doubt
that this rapidly emerging technology will become a significant
factor in the production of certain aroma chemicals and essential
oils. Ultimately, this may lessen our dependance on important agriculturally derived botanicals that suffer from periodic shortages
and price fluctuations.

The excellent paper by Armstrong shows that ethyl acetate, an
important fruit aroma compound, can be produced by the yeast Candida
utilis. He was able to increase production of ethyl acetate by manipulation of the carbohydrate source and regulation of iron levels
in the media.

Alkyl methoxy pyrazines are an important class of aroma compounds exhibiting intense green bean/green pea aroma notes.
Reineccius's group at the University of Minnesota show that mutant
strains of Pseudomonas perolens produce the isopropyl isomer to a
final level of 15 mg/L of culture. Since the threshold is 2×10^{-6}
ppm, the reported yield is substantial.

Scharpf et al extensively review the production of aroma compounds by fermentation processes. Of particular interest to those
interested in this technology is the discussion on the factors to be
considered in scale-up and the potential problems. They conclude
that aseptic fermentation is appropriate when the product has a
value of $100 to $800 per kg. They describe and reference a issued
patented procedure for the production of γ-decalactone by aseptic
fermentation using castor oil as the carbon source.

One important class of foods possessing natural flavors that
arise via enzymatic and microbial action are "fermented" dairy products. This important food group, consisting of cheeses, yoghurt,
buttermilk, sour cream and similar products is very interesting and
illustrates the importance of taste and odor in food selection.
There is no doubt that this group of fermented dairy products significantly contributes to our caloric intake and our over-all nutrition. As consumers we use these products in a multitude of ways.
We consume them directly, combine them with other foods directly or
through cooking, convert them into sauces, etc. There seems to be
no end to the creative way we utilize and consume fermented dairy
products. The lesson is clear, we use these products because we enjoy their odor and taste. In the selection process, their nutritional value is secondary when compared to their flavor. Since food
consumption is a necessary part of life, and in most societies food
selection is primarily based on hedonics, then a priori, the biological origin of aromas is an important subject.

Ho's group at Rutger's describe the use of enzymes from Candida
rugosa to convert butterfat to a series of neutral and acidic compounds possessing a Romano cheese flavor. Similar technology is
used by the food industry to produce enzyme modified cheese from
young cheddar cheese. The final product possesses a more intense
natural aroma and taste. Similar techniques could undoubtedly be
used for the production of other natural cheese flavors.

Gatfield reviews various fermentation reactions that occur in
the presence of endogenous or exogenous microbial enzymes. Among

the aroma compounds that can be generated are acetoin, acetaldehyde, ethyl acetate, pyrazines and the so-called C6 green notes.

Essential oils have been used since the beginning of recorded history as sources of aroma chemicals. Lawrence discusses how factors such as photoperiod, stress, etc. can effect oil production. He describes various essential oils that can be used as sources of specific natural flavor chemicals. Among them are leaf alcohol, benzaldehyde and tolualdehyde. This represents a technology which can be employed at the present time, since only classical isolation and separation techniques are required.

Future Applications

Can we learn to control the enzymatic pathways to produce desired chemicals in good yield and prevent undesirable reactions resulting in off-flavor production?

Hatanaka shows in detail the reaction scheme in plants whereby linolenic acid is converted via lipoxygenase and lyase enzymes to cis-3-hexenal. This is subsequently converted by other enzyme systems to leaf aldehyde and leaf alcohol. These three aroma chemicals are important "green" aroma notes in strawberries and other berries. If these enzymes were commercially available, then the production of natural cis-3-hexenal from inexpensive vegetable oil is theoretically feasible.

The work reported by Josephson and Lindsay illustrates the broad nature of aroma biogenesis. In the past, most food scientists did not consider that the desirable aroma of fresh fish was enzymatically generated after harvest. In seafood, post harvest aroma chemical formation is generally associated with negative odors. This work clearly shows that after harvest, polyunsaturated fatty acids are enzymatically converted to physiologically active compounds which are subsequently converted to volatile alcohol and carbonyl compounds that are characteristic of the odor of freshly harvested fish.

Conceptually, it is interesting to speculate on the bioconversion of inexpensive secondary metabolites to others of greater value. Along these lines, Schreier and co-workers use Botrytis cinera to convert linalool to a series of other terpenoids as well as to the furanoid and pyranoid linalool oxides. Reactions of this type are good examples of converting inexpensive, available aroma chemicals to higher valued products.

Finally, Whitaker discusses the promise of plant tissue culture. This technology was originally received by scientists with great enthusiasm. However, we now realize that much more must be learned about biochemical and genetic regulation of plant secondary metabolites before large scale production becomes realistic. There is little doubt that these goals will evenutally be achieved. As we accumulate knowledge about the biogeneration of fruit aroma, it is theoretically possible that someday we will be able to "turn on" the appropriate enzyme systems to produce a complete natural fruit flavor. This is a difficult objective, and some may consider it impossible, but the results are worth the effort. That is why symposia like this are invaluable for stimulating aroma research. It will be interesting to compare the conclusions of this symposia with a similar one five to ten years from now.

Conclusion

The present use of biotechnology consists of the production of aroma
notes and specific chemicals by fermentation; the application of improved enzymes and enzyme technology to food processing; and the application of new genetic technology for the production of improved
food crops, spices and essential oils.

   The future application of biotechnology to bioaroma production
is limited only by our own intellectual capacity to apply the technology as it emerges. Increased food crop yields with intensified
flavors, total aroma production of specific foods, production of important aroma chemicals previously unavailable and significantly
lowered costs are some of the promises of the future.

RECEIVED May 23, 1986

Author Index

Acree, T. E., 75
Albrecht, Wolfgang, 114
Andersen, R. A., 99, 193
Armstrong, David W., 254
Benda, I., 243
Bock, G., 243
Braell, P. A., 75
Bruemmer, J. H., 275
Butts, R. M., 75
Carlin, James T., 370
Chang, Stephen S., 370
Chen, Chu-Chin, 176
Croteau, Rodney, 134
Farbood, M. I., 323
Fleming, P. D., 99
Gatfield, Ian L., 310
Gooley, Paul R., 222
Guentert, M., 53, 65
Hamilton-Kemp, Thomas R., 99, 193
Hatanaka, Akikazu, 167
Hildebrand, D. F., 99
Ho, Chi-Tang, 176, 370
Hobbib, George C., 347
Huang, An-Shun, 370
Jennings, W., 53, 65
Josephson, David B., 201
Kajiwara, Tadahiko, 167
Katz, Ira, 380
Lawrence, Brian M., 363
Lee, Kuo-Chung M., 370

Lindsay, R. C., 201, 286
Liu, Su-Er, 176
Ludwig, S. Peter, 18
Lugay, Joaquin C., 11
Macku, C., 53
Marriott, R. J., 184
McIver, R. L., 266
Morris, J. A., 323
Parliment, Thomas H., 34
Reineccius, G. A., 266
Rippe, J. K., 286
Scharpf, L. G., Jr., 323
Schreier, P., 85, 243
Seitz, E. W., 323
Sekiya, Jiro, 167
Shi, Huang, 370
Smith, Sharon L., 65
Steward, Leslie A., 347
Stofberg, Jan, 2
Strauss, Christopher R., 222
Takeoka, G. R., 53, 65
Tressl, Roland, 114
Weeks, W. W., 157
Whitaker, Robert J., 347
Williams, Patrick J., 222
Wilson, Bevan, 222
Winterhalter, P., 85
Wu, Chung-May, 176
Zhou, P. G., 75

Subject Index

A

Acetaldehyde
  effect on ethyl acetate
    accumulation,
    yeast, 260,262,263t
  enzymatically produced, 313,316f
  in oranges during
    maturation, 278,279f
Acetyl-CoA, effect on ethyl acetate
  accumulation, yeast, 257,259f
Acid lability, polyols and glycosides
  in grapes, 228
Acidic components, enzyme-modified
  butterfat, 373
Actinomycetes, factors affecting
  volatile production, 333
Acyl pathways, in formation of aroma
  compounds, 1114-133

Aerobic environments, methioninase
  activity under, 293,295-296f,298f
Aerobic metabolism, in oranges during
  maturation, 278,283f
Aging
  effect on wine monoterpenes, 237
  See also Maturation, Ripening
Alcohol dehydrogenase, in citrus,
  277-280
Alcohols
  biogenesis by fruits, 115-126
  biosynthesis in passion
    fruits, 117,120f
  formed by B. cinerea, 247f
  in freshly harvested fish, 202-204
  in grapes, 222-231
  in oranges during
    maturation, 278,279f
  in passion fruits, 117, 119f

387

Production by Keith B. Belton
Indexing by Janet S. Dodd
Jacket design by Pamela Lewis

Elements typeset by Hot Type Ltd., Washington, DC
Printed and bound by Maple Press Co., York, PA

RECENT ACS BOOKS

Writing the Laboratory Notebook
By Howard M. Kanare
145 pages; clothbound ISBN 0-8412-0906-5

Phenomena in Mixed Surfactant Systems
Edited by John F. Scamehorn
ACS Symposium Series 311; 350 pp; ISBN 0-8412-0975-8

Chemistry and Function of Pectins
Edited by Marshall Fishman and Joseph Jen
ACS Symposium Series 310; 286 pp; ISBN 0-8412-0974-X

Fundamentals and Applications of Chemical Sensors
Edited by Dennis Schuetzle and Robert Hammerle
ACS Symposium Series 309; 398 pp; ISBN 0-8412-0973-1

Polymeric Reagents and Catalysts
Edited by Warren T. Ford
ACS Symposium Series 308; 296 pp; ISBN 0-8412-0972-3

Excited States and Reactive Intermediates:
Photochemistry, Photophysics, and Electrochemistry
Edited by A. B. P. Lever
ACS Symposium Series 307; 288 pp; ISBN 0-8412-0971-5

Artificial Intelligence Applications in Chemistry
Edited by Bruce A. Hohne and Thomas Pierce
ACS Symposium Series 306; 408 pp; ISBN 0-8412-0966-9

Organic Marine Geochemistry
Edited by Mary L. Sohn
ACS Symposium Series 305; 440 pp; ISBN 0-8412-0965-0

Fungicide Chemistry: Advances and Practical Applications
Edited by Maurice B. Green and Douglas A. Spilker
ACS Symposium Series 304; 184 pp; ISBN 0-8412-0963-4

Historic Textile and Paper Materials: Conservation and Characterization
Edited by Howard L. Needles and S. Haig Zeronian
Advances in Chemistry Series 212; 464 pp; ISBN 0-8412-0900-6

Multicomponent Polymer Materials
Edited by D. R. Paul and L. H. Sperling
Advances in Chemistry Series 211; 354 pp; ISBN 0-8412-0899-9

For further information and a free catalog of ACS books, contact:
American Chemical Society, Sales Office
1155 16th Street N.W., Washington, DC 20036
Telephone 800-424-6747